Applied Immunocytochemistry

Applied Immunocytochemistry

Edited by **Rachel Dell**

New York

Published by Callisto Reference,
106 Park Avenue, Suite 200,
New York, NY 10016, USA
www.callistoreference.com

Applied Immunocytochemistry
Edited by Rachel Dell

International Standard Book Number: 978-1-63239-076-9 (Hardback)

Printed in the United States of America.

Contents

Preface

This book is a concise and sophisticated introduction to applied immunocytochemistry. Traditionally, immunocytochemistry is described as a process to investigate antigens in cellular contexts through antibodies. But over time, various attributes of the process have emerged from a vast spectrum of experimental conditions. There are various methods to develop a given sample, various kinds of antibodies for applications, multiple techniques for imaging and various procedures of analyzing the information. In this book, multiple methods for carrying out every individual step of immunocytochemistry in diverse cellular contexts have been illustrated. This book provides practical and background data on various levels of immunocytochemistry and the utilization of this technology and its manifestations in relevant fields.

The researches compiled throughout the book are authentic and of high quality, combining several disciplines and from very diverse regions from around the world. Drawing on the contributions of many researchers from diverse countries, the book's objective is to provide the readers with the latest achievements in the area of research. This book will surely be a source of knowledge to all interested and researching the field.

In the end, I would like to express my deep sense of gratitude to all the authors for meeting the set deadlines in completing and submitting their research chapters. I would also like to thank the publisher for the support offered to us throughout the course of the book. Finally, I extend my sincere thanks to my family for being a constant source of inspiration and encouragement.

                                                                                    Editor

Part 1

Study of Cellular Components and Analysis of Cellular Processes by Immunocytochemistry

Immunostaining of Voltage-Gated Ion Channels in Cell Lines and Neurons – Key Concepts and Potential Pitfalls

Elke Bocksteins[1,2,*], Andrew J. Shepherd[2,*],
Durga P. Mohapatra[2] and Dirk J. Snyders[1]
[1]Department of Biomedical Sciences, University of Antwerp, Antwerp;
[2]Department of Pharmacology, University of Iowa Carver
College of Medicine, Iowa City, Iowa,
[1]Belgium;
[2]USA

1. Introduction

Widely acknowledged to be the first person to make detailed studies of microscopic objects, the pioneering work of Antonie van Leeuwenhoek (1632-1723) provided the first glimpse of neurons when he observed sections of bovine optic nerves (van Zuylen, 1981). It was over a century later when the Italian scientist Camillo Golgi (1843-1926) developed *la reazione nera* ('the black reaction'), later dubbed the Golgi stain or Golgi method, in 1873. This approach revealed a hitherto unseen world, one brought to light in stunning fashion in the decades that followed by Santiago Ramón y Cajal (1852-1934), who refined Golgi's techniques and published numerous articles detailing the fine structure of the nervous system (Ramón y Cajal, 1995). The inescapable conclusion of this work - that the nervous system is composed of discrete cellular units, or neurons, that communicate with one another in a vast network - came to be known as 'the neuron doctrine.' It is from these initial findings that the sophisticated techniques available to today's scientists are descended.

One fundamental technical aspect of Ramón y Cajal's work still holds true for many contemporary techniques; the fine structure of tissues and cells must be stabilized, or 'fixed,' before any further processing or visualization can be carried out. Fixation is a crucial step in staining cells or tissues of any type. This also applies to immunocytochemistry (ICC) – the process of labeling proteins within cells and tissues with the use of specific antibodies. Despite being initially restricted to purified serum samples of ill-defined specificity, the possibilities of this technique were immediately apparent. Provided a specific antibody is available, it is possible to visualize almost any protein within the cell.

The development of the hybridoma technique to produce monoclonal antibodies against specific proteins or a specific amino acid sequence therein spurred a great increase in the practicality and popularity of immunostaining; theoretically limitless quantities of

---

* Both authors contributed equally to this work

antibodies of identical specificity could be generated using cell culture. This has doubtlessly accelerated the rate of discovery across all biomedical science disciplines. The information that can be gleaned from immunostaining is not restricted merely to whether a protein of interest is present within a given cell or tissue region. In principle, almost all of the processes that a protein goes through during its lifetime can be explored by ICC with the use of antibodies. This includes changes in the sub-cellular localization of the protein, developmental or pathological changes in expression and/or localization patterns, association or dissociation with other proteins, even post-translational modifications such as glycosylation and phosphorylation, as well as degradation at the end of a protein's life (Misonou et al., 2006; Ogawa et al., 2008; Yus-Nájera et al., 2003).

A yet more recent development was the exploitation of fluorescent proteins, which, when 'tagged' onto any protein of interest, enable the visualization of proteins within live cells. This enabled the dynamics and the temporal characteristics of biological processes to be scrutinized like never before, and shed more light on the processes that enable cells such as neurons to establish and maintain homeostasis. Such approaches can be (and often are) used in conjunction with conventional immunostaining, both techniques complementing one another. The unique properties of fluorescent proteins have more recently been harnessed in another way, in order to explore associations between proteins at a truly molecular level. Only when two proteins with appropriate fluorescent protein tags are in extremely close proximity to one another can the process of fluorescent resonance energy transfer (FRET) occur, a process that today's scientists exploit in order to explore protein-protein interactions.

These techniques have allowed us to peer deeper into the complexities of cells than ever before, and still doubtlessly have a great deal to tell us. It is these techniques, their optimization, applications and limitations, which we intend to describe here by focusing on ion channel proteins expressed in cell lines and neurons as our principal examples. Ion channels are a functionally diverse group of proteins with a multitude of closely-related subfamily members. Furthermore, the extensive control over subcellular localization, protein-protein interaction and post-translational modifications these proteins undergo means that they represent both the difficulties and fascinating insights that ICC has to offer.

2. Cell fixation

It is self-evident that one of the key aims of ICC is to label subcellular components in a way that most closely reflects their native confirmation within the cell. This enables investigators to make judgments based on their data that are likely to be in accordance with cells and tissues in their live state. This means that all metabolic processes must be terminated as quickly as possible, whilst preserving the morphology of cells and organelles, and the distribution of proteins therein. However, fixation that is too extensive may begin to alter the structure of proteins from their native confirmation so significantly that antibodies are no longer able to bind to their target sequence in the antigen/protein (epitope), resulting in decreased staining intensity and possibly an increase in the background 'noise' of the stain. On the other hand, incomplete fixation will not halt proteolytic and other enzymatic and metabolic processes sufficiently quickly. This can also reduce staining intensity and introduce abnormal staining patterns or fixation artifacts (Spector & Goldman, 2006). Of the main fixatives generally employed in ICC (glutaraldehyde, paraformaldehyde, methanol/acetone), paraformaldehyde is the most widely used, since it provides a good balance of thorough and rapid fixation of tissue/cells to preserve cell organelle morphology and the target epitope(s).

Glutaraldehyde is usually employed in electron microscopy studies because of its ability to preserve ultrastructural detail. However, the slow rate of cellular penetration and the extensive cross-linking that results from glutaraldehyde fixation make it a poor choice for most ICC studies, since this can impede antibody penetration and binding, thereby reducing specific signal and increasing background fluorescence.

Much more rapid fixation is possible with the use of cold acetone or methanol, and as a result they are employed more frequently in ICC than glutaraldehyde. However, they still present several potential problems for ICC studies. Methanol treatment leads to precipitation of proteins, which can disrupt the localization and/or detectability of certain antigens, and the permeabilization of cells that results from the solubilization of lipids may also result in the loss of more soluble antigens. Therefore, acetone or methanol fixation is usually employed only when rapid fixation and simultaneous permeabilization of cells is required, or for studies on particularly resilient antigens (e.g. the actin cytoskeleton, microtubules, intermediate filament networks (Spector & Goldman, 2006)).

Paraformaldehyde (or, to be more accurate, simply 'formaldehyde,' once present in its monomeric form in solution) is less effective than glutaraldehyde at cross-linking proteins by forming methylene bridges, meaning that while fixation and ultrastructural preservation aren't as robust as with glutaraldehyde, antigenicity is more likely to be maintained. Like glutaraldehyde, paraformaldehyde does not permeabilize cells, meaning that, unless surface or extracellular antigens are being detected, subsequent permeabilization will be necessary. It is also for this reason that preparing fresh solutions of formaldehyde from paraformaldehyde is recommended, so that the methanol that is often present in many commercially available formaldehyde solutions does not inadvertently permeabilize the cells. For fixation of cells, formaldehyde is typically used at 3-4% in phosphate-buffered saline, pH 7.4, and applied at room temperature for 15 minutes, or at 4°C for 30 minutes. The water in which the paraformaldehyde is dissolved must first be heated to approximately 60°C and subsequently brought to alkaline pH (approximately 10.5) for the paraformaldehyde to dissolve. Once the components of the phosphate buffer have been added, the pH must then be adjusted to 7.4 with hydrochloric acid. Opinions vary regarding the shelf life of formaldehyde prepared in this way; some generate small batches that are used within 24 hours, others generate larger quantities that are stored at 4°C for several months (Williams, 1997).

Several additives may also be included in the formaldehyde solution that can better preserve cell structure, the inclusion of 4% sucrose being a prime example. It is thought that the inclusion of a non-electrolyte such as sucrose inhibits the extraction of intracellular components during fixation, a factor that can improve staining quality, particularly for membrane proteins, and especially in cells such as neurons, with a large surface area-to-volume ratio. Similar effects have also been reported with the inclusion of 1-3mM $CaCl_2$ and/or $MgCl_2$ in the fixative.

If the cross-linking that formaldehyde introduces does preclude the binding of antibodies to their target, some cross-links can be broken, restoring more of the native conformation to cellular structures. Such processes are referred to as antigen retrieval, and are mainly enzyme or heat-mediated. Incubation of specimens at temperatures of approximately 60°C in an antigen retrieval buffer (most commonly sodium citrate, pH 6.0) is performed for several minutes, although the time needed for effective retrieval varies widely from antigen to antigen and must be determined experimentally (Daneshtalab, 2010; Shi et al., 2001). In short, antigen retrieval on cultured cells should only be considered as a last resort – if at all possible, an antibody that has been validated for use specifically in ICC (and for the antigen in the species of origin) should be used in order to minimize the chances of requiring any antigen retrieval steps.

3. Blocking and permeabilization

Following fixation, but prior to incubation with primary antibodies, cells must be permeabilized in order to allow antibodies to penetrate cellular membranes. Several different detergents are often employed to this end, including Triton X-100, Tween 20 and saponin. They are usually used at concentrations between 0.1 to 0.5%, and are either administered as a separate step prior to blocking, or co-administered with the blocking step. Some protocols specify that a small amount of detergent be kept in the buffers throughout blocking and antibody incubation steps, in order to prevent non-specific interaction of antibodies in different cellular compartments. There are certain exceptions where cells need not be permeabilized; those situations where an antibody is raised against an extracellular region of a membrane protein for instance, or perhaps a component of the extracellular matrix, do not require detergent treatment. In fact, such antibodies can be used to gain insight into the level of surface expression of a particular protein - relative to the total cellular expression - and in studies of trafficking to and from the plasma membrane.

Whether or not it occurs in tandem with solubilization, samples must also be 'blocked,' the process by which all potential antibody binding sites (or 'reactive sites') in the sample are blocked by incubation with a protein with no specific affinity. If this is not carried out, the level of non-specific binding of antibodies to 'off-target' sites will increase dramatically, elevating background signal. For immunostaining of cultured cells, there are two main categories of blocking agent. Normal serum is often used, particularly because it carries immunoglobulins that will bind to reactive sites throughout the sample, reducing the likelihood that the primary antibody will do the same when it is introduced. Optimally, the serum should be sourced from the animal in which the secondary antibody was raised (i.e., when using a goat anti-mouse IgG2a secondary, goat serum should be used). This should further reduce the possibility of excessive binding of primary or secondary antibodies to off-target sites, ensuring a clean, specific staining signal. Bovine serum albumin or nonfat dry milk are also often used. Again, the principle is similar; the primary antibody 'competes' against the proteins in the blocking solution for binding to all available epitopes. Theoretically, the affinity of the antibody for its target is far greater than the affinity of the blocking protein for that same region. Furthermore, the affinity of the antibody for all other, off-target epitopes should, on average, be equal to or lower than that of the blocking protein. The result should be that the antibody binds to only those regions against which it was designed, generating the desired high signal-to-noise ratio.

which the relevant hybridoma was grown, and into which the monoclonal antibody was secreted. These preparations haven't been affinity purified, which makes them less time-consuming and costly to generate, but a larger volume has to be used relative to the blocking solution than if the antibody is purified. The volume of antibody required to achieve good staining in these situations can sometimes be very high relative to the total volume of antibody and diluent, a problem that can sometimes significantly reduce the availability of blocking agent in the solution, potentially compromising binding specificity and increasing background noise. As a precaution, any dilutions of antibodies that will reduce the final concentration of blocking agent by more than 25% should instead be made with a more concentrated blocking agent to maintain a sufficient final concentration and blocking capacity.

As each antibody generated varies, even if it is targeted against the same epitope, it is difficult to determine with any certainty what concentration of a particular antibody will yield the most satisfactory results. Typically, a good starting range for an antibody on cultured cells is 1-2µg/ml, although this varies significantly, depending on the affinity of the antibody for its epitope. This final concentration may represent a dilution of only 1:5 or 1:10 of whole serum or tissue culture supernatant, but 1:500 or 1:1000 of a purified antibody. The best way to establish good staining with a new antibody is to perform a dilution series across a broad range. This should make it possible to find a dilution that provides optimal signal of the antigen of interest with minimal background. It should also be noted that once diluted to their working concentration, the stability of antibodies is reduced (this principle of maintaining stability of proteins by keeping the total protein concentration of a sample high is the rationale behind supplementing many purified antibody samples with additional protein, usually bovine serum albumin). It is for this reason that solutions of antibodies diluted to their working concentrations should not be kept for more than 24-48 hours.

4.4 Storage of antibodies

Antibodies are of course complex protein molecules, and should be treated in the same way as any other protein sample in order to minimize its degradation and preserve its biological activity. In the majority of cases, primary antibodies should be split into small volume aliquots and stored at -20°C, whether the solution contains an antifreeze agent such as glycerol or not. This will reduce the number of freeze/thaw cycles the antibody receives, as well as minimizing the risk of contamination. However, there are several important exceptions to this rule: although rare, monoclonal antibodies of the IgG3 isotype are especially vulnerable to forming precipitates upon thawing and should be stored at 4°C. Furthermore, antibodies conjugated to enzymes should also be stored at 4°C to maintain enzyme activity. Sodium azide is often included in antibody solutions as an antimicrobial agent. If desired, it can be added to solutions that don't already contain it to a final concentration of 0.01-0.05%. However, care must be taken that it is not used with secondary antibodies conjugated to horseradish peroxidase (HRP), as sodium azide is a potent inhibitor of HRP activity (Ortiz de Montellano et al., 1988).

4.5 Secondary antibodies: fluorescent versus enzyme-based methods

Secondary antibodies are used to indirectly detect an antigen to which the primary antibody has bound. Since the vast majority of commercial antibodies are derived from a small number of species, a wide, versatile range of secondary antibodies are available to detect any primary antibody. An additional advantage of using secondary antibodies is that several secondary antibody molecules can bind to each primary antibody molecule, amplifying the signal several-fold versus a directly conjugated antibody (see section 4.6.3). Broadly speaking, there are two classes of covalently-bound tag found on secondary antibodies – fluorophores and enzymes – of which the fluorophore family is more diverse and more widely used in ICC.

Since secondary antibodies are essentially generated by immunizing one species of animal with the antibody from another, they follow the standard 'host anti-target-tag' nomenclature i.e., goat anti-mouse IgG-Alexa Fluor 488. Understanding and paying attention to this nomenclature can also help to avoid a very common pitfall when performing stainings with multiple primary antibodies. Since goats are widely used to generate secondary antibodies, any staining that happens to use a goat polyclonal *primary* antibody must not employ *any* secondary antibodies of goat origin, since any anti-goat secondary antibody applied to the sample will also bind secondary antibodies *of goat origin* present in the sample, since they are also goat IgG molecules. This will generate unexpected labeling patterns and confounding results. This can be avoided by using secondary antibodies from another species (e.g. donkey) in these cases.

In addition to secondaries with class-wide specificity, there are also isotype-specific secondary antibodies, which, depending on the extent of their purification, should only recognize one particular isotype; mouse IgG2a, for example. Affinity purification (passing the secondary through a column containing off-target immunoglobulins) and cross-adsorption (effectively filtering the secondary antibody through serum proteins from other species to remove virtually all traces of non-specific binding) are used to generate so-called 'highly cross-adsorbed' secondaries. These allow the end user to simultaneously label several different antigens in the same specimen, primary antibodies permitting. For instance, using isotype-specific secondary antibodies, each labeled with a fluorophore of a different emission wavelength, it is routinely possible to visualize labeling of a rabbit polyclonal, a mouse monoclonal IgG1, IgG2a and IgG2b antibodies, all within the same sample. This of course assumes that the appropriate excitation source and filters are available on the microscope (see section 5).

The number of fluorophore wavelengths available easily exceeds the total number of widely-used antibody classes and subtypes. Commercial suppliers have their own proprietary versions, but the emission wavelengths generally fall into 5 main classes: blue fluorophores (e.g. Alexa Fluor/DyLight 350, 405, AMCA), green (Alexa Fluor/DyLight 488, FITC), yellow-red (DyLight 550, Alexa Fluor 546, 555, Cy3, TRITC), red (DyLight 633, 650, Alexa Fluor 633, 647, Cy5) and near-IR-infrared (Dylight 680, 750, 800, Alexa Fluor 680, 750, Cy5.5).

The principles of isotype specificity also apply to those secondary antibodies conjugated to enzymes (typically horseradish peroxidase (HRP) or alkaline phosphatase (AP) and, more rarely, glucose oxidase), but the detection of antibody binding differs significantly from the fluorescent conjugate approach. In this scenario, a secondary antibody conjugated to HRP or AP (or perhaps a biotinylated secondary binding an avidin-HRP or avidin-AP conjugate) is incubated with a substrate solution which is converted into a colored precipitate in the immediate vicinity of the enzyme. These additional recruitment and binding stages mean that the amplification of the initial signal is greater than with direct conjugates, offering much greater sensitivity. In addition, the precipitates generated from the substrate solutions (or 'chromogens') are observable with a conventional visible light microscope, eliminating the need for fluorescence light sources and optics. In addition, the insoluble deposits left behind from an enzyme's activity should theoretically preclude any interaction of subsequent primary antibodies with that region, meaning that multiple primary antibodies from the same species could be detected, albeit sequentially, in the same specimen (Grube, 2004). There are, however, also some disadvantages to this approach when compared to fluorescence-based strategies. Multiple labeling is possible, since several different substrates are available for each enzyme – red, brown/black and purple colors are possible with the NovaRED, DAB/Ni and VIP substrates for HRP, respectively, as are red, blue and blue/violet with the Red, Blue and BCIP/NBT substrates for AP, but if versatility and routine triple or quadruple labeling are required, fluorescence becomes the more viable option. The additional incubation steps that the enzyme conjugate approach requires take substantially more time than incubation with a fluorophore-conjugated secondary, and there is also the issue (specifically with HRP conjugates) of endogenous peroxidases in cells and tissues which, unless 'quenched' by prior incubation with hydrogen peroxide, may contribute to high levels of background staining.

4.6 Kits and alternative strategies

Despite the vast array of antibodies now available, it is highly likely that investigators will eventually encounter a scenario where the staining that needs to be carried out doesn't appear to be compatible with the antibodies that are available. Detailed in this section are three approaches that can be used to circumvent some of these issues.

4.6.1 'Mouse-on-mouse' kits

Multiple antibody stainings are a common requirement in studies involving ICC, whether it is used to discover potential interactions between two or more proteins, to label different cell subtypes in a mixed population, or merely to confirm the specificity of one antibody relative to another. All too often, however, the popularity of mouse monoclonal antibodies and the mouse as a model organism means that investigators will be left with no option but to attempt to stain cells of mouse origin with a mouse antibody. This can generate problems when secondary detection methods are used, since a goat anti-mouse IgG secondary antibody may fail to discriminate between the primary antibody IgG molecules and the endogenous IgG molecules expressed by the cells, increasing non-specific staining. There are several commercial kits available designed for these situations that use proprietary blocking solutions or conjugation steps to reduce or eliminate this background staining. However, these so-called 'mouse-on-mouse' kits are of little use when multiple antibodies of the same isotype need to be applied to the same sample, an obstacle that the following two alternative strategies can overcome.

4.6.2 Zenon labeling

In many ways, mouse monoclonal antibody technology has become a victim of its own success; frequently there aren't antibodies available in distinct species or isotypes, which precludes specific detection of each signal. Zenon labeling, however, can overcome these limitations. In this system, the target specific antibody is incubated with fluorophore-conjugated Fab fragments specific for the Fc region of the antibody to be labeled (e.g. anti-mouse IgG1 Fab fragments conjugated to Alexa Fluor 488). Following a further incubation with a non-specific IgG (to complex any remaining unbound Fab fragments), the labeled antibody can now be applied to cells in tandem with any other primary antibodies, irrespective of their isotype. The approach is similar in principle to incubating an antibody with a covalently reactive label (and generates fluorescence comparable to that seen with such direct conjugates; see section 4.6.3), however, it affords greater flexibility in terms of switching fluorophore conjugates and no affinity purification of the sample is required post-conjugation (van Duijnhoven et al., 2005).

4.6.3 Direct conjugates

When multiple antibodies of the same isotype or species need to be applied to cells, another means of solving the issue is to switch to a directly conjugated version of the antibody, which are often commercially available (for example, instead of detecting a mouse IgG1 anti-MAP2 primary with a goat anti-mouse IgG1-Alexa Fluor 488, simply apply a mouse IgG1 anti-MAP2 conjugated to Alexa Fluor 488). There are also commercially available kits that can covalently label even relatively small amounts of antibody with a fluorophore, as little as 10-20μg. These approaches can help circumvent issues of isotype availability, but there are some disadvantages. Aside from the additional expense and the limitation of only being able to visualize that antigen in one wavelength, the signal from a direct conjugate will not usually be as intense as with conventional secondary detection, since there hasn't been an additional round of signal amplification (see section 4.5). This means that a higher concentration of antibody may be required to generate signals of sufficient strength, but this may also boost non-specific staining which poses problems, especially if the goal is to detect low abundance proteins. Biotinylated primary antibodies can help overcome some of these amplification issues – subsequent incubation of a sample labeled with a biotinylated primary with an avidin or streptavidin-fluorophore conjugate will boost the signal relative to a fluorophore-conjugated primary, but there are several issues with this system also. Correct blocking of avidin-binding sites using an avidin/biotin blocking kit is almost always required in these circumstances, and the blocking procedure may need to be further altered, since blocking buffers containing nonfat dry milk contain endogenous biotin, a factor that could prevent effective blocking and/or labeling.

5. Fluorescence microscopy

Fluorescent molecules are a useful and convenient means of visualizing protein(s)/peptide(s) of interest within a cell and therefore are widely used in ICC. However, fluorescence-based stainings must be interpreted with caution, since the use of fluorescent molecules – especially when using multiple fluorophores together – has a few disadvantages.

5.1 Bleed-through

One of the major pitfalls with fluorescence-based stainings is bleed-through. Bleed-through occurs when the emission wavelength of one fluorophore is detected in the emission detection channel of another fluorophore (Fig. 1). This can happen in either direction, i.e.

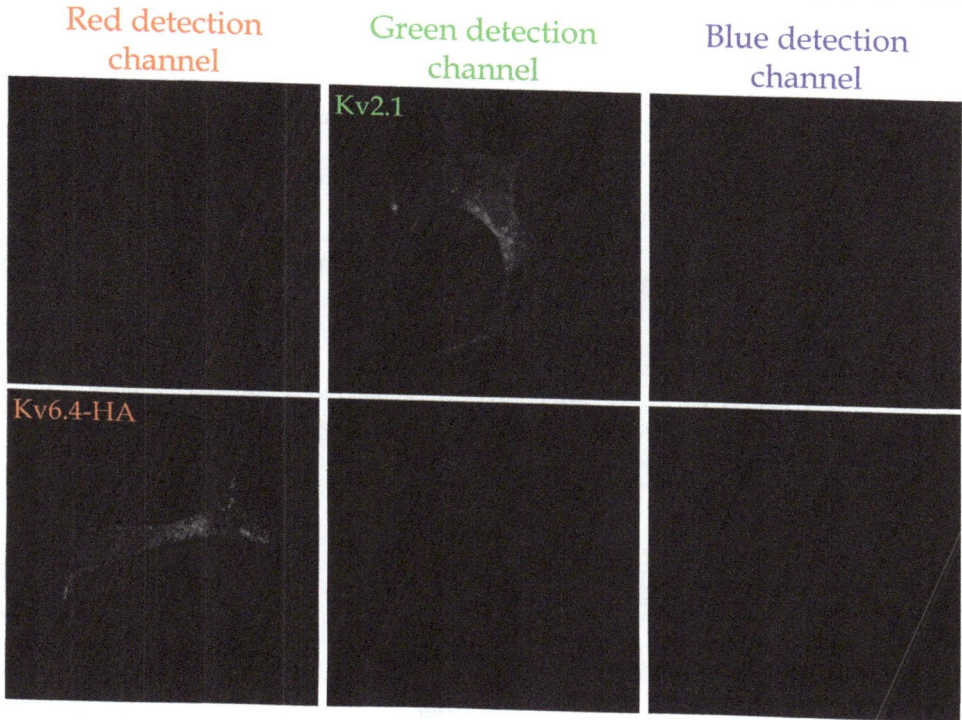

Fig. 1. Bleed-through of the fluorescent signal with immunofluorescence-based detection of voltage-gated K⁺ (Kv) channel subunits. In the upper row, the expression of Kv2.1 in a HEK293 cell has been fluorescently detected by incubating with a Kv2.1-specific antibody (1μg/ml; K89 from NeuroMab, UC Davis, CA, USA) followed by a green Alexa Fluor 488 labeled goat anti-mouse IgG1 secondary antibody (1/1600 dilution, Invitrogen, San Diego, CA, USA). The lower row represents a HEK293 cell expressing HA-labeled Kv6.4, which was detected by incubating with a HA-specific antibody (4 μg/ml; 12CA5 from Roche Diagnostics, Basel, Switzerland) followed by a red Alexa Fluor 555 labeled goat anti-mouse IgG2b secondary antibody (1/1600 dilution, Invitrogen). In each row, the left, middle and right image represents the red, green and blue detection channel, respectively. Note the slight bleed-through of the green fluorescent Kv2.1 signal in the blue detection channel (upper row) and the more pronounced bleed-through of the red fluorescent Kv6.4-HA signal in the green detection channel (lower row).

bleed-through from a shorter wavelength into a longer wavelength's detection channel and vice versa. Due to bleed-through, results can be interpreted incorrectly; it can be mistakenly believed that the protein of interest was detected in its wavelength detection channel, when in reality only the emission of a fluorophore used to observe another protein was detected. Bleed-through can be minimized by using fluorophores where the emission wavelengths are widely separated from each other; when one protein of interest is detected at a shorter wavelength (350 nm for example) and the other protein of interest is investigated in a longer wavelength detection channel (650 nm for example), bleed-through of emission of one wavelength in the other detection channel will not occur. However, detecting proteins in those shorter and longer wavelengths is only possible when the detected protein is present in abundance and the antibodies used are providing a strong signal, since weaker fluorescent signals are more difficult to observe in longer wavelengths such as 650 nm. In addition, the far red and infrared detection channels (i.e. 650 nm and above) are not especially common detection channels on most conventional fluorescent microscopes. Although bleed-through can happen in either direction, bleed-through usually occurs from a lower emission wavelength detection channel into a higher emission wavelength detection channel. Therefore, it may be better to start imaging with higher wavelengths and continue imaging toward lower wavelengths when multiple fluorphores are used to minimize bleed-through. Bleed-through can also be minimized by minimizing the exposure time of the fluorophore to the lowest necessary level; minimizing the exposure time would also decrease the excitation of the fluorophore causing the bleed-through. If possible, the specificity of the emission spectrum reaching the camera/eyepieces can be increased by employing bandpass filters specific for the fluorophore in question rather than the less specific and less costly short or long-pass filters.

5.2 Background and non-specific signals

Another major pitfall of fluorescence-based staining is background fluorescence. This could be caused by autofluorescence of the sample, non-specific binding of the antibody in the case of immunostainings and/or non-specific excitation of another fluorophore by the selected excitation wavelength when multiple fluorophores are used. Such background fluorescence is a particular problem when the fluorescent signal of interest is relatively weak. With stronger fluorescent signals of interest the microscope settings can easily be changed (i.e. lower laser strength and lower gain) to minimize background fluorescence. However, with a weaker fluorescent signal, reducing the microscope settings will mask the signal of interest. In this case trying to obtain a stronger fluorescent signal of interest is the only way to discriminate between background fluorescence and the fluorescent signal of interest. An increased fluorescent signal of interest can be achieved by using a primary antibody that exhibits less non-specific binding and/or provides a stronger signal by having a stronger binding affinity to the protein of interest and/or by using another detection channel (i.e. fluorescent signals can be easier to detect in shorter wavelengths). Since secondary antibodies can also exhibit non-specific binding, secondary antibodies can also contribute to background fluorescence. To determine whether secondary antibodies are also causing background fluorescence, the sample can be incubated with the secondary antibody alone; any remaining fluorescence is by definition background fluorescence. To avoid such background fluorescence, it may help to optimize the dilution of the secondary antibody, or switch to an alternative secondary antibody.

5.3 Discerning the signals

Due to limitations in microscope resolution, it is difficult to discern the presence of a few proteins at the plasma membrane (PM) when the majority resides in the endoplasmic reticulum (ER). This could be solved by making z-stack images on a confocal microscope. These are generated by sequentially focusing on a thin horizontal section of the cell, and generating a final composite of the stack in each image. However, the cost of a confocal microscope is almost five times a conventional microscope and is therefore not always an option. Another way to determine the presence of a few proteins at the PM when the majority is localized in the ER is by performing surface staining prior to cell permeabilization. However, surface staining can also be a challenge; just a few thousand copies of a (trans)membrane protein such as ion channels may still have significant functional contributions to the cell membrane properties, but this level of expression may be difficult to detect using immuno-based surface staining.

5.4 Focusing level

To prevent misinterpretation of the results using fluorescence microscopy it is also important to know the level within the cell at which the microscope is focused. For example, the voltage-gated K+ (Kv) channel Kv2.1 demonstrates a clustered membrane staining pattern both natively in neurons and in heterologous expression systems (Mohapatra et al., 2008; Scannevin et al., 1996; Vacher et al., 2008). However, this is only visible when focused on the cell membrane (Fig. 2, left). When the focus lays more on intracellular cell compartments/organelles, these Kv2.1 clusters are no longer detectable and the Kv2.1 membrane localization is only visible as a colored 'ring' around the edge of the cell (Fig. 2, right).

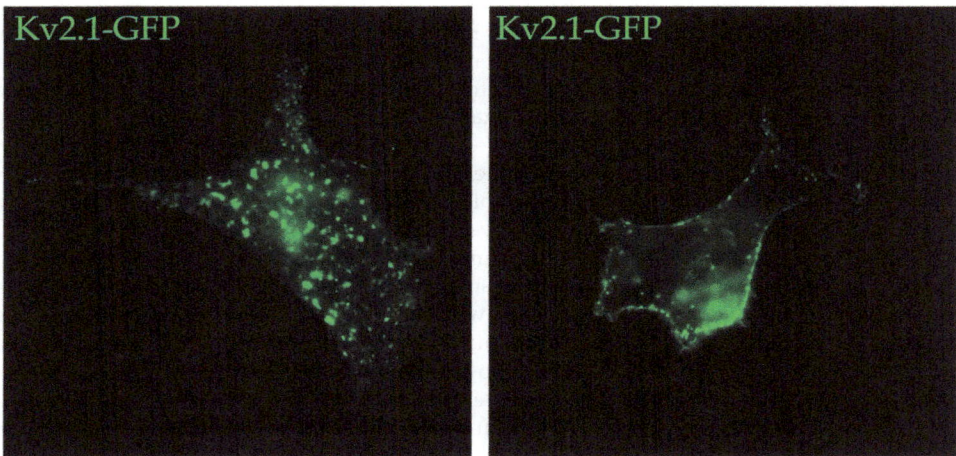

Fig. 2. Visualization of the plasma membrane (PM) localization of the voltage-gated K+ (Kv) subunit Kv2.1 depends on the level in the cell at which the microscope is focused. Both left and right images show a representative HEK293 cell expressing GFP-labeled Kv2.1. When the focus in the cell is on the PM, the clustered PM localization of Kv2.1 is readily detectable (left image) while the Kv2.1 PM localization is only noticeable as a colored 'ring' around the edge of the cell when the focal plane lies deeper in the cell (right image).

5.5 Bleaching and weakening of fluorescent signals

A last point that has to be taken into account when using fluorescent molecules to detect proteins of interest is bleaching. Fluorophores are highly light sensitive and exposure of those fluorescent molecules to light cause bleaching of the fluorophore. Because of this, weak signals could disappear before detection, resulting in misinterpretation of the results. To avoid this, exposure of the fluorophores to light has to be minimized which includes storing the fluorophores in a dark place, covering the samples during incubation with fluorophore conjugates and minimizing the exposure of the fluorophores to light when images are taken.

6. Alternative approaches

Because the use of antibodies to investigate the (co-)localization of proteins experimentally is usually not that easy as it is in theory, it is sometimes more convenient to use genetically-encoded fluorescent proteins to define the (sub-)cellular expression of the protein(s)/peptide(s) of interest and to define specific cellular compartments/organelles.

6.1 Genetically-encoded labels

To examine the (co-)localization of one or more proteins of interest in cells, those proteins can be labeled with a genetically-encoded fluorescent label, such as green fluorescent protein (GFP) or its yellow (YFP), cyan (CFP), blue (BFP) or red (RFP) variants, in which a mutation in the GFP chromophore is introduced to change the color. To obtain such genetically-encoded fluorescent labeled proteins, the cDNA of the fluorescent label and that of the protein of interest have to be cloned in frame with each other.

The use of genetically-encoded protein labels has a few major advantages compared to the use of immunofluorescent detection of the protein of interest. Using genetically-encoded labeled proteins allows experiments to be performed in live cells, which eliminates possible artifacts that could occur as a result of cell fixation. In addition, this opens the opportunity to perform live cell imaging - in which changes in a protein's localization can be tracked in real-time within the same cell. The use of genetically-encoded labeled proteins also permits less protein to be expressed in order to be able to visualize the protein of interest in cells, since the fluorescent signal of genetically-encoded labels is stronger than that of immunofluorescence-based stainings. For example, at least 1 µg of the plasmid containing the Kv channel Kv2.1 cDNA has to be transfected to observe good clustered plasma membrane staining when antibodies are used while only 100 ng GFP-labeled Kv2.1 has to be transfected to obtain a similar Kv2.1 plasma membrane staining. Because less protein has to be expressed, possible artifacts caused by an overload of the cell transcription/translational machinery are minimized. Using genetically-encoded labeled proteins also eliminate non-specific/background staining caused by non-specific binding of the used antibodies in immunofluorescence-based stainings.

As with any experimental technique, the use of genetically-encoded labeled proteins also has some disadvantages which have to be taken into account. GFP and its variants are fairly large proteins (~ 30 kDa) which are added to the protein of interest. This could result in changes in the native properties of the protein or in native protein-protein interactions. Therefore, one always has to verify that the properties of the fluorescently-labeled fusion protein have not significantly been changed compared to the non-labeled protein. In the case of ion channels, this can be easily determined based on the electrophysiological properties of the channels using the patch clamp technique. Even if the native properties of a genetically-encoded fluorescent-labeled protein have not been changed, such labeled proteins cannot readily be used to investigate the localization of native proteins in native cells (such as neurons), since those genetically-encoded labeled proteins always have to be transfected into cells or introduced into the animal (i.e. GFP knock-in mice).

6.2 Genetically-encoded marker proteins to identify cellular compartments

In addition to the fluorescent labels used to localize a protein of interest in live cells, different cellular compartments can also be identified by transfecting genetically-encoded fluorescent proteins. Those proteins are so modified that they are targeted to a certain cellular compartment using established signal sequences.

DsRED-ER is a genetically-encoded fluorescent protein that stains the endoplasmic reticulum (ER) red when expressed in cells. This ER marker was produced by cloning the first 17 amino acids of calreticulin (the ER signal sequence) in frame with the DsRED sequence and adding the ER retention signal KDEL behind the DsRED sequence (Ottschytsch et al., 2002). Co-transfecting this ER marker with a protein of interest into cells can reveal whether this protein of interest is located in the ER or not, as demonstrated in figure 3. Co-transfection of DsRED-ER with the GFP-labeled silent Kv subunit Kv6.4 demonstrated a maximal overlap of the green GFP staining pattern with the red DsRED-ER staining pattern, indicating that this Kv6.4-GFP subunit is retained in the ER. In contrast, the green staining pattern of Kv2.1-GFP shows minimal overlap with the DsRED-ER staining, indicating that this Kv2.1 channel is not localized in the ER.

The genetically-encoded fluorescent protein mtGFP labels the mitochondria green when transfected into cells. This mitochondrial marker was obtained by consecutively subcloning the sequence encoding the N-terminal 31 amino acids of the precursor of subunit VIII of cytochrome c oxidase (COX8) which is an endogenous mitochondrial transmembrane protein and therefore acts as the mitochondrial targeting sequence, the sequence encoding the hemagglutinin HA1 epitope and the sequence encoding GFP in frame with each other (Rizutto et al., 1995). mtGFP can be used to visualize the mitochondria and to investigate whether a protein of interest is localized to the mitochondria or not. Figure 4 demonstrates this; no overlap could be observed between the red clustered membrane staining pattern of Kv2.1-RFP with the green staining pattern of mtGFP, indicating that this Kv channel is not located in the mitochondria. Mitochondria can also be visualized with the use of mtPericam that labels the mitochondria green after transfecting into cells. In addition, mtPericam, due to its Ca^{2+} sensitivity, can be used to investigate changes in the intracellular mitochondrial Ca^{2+} concentration; its fluorescence intensity changes with changes in intracellular mitochondrial Ca^{2+} concentrations. mtPericam is a chimeric protein consisting of (consecutively): the N-terminal 12 amino acid presequence of subunit IV of cytochrome c oxidase (COX4), M13 which is a 26 amino acid-long peptide derived from the calmodulin (CaM)-binding region of the skeletal muscle light-chain kinase, circularly permuted GFP (cpGFP) in which the N-and C-terminus of GFP are linked to each other, and CaM (Nagai et al., 2001).

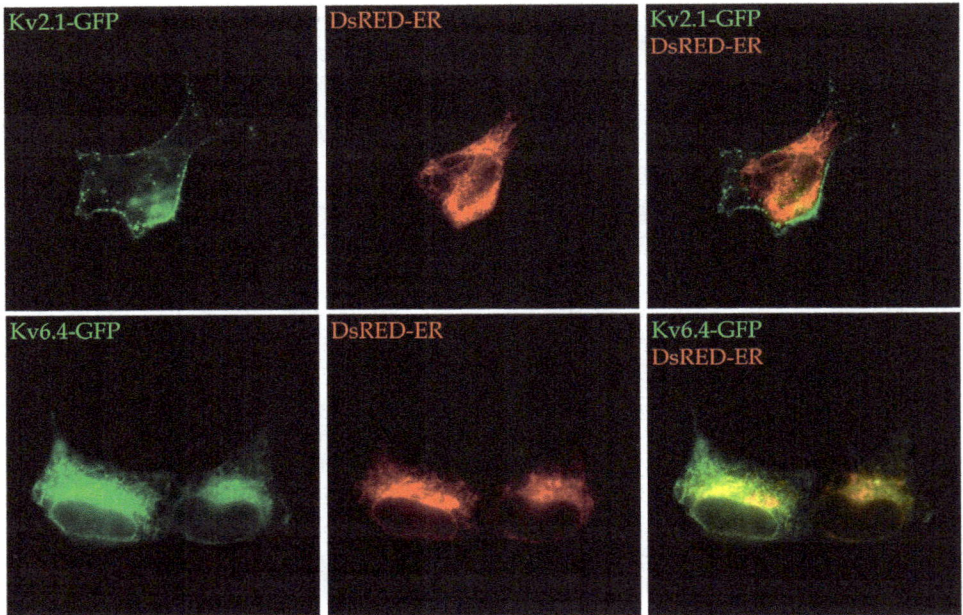

Fig. 3. Localization of GFP-labeled voltage-gated K⁺ (Kv) channels in the endoplasmatic reticulum (ER) using the ER marker DsRED-ER. The upper and lower rows represent typical images of GFP-labeled Kv2.1 and Kv6.4 subunits in HEK293 cells. The left, middle and right image represent the green GFP fluorescence, the red DsRED-ER fluorescence and the overlay of both fluorescent signals, respectively. The green fluorescent staining pattern of Kv2.1-GFP demonstrates minimal overlap with the red DsRED-ER staining pattern, indicating that Kv2.1-GFP is not located in the ER. Kv6.4-GFP, on the other hand, is localized almost entirely in the ER which is noticeable in the maximal overlap of the green Kv6.4-GFP staining pattern with the red DsRED-ER staining pattern.

Lck-mCherry is a genetically-encoded fluorescent protein that stains the PM red after transfecting into cells. This PM marker is obtained by attaching the myristoylation/palmitylation motif from the Lck tyrosine kinase to the N-terminus of mCherry (Naumann et al., 2010). After co-transfecting Lck-mCherry with a protein of interest, it can be determined whether this protein of interest is localized at the PM or not. In figure 5, the PM localization of transiently transfected Kv2.1-GFP in HEK293 cells is confirmed by the maximal overlap of its green PM staining pattern with the red staining pattern of the PM marker Lck-mCherry. In addition, Kv6.4-GFP is not located at the PM since no overlap between the green staining pattern of Kv6.4-GFP and the red staining pattern of Lck-mCherry could be detected.

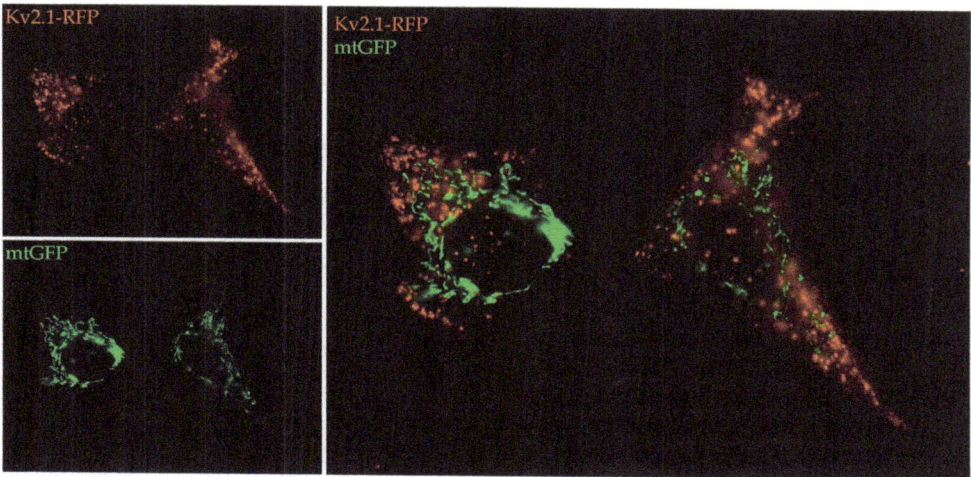

Fig. 4. Localization of the RFP-labeled voltage-gated K⁺ (Kv) channel Kv2.1 in comparison to the mitochondria by using the mitochondrial marker mtGFP. Shown are representative HEK293 cells expressing Kv2.1-RFP. The left upper and left lower images represent the red RFP fluorescence and the green mtGFP fluorescence, respectively. The image on the right represents the overlay of both fluorescent signals. Note that Kv2.1-RFP (consistent with its efficient trafficking to the plasma membrane) is not localized in the mitochondria since no overlap between the red Kv2.1-RFP staining pattern and the green mtGFP staining pattern is observed.

The above-mentioned genetically-encoded marker proteins for cellular compartments can be used in both heterologous expression systems such as HEK293 cells as well as native cells such as neurons by transfecting those marker proteins in the cells of interest. However, neurons are not as amenable to transfection as heterologous cells, and transfection could also change the native properties of the neuron. For this reason, using antibodies raised against endogenous marker proteins of a certain cellular compartment is a useful alternative that can be used to differentiate cellular compartments in neurons. In this case antibodies raised against the cis-Golgi matrix protein GM130, early endosome-associated protein EEA1, Calnexin and E-Cadherin could be used to indicate Golgi (Nakamura et al., 1995), endosomes (Mu et al., 1995), ER (Wada et al., 1991) and PM (Volk & Geiger, 1984), respectively.

In addition to the genetically-encoded cellular compartment/organelles markers, there also exist phenotypic markers to differentiate cell types. For example, using antibodies raised against the endogenously expressed microtubule-associated protein 2B (MAP2B) and glial fibrillary acidic protein (GFAP) allow investigators to determine whether a protein of interest is expressed in neurons (Izant & McIntosh, 1980) or in glial cells (Raff et al., 1979), respectively, since it is not always that easy to discriminate neuronal from glial cells based on their morphological properties alone. This is demonstrated in figure 6; Kv2.1 (represented by green fluorescence) is expressed in rat hippocampal neurons (marked by the blue staining pattern of MAP2B) and not in glial cells (marked by the red staining pattern of GFAP).

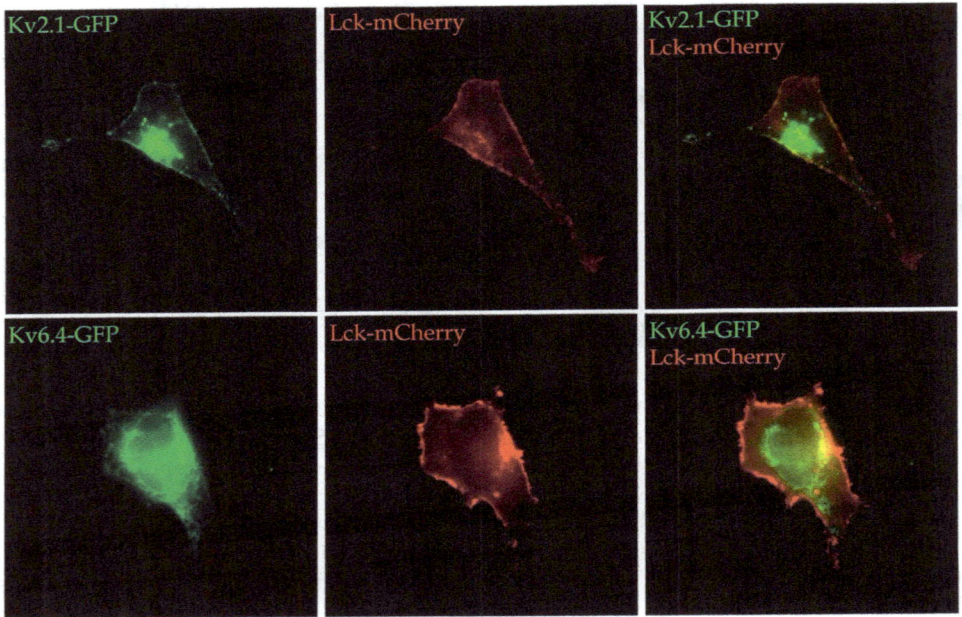

Fig. 5. Localization of GFP-labeled voltage-gated K+ (Kv) channels at the plasma membrane (PM) in HEK293 cells using the PM marker Lck-mCherry. The upper and lower rows represent typical images of GFP-labeled Kv2.1 and Kv6.4 subunits in HEK293 cells. The left, middle and right images on both rows represent the green GFP fluorescence, the red Lck-mCherry fluorescence and the overlay of both fluorescent signals, respectively. The green fluorescent signal of Kv2.1-GFP demonstrates substantial overlap with the red Lck-mCherry signal, indicating that Kv2.1-GFP is located at the PM. Kv6.4-GFP is not localized at the PM, which is exemplified by the minimal overlap of the green Kv6.4-GFP signal with the red Lck-mCherry signal.

6.3 Fluorescence Resonance Energy Transfer (FRET)

ICC can tell us about co-localization of proteins/peptides of interest due to the overlap of different fluorescent staining patterns, but it cannot tell us if both proteins are merely adjacent to one another, or if their physical proximity is the result of a functional interaction. Fluorescence resonance energy transfer (FRET) can be used to determine both co-localization and actual interaction between the proteins of interest (Kenworthy, 2001).

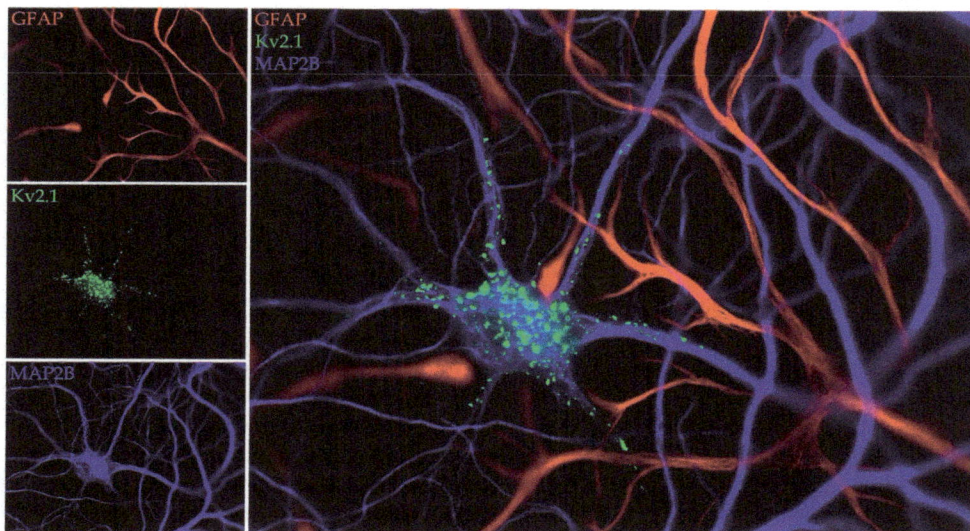

Fig. 6. Localization of the voltage-gated K⁺ (Kv) channel Kv2.1 in cultured rat hippocampal neurons using neuronal and glial cell marker antibodies. In the left column, the upper, middle and lower images represent the red GFAP fluorescence, the green Kv2.1 fluorescence and the blue MAP2B fluorescence, respectively. The image on the right represents the overlay of the three fluorescent signals. Note that Kv2.1 expression is confined to the blue hippocampal neuron and absent from the red glial cell processes.

FRET, also known as Förster resonance energy transfer, is the radiation-free transmission of energy from a donor molecule, which is the chromophore that initially absorbs the energy, to an acceptor molecule, the chromophore to which the energy is subsequently transferred (Förster, 1948). FRET can only occur between two chromophores when the following criteria have been fulfilled: (i) the donor and acceptor molecule must be situated in a range within 10 Å (1 nm) and 100 Å (10 nm) of each other since closer proximity (< 10 Å) will allow several other ways of energy transfer while more distance (> 100 Å) will prevent the radiation-free energy transmission, (ii) the donor emission spectrum has to have a certain spectral overlap with the acceptor absorption spectrum and (iii) the donor emission dipole moment has to be almost parallel with the acceptor absorption dipole moment.

When FRET occurs, light emitted by the acceptor will be observed after excitation of the sample with the donor excitation wavelength. However, bleed-through of donor emission light in the acceptor emission detection channel, as well as direct excitation of the acceptor chromophore by the donor excitation wavelength through to overlap in the donor and acceptor excitation spectra could result in detecting acceptor emission light even when no

FRET actually takes place. Therefore, it is better to determine the occurrence of FRET based on changes in the donor fluorescence. FRET results in both a decrease of the donor fluorescence lifetime and the donor fluorescence intensity, the latter of which is more convenient to determine.

Based on the donor fluorescence intensity, the efficiency with which FRET occurs can be determined with following equation:

$$FRET = \left(1 - \frac{f_{DA} - f_{background}}{f_{D} - f_{background}}\right) x \frac{1}{paired_{DA}} \tag{1}$$

In which f_{DA} represents the fluorescence intensity of the donor in the presence of the acceptor, f_D the fluorescence intensity of the donor in the absence of the acceptor, $f_{background}$ the background fluorescence intensity, and $paired_{DA}$ the fraction of paired donor-acceptor chromophores. The donor fluorescence intensity in the presence of the acceptor (f_{DA}) is determined by recording the donor emission light after exciting the sample with the donor excitation wavelength. By eliminating the acceptor chromophore which is obtained by bleaching the acceptor molecule with full power acceptor excitation wavelength laser light, the donor fluorescence intensity in the absence of the acceptor (f_D) is obtained. If FRET occurs, eliminating the acceptor chromophore by bleaching will result in an increase of the donor fluorescence intensity since the donor will no longer be able to transfer its energy to the acceptor molecule (Fig. 7). The background fluorescence ($f_{background}$) is determined by recording the donor emission light after bleaching of the donor chromophore with full power donor excitation wavelength laser light. When the fraction paired donor-acceptor chromophores ($paired_{DA}$) cannot be determined precisely (which is mostly the case when both chromophores are introduced in different proteins of which the interaction is tested) the acceptor chromophore has to be "over expressed" in comparison to the donor chromophore to ensure that this fraction is 1 and that the FRET efficiency can be determined adequately.

To prevent misinterpretation of the obtained results, two ratios have to be taken into account when determining FRET efficiencies; both the donor:acceptor intensity ratio and the background:signal intensity ratio have to be high enough. If the donor:acceptor intensity ratio is too small (<1), FRET efficiencies will be underestimated, while a high background:signal intensity ratio will overestimate the effective FRET efficiency.

Not all possible chromophore pairs can be used to perform FRET experiments. A good FRET donor-acceptor pair must fulfill the following criteria: (i) the acceptor and donor excitation spectra have to be separated enough so that the acceptor chromophore has been excited minimally after excitation of the donor molecule, (ii) the acceptor and donor emission spectra have to be separated sufficiently such that the emission spectra of the donor is detected minimally at the acceptor emission detection wavelength and (iii) the donor emission and acceptor excitation spectra have to demonstrate a maximal overlap with each other so that the acceptor chromophore is excited by the donor's emission wavelength after being excited with its own excitation wavelength. Due to these criteria, the optimal excitation and emission wavelength of both donor and acceptor molecule is not always at the maximum of their excitation and emission spectra.

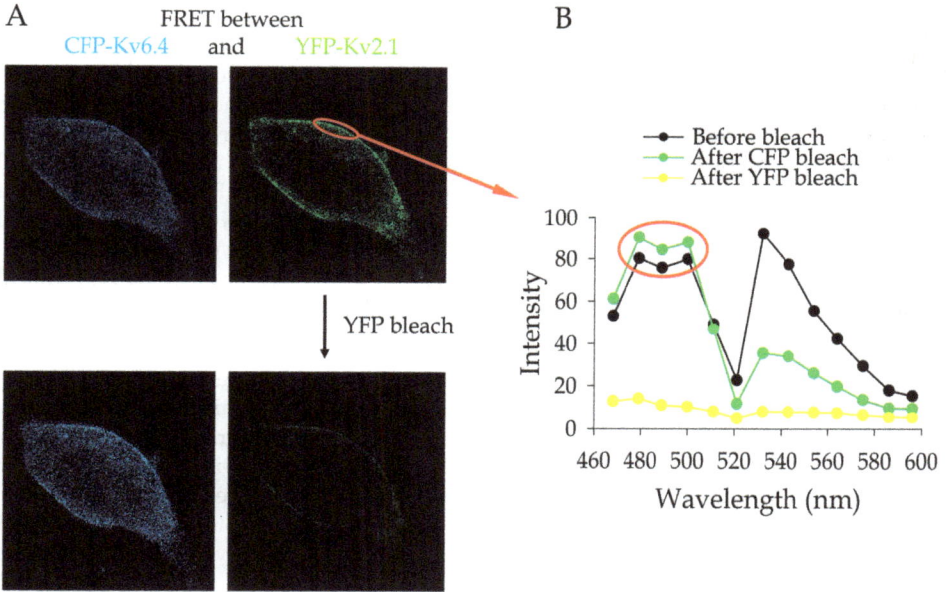

Fig. 7. Interaction between voltage-gated K+ (Kv) subunits determined by FRET using CFP and YFP-labeled Kv channel constructs. (A) Representative cell expressing both N-terminally labeled CFP-Kv6.4 and YFP-Kv2.1 subunits. The panels on the left and right show the CFP and YFP emission light after excitation with the CFP excitation wavelength (in this case 458 nm), respectively, while the top and bottom panels represent CFP and YFP emissions before and after bleaching of YFP in the whole cell, respectively. If FRET occurs, the CFP emission is underestimated since CFP transfers its emission light onto YFP which is relieved by eliminating YFP by bleaching with full power YFP excitation wavelength laser light (514 nm in this case); note the increased CFP intensity and the decreased YFP emission after YFP bleach. (B) Recorded emission spectra of the indicated region in panel A (red oval) before (black line) and after (green line) YFP bleaching. Note the increase in intensity in the CFP detection wavelengths (indicated by the red circle) and the significant decrease in the YFP detection wavelength intensity after YFP is bleached in the whole cell. Background signal (yellow line) was determined after fully bleaching CFP.

The donor-acceptor pair CFP-YFP fulfills these criteria reasonably and is therefore often used as a FRET donor-acceptor pair. Another good FRET donor-acceptor pair is the donor-acceptor pair BFP-GFP (Pollok & Heim, 1999). In addition to the genetically-encoded chromophores, organic cyanine dyes such as Cy3 and Cy5 which act as donor and acceptor chromophore, respectively, are also often used to perform FRET experiments. Those dyes have some advantages compared to the genetically-encoded chromophores (i.e. less spectral overlap between the donor emission and acceptor excitation spectrum is required and larger distances between both chromophores are possible) but these organic cyanine dyes have to be chemically linked to the proteins of interest which is a great disadvantage.

The position in which the acceptor/donor chromophore is introduced into the protein of interest is very important when performing FRET experiments. In addition to the fact that these chromophores could change the properties of the native protein significantly, it is also possible that those chromophores are introduced into the proteins at positions where the distance between both chromophores is too large to observe FRET. For example, when the interaction between voltage-gated K^+ subunits is investigated by FRET experiments, the subunits have to be labeled N-terminally and not C-terminally since no FRET could be observed when those subunits are labeled C-terminally although those subunits physically interact with each other in a functional tetrameric Kv channel (Bocksteins et al., 2009; Kobrinsky et al., 2006).

In addition to using FRET to investigate whether two proteins of interest only co-localize or actually interact with each other, FRET can also be used to ascertain whether the lack of functional membrane proteins at the PM is due to a trafficking or tetramerization deficiency (Bocksteins et al., 2009). For this purpose, the FRET efficiency in both a region that represents the PM and a region that represents the ER has to be determined. When a trafficking deficiency causes the lack of functional membrane proteins at the PM, FRET could still be observed in the ER while a tetramerization deficiency would also eliminate FRET in the ER. With this approach, it has been demonstrated that the reduced PM expression of mutant Kv2.1 channels was due to a tetramerization deficiency, since the observed reduced current amplitude of those mutant channels was accompanied by a significant reduction of FRET in the ER (Bocksteins et al., 2009).

7. Conclusion

ICC is a powerful, versatile tool for furthering our understanding of the expression, localization and interactions of proteins at the cellular level. This power has only been augmented by more recent developments in molecular biology, where insights into the behavior of proteins in live cells, or interactions on the atomic scale can now be investigated. Unfortunately, there are many ways in which the results of ICC can be sub-optimal or completely negative if attention isn't paid to certain key concepts, and an understanding of the principles involved isn't developed. Taking voltage-gated ion channels and neurons as our examples, we have explained the core principles behind the most widely-used techniques in ICC: how to correctly fix cells so that the structure of even complex transmembrane proteins will be preserved faithfully; the processes of blocking and permeabilization and the ways in which they can be modified to ensure the best possible results; the importance of the difference between monoclonal and polyclonal antibodies and how the signals from antibodies of interest should be validated; as well as the dilution and storage conditions that will minimize the loss of antibody activity. Secondary antibody options were detailed, explaining the strengths and weaknesses associated with the common approaches to detecting primary antibody signal. Alternative strategies, which can enable staining that wasn't previously possible, were outlined, as were the limitations and errors that are frequently encountered in fluorescence microscopy. Finally, we discussed approaches that can complement antibody-based studies, namely the transfection of fluorescent-labeled constructs targeted to specific sub-cellular compartments, and the use of FRET to discover if two proteins that appear in close proximity are physically interacting or not. The combined powers of these techniques still have a great deal to teach us across the life sciences as a whole, and will continue to be an integral part of studies therein for the foreseeable future.

8. Acknowledgments

We thank Dr. Yuriy Usachev and Dr. Stefan Strack, Department of Pharmacology, The University of Iowa Carver College of Medicine for kindly providing us with the mammalian expression plasmids containing mtGFP and Lck-mCherry cDNA, respectively.

9. References

Bocksteins, E., Labro, A.J., Mayeur, E., Bruyns, T., Timmermans, J.P., Adriaensen, D. & Snyders, D.J. (2009). Conserved negative charges in the N-terminal tetramerization domain mediate efficient assembly of Kv2.1 and Kv2.1/Kv6.4 channels. *The Journal of Biological Chemistry*, Vol. 284, No. 46 (November 2009), pp.31625-31634, ISSN 0021-9258.

Burry, R.W., (2010). *Immunocytochemsitry: A Practical Guide for Biomedical Research*. Springer, ISBN 978-1441913036, New York, New York.

Cambrioso, A. & Keating, P. (1992). Between Fact and Technique: The Beginnings of Hybridoma Technology. *Journal of the History of Biology*, Vol. 25, No. 2 (Summer 1992), pp. 175-230, ISSN 0022-5010.

Daneshtalab, N., Doré, J.J. & Smeda, J.S. (2010). Troubleshooting tissue specificity and antibody selection: Procedures in immunohistochemical studies. *Journal of Pharmacological and Toxicological Methods*, Vol. 61, No. 2 (March-April 2010), pp. 127-135, ISSN 1056-8719.

Förster, T. (1948). Zwischenmolekulare Energiewanderung und Fluoreszenz. *Annalen der Physik*, Vol. 437, No.1-2. pp. 55-75. ISSN 1521-3889.

Grube, D. (2004) Constants and variables in immunohistochemistry. *Archives of Histology and Cytology*, Vol. 67, No. 2 (June 2004), pp. 115-134. ISSN 0914-9465.

Izant, J.G. & McIntosh, J.R. (1980). Microtubule-associated proteins: a monoclonal antibody to MAP2 binds to differentiated neurons. *Proceedings of the National Academy of Sciences of the USA*, Vol. 77, No.8 (August 1980), pp. 4741-4745, ISSN 0027-8424.

Kenworthy, A.K. (2001). Imaging protein-protein interactions using fluorescence resonance energy transfer microscopy. *Methods*, Vol. 24, No. 3 (July 2001). pp. 289-296, ISSN 1046-2023.

Kobrinsky, E., Stevens, L., Kazmi, Y., Wray, D. & Soldatov, N.M. (2006). Molecular rearrangements of the Kv2.1 potassium channel termini associated with voltage gating. *The Journal of Biological Chemistry*, Vol. 281, No. 28 (July 2006), pp. 19233-19240, ISSN 0021-9258.

Misonou, H., Menegola, M., Buchwalder, L., Park, E.W., Meredith, A., Rhodes, K.J., Aldrich, R.W. & Trimmer, J.S. (2006). Immunolocalization of the Ca2+-activated K+ channel Slo1 in axons and nerve terminals of mammalian brain and cultured neurons. *The Journal of Comparative Neurology*, Vol. 496, No. 3 (May 2006), pp. 289-302, ISSN 1096-9861.

Mohapatra, D.P., Siino, D.F. & Trimmer, J.S. (2008). Interdomain cytoplasmic interactions govern the intracellular trafficking, gating, and modulation of the Kv2.1 channel. *The Journal of Neuroscience*, Vol. 28, No. 19 (May 2008), pp. 4982-4994, ISSN 0270-6474.

Mu, F.T., Callaghan, J.M., Steele-Mortimer, O., Stenmark, H., Parton, R.G., Campbell, P.L., McCluskey, J., Yeo, J.P., Tock, E.P. & Toh, B.H. (1995). EEA1, an early endosome-associated protein. EEA1 is a conserved alpha-helical peripheral membrane protein flanked by cysteine "fingers" and contains a calmodulin-binding IQ motif. *The Journal of Biological Chemistry*, Vol. 270, No. 22 (June 1995), pp. 13503-13511, ISSN 0021-9258.

Nagai, T., Sawano, A., Park, E.S. & Miyawaki, A. (2001). Circularly permuted green fluorescent proteins engineered to sense Ca2+. *Proceedings of the National Academy of Sciences of the USA*, Vol. 98, No. 6 (March 2001), pp. 3197-3202, ISSN 0027-8424.

Nakamura, N., Rabouille, C., Watson, R., Nilsson, T., Hui, N., Slusarewicz, P., Kries, T.E. & Warren, G. (1995). Characterization of a cis-Glogi matrix protein, GM130. *The Journal of Cell Biology*, Vol. 131, No. 6 pt. 2 (December 1995), pp. 1715-1726, ISSN 0021-9525.

Naumann, U., Cameroni, E., Pruenster, M., Mahabaleshwar, H., Raz, E., Zerwes, H.G., Rot, A. & Thelen M. (2010). CXCR7 functions as a scavenger for CXCL12 and CXCL11. *PLoS One*, Vol. 5, No. 2 (February 2010), pp. e9175.

Ogawa, Y., Horresh, I., Trimmer, J.S., Bredt, D.S., Peles, E., & Rasband, M.N. (2008). Postsynaptic density-93 clusters Kv1 channels at axon initial segments independently of Caspr2. *The Journal of Neuroscience*, Vol. 28, No. 22 (May 2008), pp. 5731-5739, ISSN 0270-6474.

Ortiz de Montellano P.R., David S.K., Ator, M.A., & Tew, D. (1988). Mechanism-based inactivation of horseradish peroxidase by sodium azide. Formation of meso-azidoprotoporphyrin IX. *Biochemistry*, Vol. 27, No. 15 (July 1988), pp. 5470-5476, ISSN 0006-2960.

Ottschytsch, N., Raes, A., Van Hoorick, D. & Snyders, D.J. (2002). Obligatory heterotetramerization of three previously uncharacterized Kv channel alpha-subunits identified in the human genome. *Proceedings of the National Academy of Sciences of the USA*, Vol. 99, No. 12 (June 2002), pp. 7986-7991, ISSN 0027-8424.

Pollok, B.A. & Heim, R. (1999). Using GFP in FRET-based applications. *Trends in Cell Biology*, Vol. 9, No. 2 (February 1999), pp. 57-60, 0962-8924.

Raff, M.C., Fields, K.L., Hakomori, S.I., Mirsky, R., Pruss, R.M. & Winter, J. (1979). Cell-type-specific markers for distinguishing and studying neurons and the major classes of glial cells in culture. *Brain Research*, Vol. 174, No. 2 (October 1979), pp.283-308, ISSN 0006-8993.

Ramon y Cajal, S., (1995). *Histology of the Nervous System of Man and Vertebrates (History of Neuroscience, No 6)*. Oxford University Press, ISBN 978-0195074017, New York, New York.

Rhodes, K.J. & Trimmer, J.S. (2006). Antibodies as valuable neuroscience research tools versus reagents of mass distraction. *The Journal of Neuroscience*, Vol. 26, No. 31 (August 2006), pp. 8017-8020, ISSN 0270-6474.

Rizutto, R., Brini, M., Pizzo, P., Murgia, M. & Pozzan, T. (1995). Chimeric green fluorescent protein as a tool for visualizing subcellular organelles in living cells. *Current Biology*, Vol. 5, No. 6, (June 1995), pp. 635-642, ISSN 0960-9822.

Saper, C.B. (2005). An open letter to our readers on the use of antibodies. *The Journal of Comparative Neurology*, Vol. 493, No. 4 (December 2005), pp. 477-478, ISSN 1096-9861.

Scannevin, R.H., Murakoshi, H., Rhodes, K.J. & Trimmer, J.S. (1996). Identification of a cytoplasmic domain important in the polarized expression and clustering of the Kv2.1 K+ channel. *The Journal of Cell Biology*, Vol. 135, No. 6, Part 1 (December 1996), pp. 1619-1632, ISSN 0021-9525.

Shi, S.R., Cote, R.J. & Taylor, C.R. (2001). Antigen retrieval techniques: current perspectives. *The Journal of Histochemistry and Cytochemistry*, Vol. 49, No. 8, (August 2001), pp. 931-937, ISSN 0022-1554.

Spector, D.L. & Goldman, R.D. (Eds.). (2006). *Basic methods in Microscopy: Protocols and Concepts from Cells: A Laboratory Manual*. Cold Spring Harbor Laboratory Press, ISBN 087969751-2, Cold Spring Harbor, New York.

Vacher, H., Mohapatra, D.P. & Trimmer, J.S. (2008). Localization and targeting of voltage-dependent ion channels in mammalian central neurons. *Physiological Reviews*, Vol. 88, No. 4 (October 2008), pp. 1407-1447, ISSN 0031-9333.

van Duijnhoven, M.W., van de Kerkhof, P.C., Pasch, M.C., Muys, L. & van Erp, P.E. (2005). The combination of the Zenon labeling technique and microscopic image analysis to study cell populations in normal and psoriatic epidermis. *Journal of Cutaneous Pathology*, Vol. 32, No. 3 (March 2005), pp. 212-219, ISSN 1600-0560.

Van Zuylen, J. (1981). The microscopes of Antoni van Leeuwenhoek. *Journal of Microscopy*, Vol. 121, No. 3 (March 1981), pp. 309-328, ISSN 0022-2720.

Volk, T. & Geiger, B. (1984). A 135-kd membrane protein of intercellular adherens junctions. *EMBO J*, Vol. 3, No. 10 (October 1984), pp. 2249-2260, ISSN 0261-4189.

Wada, I., Rindress, D., Cameron, P.H., Ou, W.J., Doherty, J.J. 2nd, Louvard, D., Bell, A.W., Dignard, D., Thomas, D.Y. & Bergeron, J.J. (1991). SSR alpha and associated calnexin are major calcium binding proteins of the endoplasmic reticulum membrane. *The Journal of Biological Chemistry*, Vol. 266, No. 29 (October 1991), pp. 19599-19610, ISSN 0021-9258.

Williams, J.H., Mepham, B.L. & Wright, D.H. (1997). Tissue preparation for immunocytochemistry. *Journal of Clinical Pathology*, Vol. 50, No. 5, (May 1997), pp. 422-428, ISSN 0021-9746.

Yus-Nájera, E., Muñoz, A., Salvador, N., Jensen, B.S., Rasmussen, H.B., & Defelipe, J., Villarroel, A. (2003). Localization of KCNQ5 in the normal and epileptic human temporal neocortex and hippocampal formation. *Neuroscience*, Vol. 120, No. 2 (February 2003), pp. 353-364, ISSN 0306-4522.

Optimizing Multiple Immunostaining of Neural Tissue

Araceli Diez-Fraile[1], Nico Van Hecke[1],
Christopher J. Guérin[2,3] and Katharina D'Herde[1]
[1]Department of Basic Medical Sciences, Ghent University, Ghent
[2]Department for Molecular Biomedical Research, VIB, Ghent
[3]Department of Biomedical Molecular Biology, Ghent University, Ghent
Belgium

1. Introduction

This chapter will discuss the optimization of a multiple immunostaining method for studying the distribution of calcium binding proteins in neural tissue. We discuss the different possibilities for brain collection and tissue preservation. Then we present a basic protocol for a multicolor immunostaining technique.

Interference from autofluorescence is one of the major problems in immunofluorescence analysis of neural tissue. This difficulty is experienced even with the use of laser-scanning microscopy, which can eliminate out of focus light but not the undesirable tissue-specific autofluorescence that is in focus. Therefore, we will center our discussion on different strategies to minimize neural tissue autofluorescence. The two greatest sources of autofluorescence discussed here are those induced by aldehyde fixation and by the fluorescent pigment lipofuscin, which accumulates with age in the cytoplasm of cells of the central nervous system. Finally, it is important to determine whether any treatments used to minimize autofluorescence might quench specific fluorescent compounds used to label the proteins of interest. We present the results of multi-label fluorescence immunostaining of cryostat sections of 20 μm of the human inferior colliculus. We conclude that the use of chemical agents to reduce autofluorescence and quench lipofuscin-specific fluorescence helps to maximize the fluorescent signal-to-noise ratio in immunocytochemical studies of fixed neural tissues.

2. Antigen preservation of human brain material

Preservation of antigens in brain is one of the major methodological problems in the immunohistochemical studies of this tissue. Various premortem events have a negative effect on antigen preservation in human brain tissue, such as prolonged agonal state, hypoxia, acidosis, fever and seizures. Factors related to postmortem events are the time between death and fixation (postmortem delay), temperature variations in cadaver storage, freshness and composition of the fixative solutions.

Postmortem delay is the most important contributor to the quality difference of immuno-stained neural tissues between experimentally controlled animal models and human (Hilbig et al., 2004). Indeed, optimal protocols for rapidly fixing central nervous tissue, such as intravascular infusion of fixative into deeply anesthetized animals (Selever et al., 2011), are essential for prevention of enzymatic degradation and antigen loss (Martin & O'Callaghan, 1995). However, human tissues are usually obtained only several hours after death. Fresh human brain tissue is sometimes obtained during surgery, but such tissue is usually pathologically altered and cannot be used to draw conclusions about normal physiology. Brain banks remain the best source of high quality postmortem human brain tissue for neuropathological studies and related research (McKee, 1999; Waldvogel et al., 2006; Ravid & Grinberg, 2008; Waldvogel et al., 2008).

For qualitative immunohistochemical analysis, a postmortem delay longer than one day is usually acceptable, but two crucial controls are needed if protein expression levels are to be estimated from immunofluorescence results. Because different proteins degrade at different rates following death (De Groot et al., 1995; Fodor et al., 2002; Hilbig et al., 2004) it is advised to biochemically assess the degredation profile of the antigen(s) of interest prior to immunochemical studies. Indeed, different antigens are affected differently by postmortem catabolic processes (De Groot et al., 1995; Fodor et al., 2002; Hilbig et al., 2004). Second, the specimens must be rigorously matched, e.g., for age, sex, postmortem delay and fixation time.

The most common method for preserving antigen for fluorescence microscopy is immersion in cross-linking agents, such as formaldehyde and glutaraldehyde. In general, 4% formaldehyde in PBS (pH 7.4) made from freshly depolymerized paraformaldehyde is preferred, because treatment with glutaraldehyde may mask amine-containing epitopes, making immunostaining impossible (Willingham, 1999). It is sufficient to incubate small tissue pieces in the fixative overnight at 4°C. For minimal effects on antibody binding, brain material should be fixed as short a time as possible consistent with good preservation. Fixation for not more than two weeks with refreshment of the fixative every third day has been shown to be successful in immunohistochemical studies (Romijn et al., 1999). Freshly made paraformaldehyde induces a lower background autofluorescence than glutaraldehyde or pre-prepared formalin. How fixative-induced autofluorescence can be reduced is the subject of section 6.1.

3. Storage and sectioning of tissue

Cryoprotection is used to prevent freezing artifacts and loss of tissue architecture. Once properly fixed, the brain tissue is immersed in a graded series of sucrose solutions (10%, 20% and 30% in PBS pH 7.4) at 4°C, until the tissue sinks to the bottom of the container. The sucrose acts as a partial dehydrating substance that protects the tissue against freezing artifacts caused by ice crystal formation when the tissue is frozen (Tokuyasu, 1973). Morphology and immunoreactivity are generally preserved for up to six months when the brain tissue is stored at 4°C in buffered 30% sucrose (supplemented with 0.05% NaN_3 to prevent growth of bacteria) (Romijn et al., 1999). For longer-term storage, freezing in a 30% sucrose solution in PBS at –80°C has been shown to reasonably preserve the hypothalamus and to cause only minimal freeze artifacts (Romijn et al., 1999). Gathering even a small number of human brain tissue samples usually takes weeks or months. Therefore, to perform a reliable, comparative immunocytochemical study, the brain tissue usually needs to be stored frozen.

Before sectioning, the embedded tissue blocks are equilibrated at –20°C for at least one hour. Section thickness is important. Sections that are too thin may be very fragile and tear when handled in long protocols, whereas thick sections have a higher intensity of autofluorescence (Del Castillo et al., 1989). After sectioning, the tissues are rehydrated for at least 40 minutes in TBS or PBS buffer (pH 7.4) before further treatment. Rehydration is done on floating tissue sections or on sections adhered to coated slides.

4. Multi-label fluorescence immunostaining

Several antigens can be simultaneously detected in the same tissue section by incubating unconjugated primary antibodies of different origin, followed by labeling with fluorophore-conjugated secondary antibodies against the primary antibodies. This strategy can be more sensitive than directly labeled primary antibodies due to signal amplification through multiple secondary antibody reactions with different antigenic sites on the primary antibody. Higher signal-to-noise ratios can also be achieved through the use of new generation fluorophores such as AlexaFluor and Dylight, which are brighter and more photostable than their older counterparts (e.g. FITC, TRITC). Various controls should be run to verify the specificity of the antibodies in order to avoid false-positive results due to non-specific binding or to recognition of epitopes shared by different molecules. Negative controls should include incubation in non-immune mouse IgG or normal rabbit serum instead of the primary antibody, omission of the primary and/or the secondary antibody and pre-adsorption of the antibody with an excess of purified antigen. Controls can also be sections from other tissues or brain regions known to contain (positive control) or to lack (negative control) the protein of interest. Staining is deemed specific only if both negative and positive controls give unequivocal results. Neurons and other cells present in neural tissue can also be visualized by combining the previous protocol with neuron specific reagents such as Neurotrace™ fluorescent Nissl stains and DAPI or Hoechst 33258 DNA stains.

When primary antibodies from different species are not available, methods for multiple immunolabeling with primary antibodies originating from the same host species can be used (Buchwalow et al., 2005; Frisch et al., 2011). Moreover, when greater sensitivity is required amplification methods such as the tyramide signal amplification method or avidin-biotin complexing can be employed (Brouns et al., 2002; Tóth & Mezey, 2007; Bratthauer, 2010).

In Section 2 we have mentioned the importance of using rigorously matching brains in quantitative studies. In addition, the matching samples need to be processed and evaluated simultaneously using the same solutions. In this way, large differences in the intensity between conditions can be shown. However, the results should be confirmed by truly quantitative methods, such as western blotting technique.

5. Confocal laser scanning microscopy

The use of confocal fluorescence microscopy to analyze multiple immunolabeled sections has several advantages over conventional methods. First, confocal microscopy can filter out-of-focus light to yield sharp, high-contrast images of thick samples. In addition, by collecting 2-D images from different planes in the tissue, a volume of tissue can be reconstructed in 3-D. Another advantage is that since most systems are equipped with multiple lasers of different wavelengths, they can excite a set of fluorescent probes spanning nearly the full range of visible wavelengths, allowing multiprobe imaging of the specimen (Dailey et al., 1999).

Nevertheless, confocal microscopy will not eliminate non-specific background fluorescence, which is the main problem in the immunohistochemical analysis of adult human neural tissue. Indeed, imaging of neural tissue is usually not satisfactory due to the broad emission spectrum of the autofluorescence which can overlap the specific fluorescent labels. An example of autofluorescence in a human brain tissue sample is shown in Figure 1. Therefore, the protocol should be optimized for multiple fluorescent immunolabeling by treating the samples as described in section 6 in order to improve the visualization of the fluorochromes of interest.

6. Non-specific background fluorescence or autofluorescence

Autofluorescence is a common problem in aged human brain tissue and when present, it decreases the signal-to-noise ratio of specific labeling. The main causes of autofluorescence in the central nervous system are artifacts due to fixation and endogenous causes such as the fluorescent pigment lipofuscin, which accumulates as cells age.

6.1 Evaluation of fixative-induced fluorescence

The most common method to fix tissue for fluorescence microscopy is immersion in cross-linking agents, such as formaldehyde and glutaraldehyde. The aldehydes form covalent bonds between adjacent amine-containing groups through a Schiff acid-base reaction, which leads to formation of methylene bridges and other types of links (Collins & Goldsmith, 1981; Fox et al., 1985). However, unreacted aldehyde groups fluoresce efficiently at the same wavelengths similar to the fluorescent probes employed for immunofluorescence assays. Since glutaraldehyde possesses two functional groups per molecule while formaldehyde has only one, background autofluorescence is more problematic for tissues fixed with glutaraldehyde. To keep the background autofluorescence as low as possible in immunocytochemistry, the preferred buffer for fixing tissue is 4% formaldehyde in PBS (pH 7.4) made from freshly depolymerized paraformaldehyde. Fixative-related autofluorescence is visualized as fairly uniform broad spectrum fluorescence and may be brighter in some cells than in others, depending on the presence of biogenic amines (Clancy & Cauller, 1998).

Fixative-induced autofluorescence may be circumvented by quenching unreacted aldehydes. A common method is to pretreat fixed sections with reducing agents, such as freshly prepared sodium borohydride ($NaBH_4$; 1% w/v) in PBS for 10–20 min at 4°C (Willingham 1983; Beisker et al., 1987; Tagliaferro et al., 1997; Figure 2) to reduce free aldehyde groups to alcohols. Care should be taken when using $NaBH_4$ solution because it is very caustic and can cause explosions due to release of hydrogen. Alternatively, fixed sections may be treated with exogenous amine-containing reagents, such as: 50 mM NH_4Cl in PBS for 5–10 min, or in 100–300 mM glycine in Tris buffer pH 7.4 or PBS for 1–2 h at 4°C (Callis 2010; Ngwenya et al., 2005). The best method should be determined empirically. These techniques can be used alone or sequentially. If the tissue is fragile, the glycine method is advised.

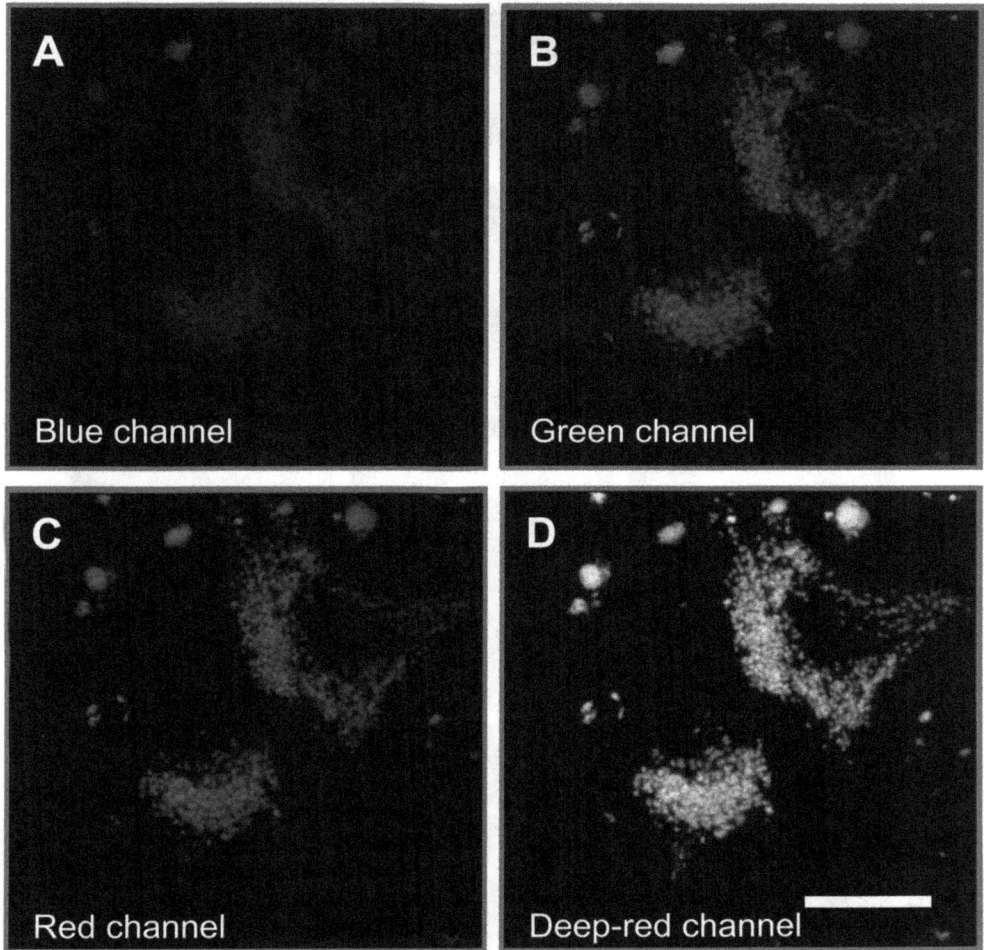

Fig. 1. Punctate autofluorescent structures in a 20-μm sagittal section from the human inferior colliculus exposed to the same laser power for the blue (A), green (B), red (C) and deep-red (D) channels. The lipofuscin-related autofluorescence is less visible in the blue channel and strongest in the deep-red channel. The wide overlap in the blue, green, red and deep-red channels supports previous studies showing that lipofuscin has broad spectral emission (Kikugawa et al., 1997; Schnell et al., 1999; Zhang et al., 2010). Magnification = 63x; scale bar = 20 μm

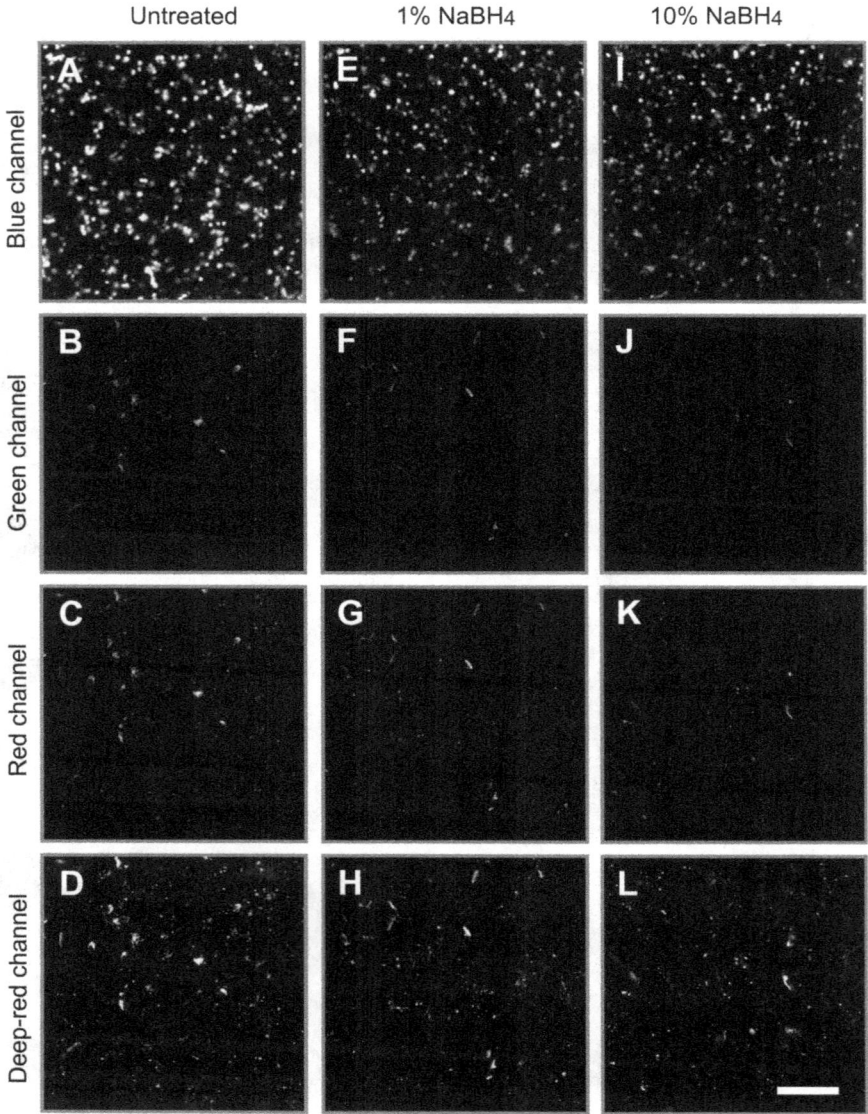

Fig. 2. Effect of incubation under control conditions (A–D), 1% (E–H) or 10% sodium borohydride (I–L) on the background fluorescence in 20-μm consecutive sections of the human inferior colliculus stained with DAPI and visualized in the blue (A, E, I), green (B,F,J), red (C, G, K), and deep-red (D, H, L) channels by confocal microscopy. Incubation in NaBH₄ resulted in a small decrease of the background autofluorescence in each channel, indicating that the un-reduced aldehydes induced by tissue fixation (4% paraformaldehyde in PBS pH 7.4 for two weeks at 4°C) were not substantially contributing to non-specific background. Magnification 20x; scale bar = 100 μm

6.2 Reduction of lipofuscin autofluorescence

Lipofuscin is a ubiquitous fluorescent pigment that accumulates with age in long-lived cells, such as neurons and cardiac myocytes (Brizzee et al., 1974). The topographical and temporal patterns of lipofuscin production vary throughout the brain (Mann & Yates, 1987; Peters et al., 1991). However, the development of lipofuscin generally begins during infancy, from the age of three or four months (Porta, 2002), becomes obvious in human neurons at the age of nine years (Sulzer et al., 2008), and is present in most cortical neurons and glial cells by the age of thirty years (Benavides et al., 2002). This pigment may occupy up to 75% of the perikaryon of some neurons in the aged brain (Terman & Brunk, 1998). Lipofuscin in neurons appears to be derived primarily from the breakdown of autophagocytosed mitochondria into a polymer of oxidatively modified protein residues bridged by acid and lipid residues (Terman & Brunk, 2004, 2006). Mature large lipofuscin granules are found mainly in the perinuclear area but may occasionally be present in the perikaryon, dendrites, axons and even presynaptic areas (Riga et al., 2006).

Under light microscopy, lipofuscin appears yellow-brown or translucent. Under the electron microscope, it appears as a homogeneously dark, irregular mass often containing small lipid droplets, and is surrounded by a single membrane (Harman, 1989). When visualized by a fluorescence or laser-scanning microscope, lipofuscin appears as tiny, punctate intracellular structures that are strongly fluorescent over a wide spectral range (excitation 350–580 nm; emission 400–650 nm). The excitation maximum, about 400 nm, yields a broad emission spectra for lipofuscin granules between 530 and 650 nm, which overlaps with the commonly used fluorophores (Hyden & Lidstrom, 1950; Eldred et al., 1982; Katz et al., 1984; Porta, 2002; Haralampus-Grynaviski et al., 2003; Double et al., 2008; Figure 1) and thereby interferes with specific immunofluorescent signals. When the fluorescence of antibody-fluorochrome complexes is weak, it is important to unequivocally discriminate between nonspecific (lipofuscin) and specific fluorescence. Four distinct lipofuscin fluorescence spectra have been reported in a sodium dodecyl sulfate extract of rat kidneys: blue fluorescence with excitation/emission maxima at 360/440 nm, respectively, greenish fluorescence at 460/530 nm, and yellow fluorescence at 400/620 nm and 580/620 nm. The yellow fluorescence at 400/620 nm accumulates with aging, and its excitation and emission maxima are close to those observed histochemically under a fluorescence microscope (Kikugawa et al., 1997).

Treatment of neural tissue with the lipophilic dye Sudan Black B and CuSO$_4$ have been consistently reported to be useful in quenching the yellow 400/620 nm fluorescence of the lipofuscin to very low levels without substantially affecting the specific fluorescence of the fluorochromes coupled to secondary antibodies (Kikugawa et al., 1997; Brehmer et al., 2004; Schnell et al., 1999; Romijn et al., 1999; Oliveira et al., 2010; Zhang et al., 2010). Moreover, the Autofluorescence Eliminator Reagent from Chemicon International reduces lipofuscin autofluorescence. This reagent consists of a solution of Sudan Black (and possibly other components) in ethanol. One drawback of treating tissues to reduce lipofuscin-related autofluorescence is reduction of the fluorescent signal from the introduced fluorophore. Therefore, it is important to check to what extent the treatment affects the fluorescent signal and choose the method accordingly.

An example of human serial sections of the inferior colliculus treated with Sudan Black B and CuSO$_4$ is shown in Figure 3. Sudan Black B treatment was done according to Kiernan

(1981), as cited in Romijn et al. (1999), who prescribed dissolving the dye by stirring 0.3% Sudan Black B in 70% ethanol at room temperature in the dark for two hours and leaving it standing overnight before filtering it, after which it can be stored in the dark at 4°C for at least two months. After immunolabeling the sections, the solution is applied for 10 minutes at room temperature and then the sections are rinsed rapidly eight times in TBS. Another protocol consists of treating the tissue with $CuSO_4$ (Schnell et al., 1999). After immunocytochemistry, the sections are taken out of TBS and dipped briefly in distilled water, followed by treatment with 1 mM $CuSO_4$ in 50 nM ammonium acetate buffer (pH 5.0) for 10–30 min, a brief dip in distilled water, and a return to TBS. The samples were coverslipped using 25 mg/ml DABCO dissolved in Mowiol^R 4-88 as mounting media. Application of $CuSO_4$ for 10 minutes after immunohistochemistry substantially reduced tissue autofluorescence while preserving the specific fluorochrome signals. The overall background fluorescence increased substantially when the samples were treated with Sudan Black B, especially when using the 543-nm laser (Figure 3, panel K). The increase in background fluorescence in the samples treated with Sudan Black B could be related to the DABCO antifading agent in the tissue mounting media, as previously observed by Romijn et al., 1999. The commercial product Vectashield and the antifading agent p-phenylene-diamine dissolved in glycerol also react similarly (Romijn et al., 1999). If Sudan Black B is used to decrease the lipofuscin-induced autofluorescence, Tris-buffered glycerol at a pH of 7.0–7.6 (not higher) is recommended as the best mounting medium.

7. Blocking non-specific binding of antibodies

The most commonly used method to reduce background staining due to non-immunological binding of the specific immune sera by hydrophobic interaction is to use a protein blocking solution before application of the primary antibody. The solution should contain proteins similar to those present in the secondary antibody but not resembling those in the primary antibody. Bovine serum albumin (BSA) is probably the most widely used blocking agent for reducing non-specific binding due to hydrophobic interaction, but non-fat dry milk, casein, fish skin gelatin, horse serum and goat serum have also been recommended (Duhamel & Johnson, 1985; Kaur et al., 2002). However, blocking agents can also cause problems. A decrease in Neurotrace™ fluorescence has been reported after incubating the samples in the presence of bovine serum albumin, non fat dried milk or horse serum (reported in the manufacturer's web page http://probes.invitrogen.com/media/pis/mp21480.pdf). An alternative to these reagents is fish skin gelatin. However, in our experience, bovine serum albumin (3%) or fish skin gelatin (2%) quenched the fluorescence of Neurotrace™ at 1/200 dilution, which made it necessary to use a higher concentration of Neurotrace™ (1/100) to achieve adequate staining in 20-μm sagittal sections of the human inferior colliculus (data not shown). Doubling of the deep-red Neurotrace™ concentration yielded correct fluorescence staining when blocking was done with BSA or FSG for the deep-red fluorochrome. No substantial differences in fluorescence intensity were observed when BSA or FSG was used for blocking for the other fluorochromes employed, i.e. AF488, AF586, and DAPI (Figure 4). Thus, blocking in a neutral protein solution such as BSA or FSG decreases background of primary and secondary antibodies successfully, but requires a higher concentration of Neurotrace™ for optimal staining. In previous experiments (data not shown) higher concentrations of blocking proteins also produced increased background staining, so this should be optimized

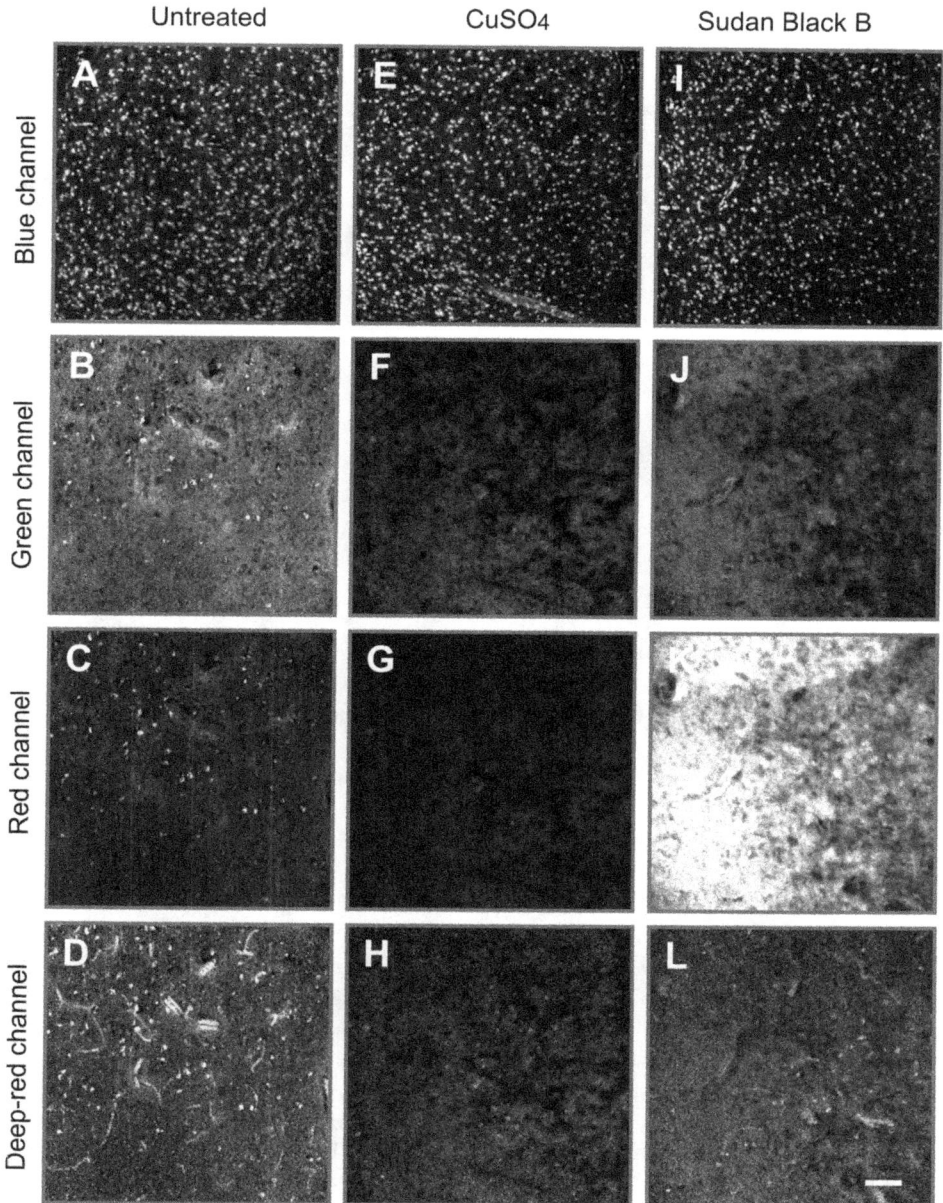

Fig. 3. Lipofuscin autofluorescence in sections of the human inferior colliculus under control conditions (A–D) and after treatment with 1 mM CuSO₄ in 50 mM ammonium acetate buffer (pH 5.0) (E–H) or 0.3% Sudan Black B in 70% ethanol (I–L) and visualized in the blue (A, E, I), green (B,F,J), red (C, G, K), and deep-red (D, H, L) channels by confocal microscopy. Magnification 20x; scale bar = 100 μm.

for different tissue types. In general beginning with 0.5% blocking protein is recommended and testing higher concentrations if necessary to block non-specific protein absorption.

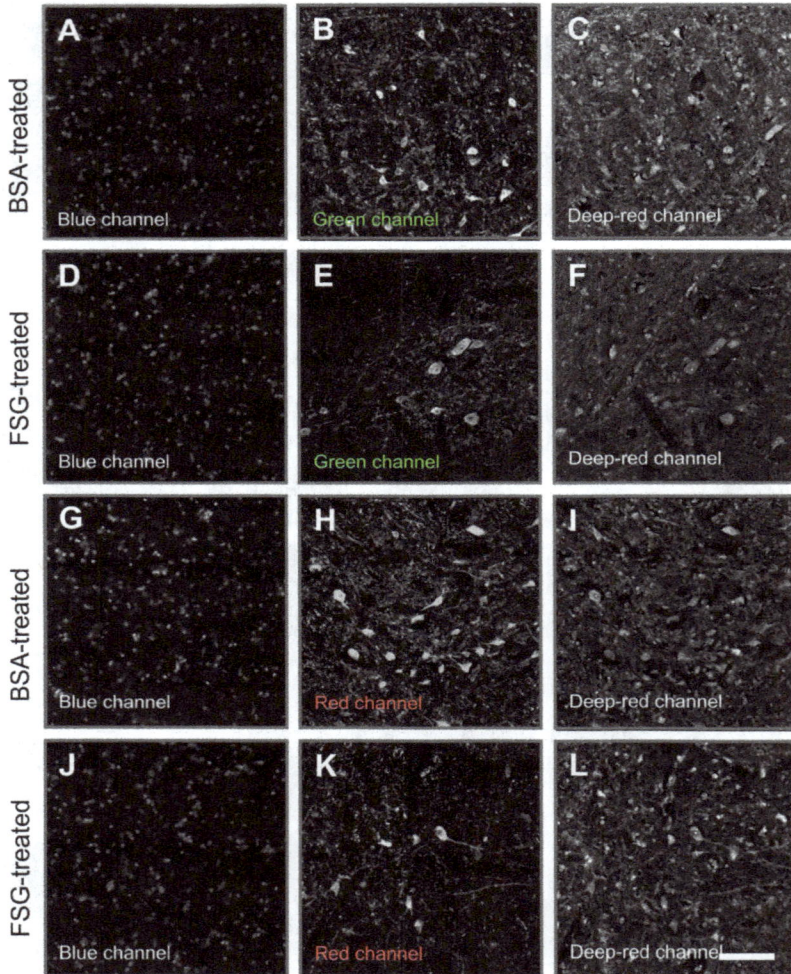

Fig. 4. Effect of bovine serum albumin (BSA) (A–C; G–I) and fish skin gelatin (FSG) (D–F; J–L) on the fluorescence intensity as visualized by confocal microscopy in the blue (A, D, G, J), green (B,E), red (H, K), and deep-red (C, F, I, L) channels. The sections were blocked with 2% BSA or 3% FSG before incubation with mouse anti-parvalbumin (2.2 µg/ml; P3088, Sigma) and appropriate secondary antibodies labeled with the fluorochrome AlexaFluor 488 (15 µg/ml; A11059, Invitrogen) or AlexaFluor 568 (15 µg/ml; A10037, Invitrogen). The samples were subsequently stained with Neurotrace™ (1/100 dilution) and DAPI (1 µg/ml). Lipofuscin autofluorescence was then reduced with $CuSO_4$ as previously reported (see section 6.2.). No differences were observed in the intensity of the various fluorochromes when blocked with BSA or FSG. Magnification 20x; scale bar = 100 µm

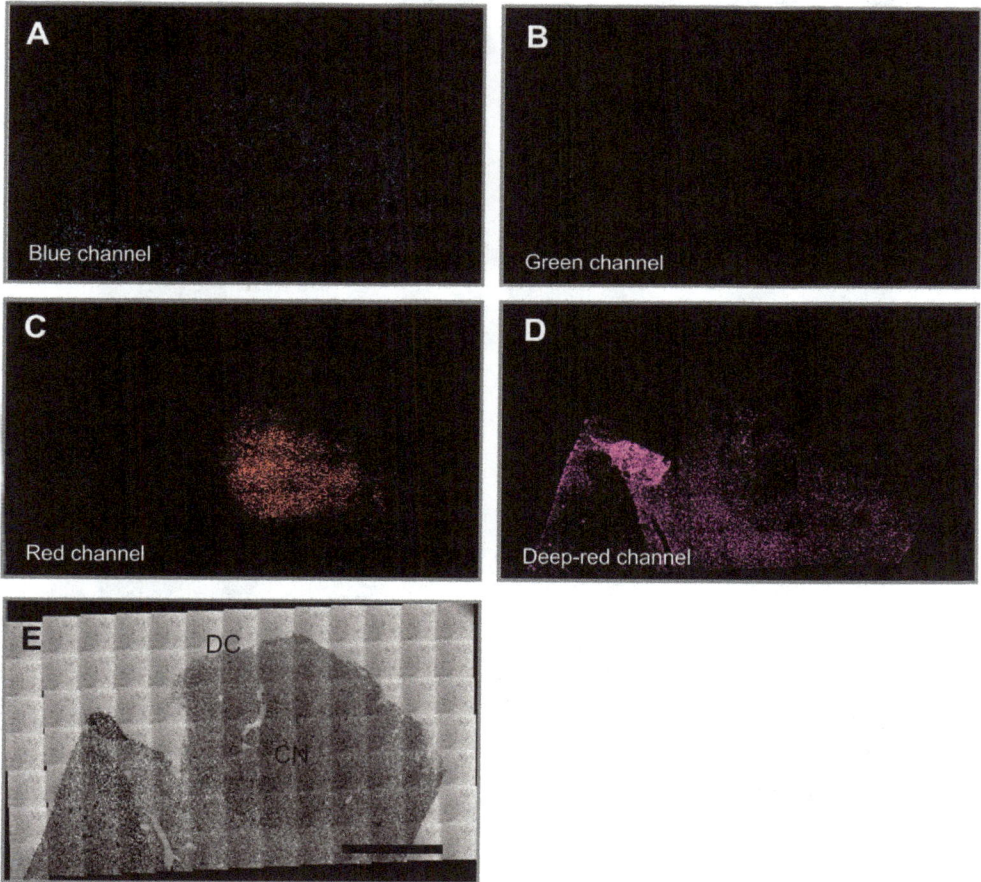

Fig. 5. Two dimensional reconstruction of confocal micrographs of a 20-μm sagittal section of the human inferior colliculus. The section was stained for the nuclei with DAPI (A), for calretinin with AlexaFluor 488 (B), for parvalbumin with AlexaFluor 568 (C), and for neurons with deep-red Neurotrace™ (D). A light microscopy image shows an overview of the inferior colliculus (E). Parvalbumin positive cells are located at the central nucleus (CN) of the inferior colliculus while calretinin positive cells are less abundant and located at the dorsal cortex (DC) as previously reported (Sharma et al., 2009; Tardif et al., 2003). Magnification 20x; bar= 2 mm

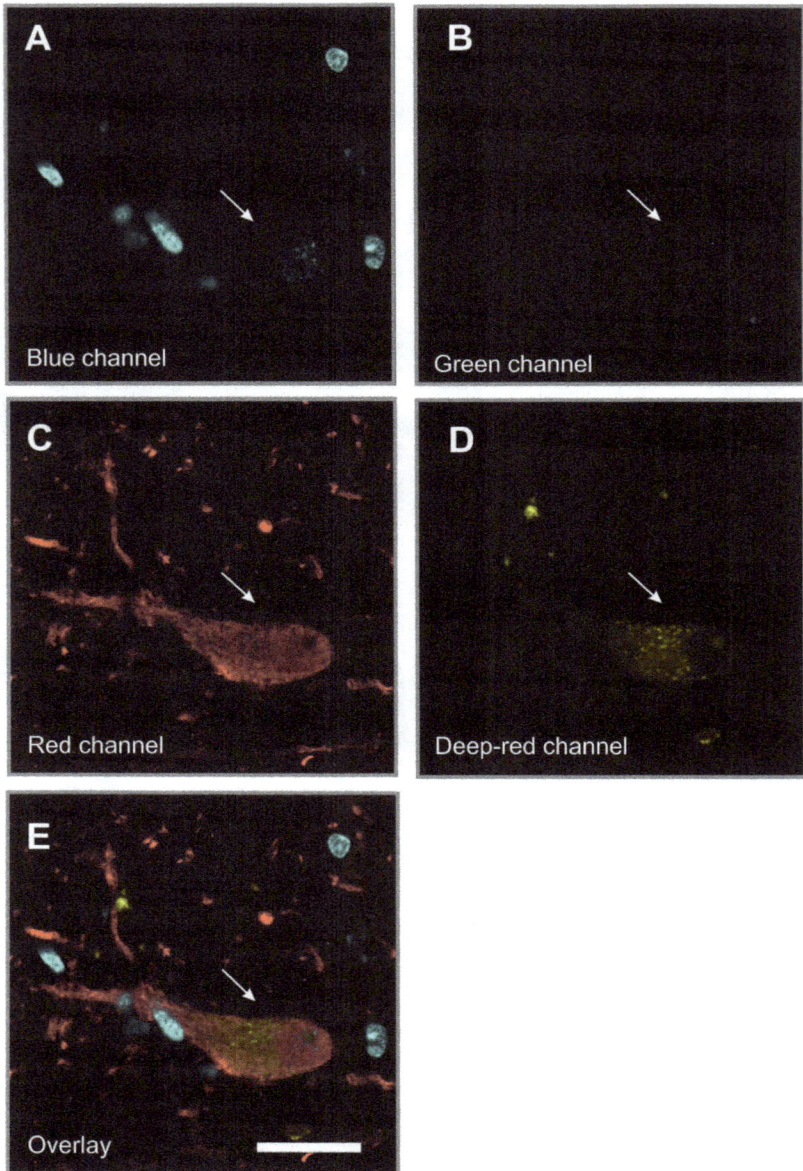

Fig. 6. A parvalbumin positive cell (arrow) located at the central nucleus of the human inferior colliculus visualized by confocal microscopy. The 20-µm sagittal section of the human inferior colliculus was stained for the nuclei with DAPI (A), for calretinin with AlexaFluor 488 (B), for parvalbumin with AlexaFluor 568 (C), and for neurons with deep-red NeurotraceTM (D). Panel E shows the overlay of panels A, B, C and D. Magnification 63x; bar= 20 µm

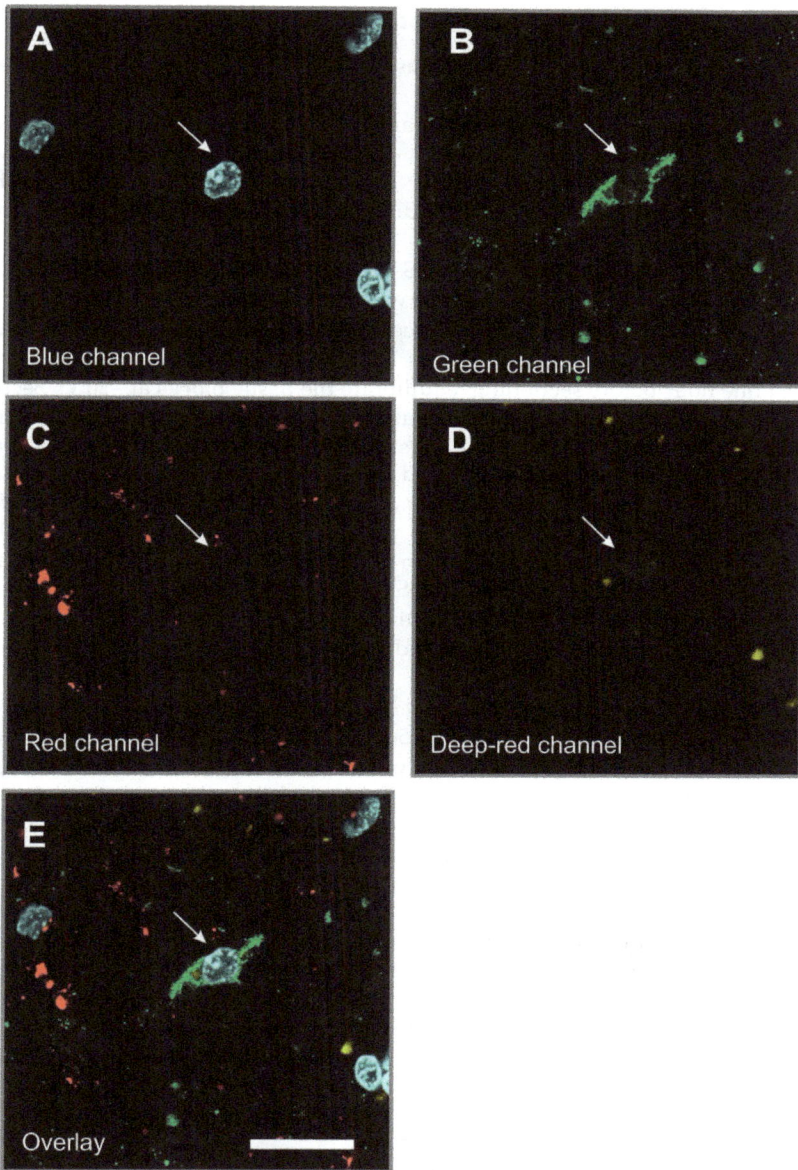

Fig. 7. A calretinin positive cell (arrow) located at the dorsal cortex of the human inferior colliculus visualized by confocal microscopy. The 20-μm sagittal section of the human inferior colliculus was stained for the nuclei with DAPI (A), for calretinin with AlexaFluor 488 (B), for parvalbumin with AlexaFluor 568 (C), and for neurons with deep-red Neurotrace™ (D). Panel E shows the overlay of panels A, B, C and D. Magnification 63x; bar= 20 μm.

8. Basic protocol for multiple immunostaining of neural tissue

- Obtain human brain tissue as quickly as possible after death (less than one day is preferred).
- Fix the tissue with freshly prepared 4% paraformaldehyde in phosphate buffered saline (PBS; pH 7.4) for not more than two weeks at 4°C in the dark. Refresh the fixative every three days. Overnight fixation for small (<1 cm^3) tissue pieces is sufficient. Freshness of the fixative and the incubation conditions are of great importance to avoid autofluorescence.
- Immerse the tissue in 10% sucrose in PBS (pH 7.4) at 4°C until the tissue sinks. Repeat the procedure with 20% sucrose and then with 30% sucrose.
- Freeze the tissue in a cryo-embedding medium such as OCT and store it in an air tight container at –80°C.
- At least one hour before sectioning, equilibrate the tissue blocks at –20°C. Sections are cut, rehydrated and washed in Tris buffered saline (TBS) for at least 40 minutes immediately before any treatment. TBS is used for all washes and dilutions, but other buffers such as PBS may also be used. To minimize autofluorescence, make the sections as thin as possible; i.e. 15-20 μm. Wash the floating sections in TBS for 3 x 5 minutes.
- Permeabilize the sections by immersion in 0.5% Triton X-100 in TBS for 30 min at 4°C in the dark. Wash the floating sections in TBS for 3 x 5 minutes.
- For blocking, incubate the sections in a mixture of 0.5% serum from the species that the secondary antibody is originated from and 1% fish skin gelatin (or 0.5% BSA) in TBS for one hour at room temperature.
- To reduce fixative-induced autofluorescence, if necessary, pretreat the fixed sections with 1% NaBH$_4$ in PBS for 10–20 min, or 50 mM NH$_4$Cl in PBS for 5–10 min, or 100–300 mM glycine in TBS or PBS pH 7.4 for one or two hours at 4°C. If the tissue is fragile, it is better to use glycine.
- Blot excess blocking solution from sections and incubate overnight at 4°C with a mixture of appropriately diluted primary antibodies in TBS containing 1% FSG (or 0.5% BSA).
- Wash sections in 1% FSG (or 0.5% BSA) in TBS for 4x15 min. Incubate sections overnight at 4°C with a mixture of fluorophore-conjugated secondary antibodies raised against the corresponding species of primary antibodies in 1% FSG (or 0.5% BSA). Adding 0.5% Tween-20 into the secondary antibody dilution aids to reduce nonspecific staining (Sun et al., 2009). Wash sections in TBS for 4 x 15 min.
- Counterstain neurons if necessary, e.g. with 1/100 dilution of deep-red Neurotrace™ (Invitrogen), for 30 min at room temperature followed by rinsing for 2 h in TBS. Then, counterstain nuclei if necessary, e.g., with 1 μg/ml DAPI in TBS for 3 min at room temperature. Wash sections in TBS for 3x5 min.
- Reduce lipofuscin-induced autofluorescence by rinsing briefly with distilled water, then incubating the samples for 10–60 min with a solution of 1 mM CuSO$_4$ in 50 mM ammonium acetate buffer (pH 5). Rinse quickly in distilled water and wash eight times in TBS. Alternatively, the samples can be treated for 10 min with 0.3% Sudan Black B in 70% ethanol.
- Place the sections on a slide and mount with a mixture of glycerol/PBS (9:1, v/v) supplemented with an antifading agent, e.g. 25 mg/ml 1,4-diaza-bicyclo[2.2.2]octane

(DABCO) or 0.5% n-propylgallate (pH 7.4). Do not use anti-fade mounting media on samples treated with Sudan Black B.

- An example of human brain tissue multi-labeled as described in this section is shown in Figures 5-7. Briefly, the inferior colliculi were fixed with 4% paraformaldehyde, sectioned at 20 μm, incubated in a pooled primary antibody solution containing mouse anti-parvalbumin monoclonal antibody (2.2 μg/ml; P3088, Sigma-Aldrich, Bornem, Belgium) and rabbit anti-calretinin polyclonal antibody (0.8 μg/ml; C7479, Sigma-Aldrich) diluted in 1% FSG, followed by incubation in AlexaFluor 488 donkey-anti-mouse and AlexaFluor 568 donkey anti-rabbit from Invitrogen (15 μg/ml). The figures also show samples counterstained with deep-red Neurotrace™ (1/100 dilution) and DAPI (1 μg/ml) before treatment with $CuSO_4$ for 10 minutes as described in this section. While lipofuscin-induced autofluorescence is no longer observed in the blue, green and red channel, it is still an issue in the deep-red channel. The remaining autofluorescence in the deep-red channel is possibly related to the weak fluorescent signal detected upon staining with deep-red Neurotrace™, especially in small neurons of the dorsal cortex.

9. Conclusion

In this chapter we discuss multiple-label fluorescence confocal microscopy and methods to reduce autofluorescence in sections of the human central nervous system. We also evaluate the quenching of fluorescence by non-specific protein adsorption. We conclude that treatment of neural tissues with reagents to reduce endogenous autofluorescence is helpful in improving specific signal-to-noise ratios. Sections processed in this way are suitable for examining the complex structure and organization of human brain tissue.

10. Acknowledgments

The authors thank Prof. Piette and Prof. Cuvelier (Forensic Institute and Department of Pathology, respectively, University of Ghent, Belgium), for providing the human inferior colliculi. They also acknowledge Dr. Amin Bredan for editing, Barbara De Bondt, Dominique Jacobus, and Stefanie Mortier for technical assistance with the immunohistological sections and Eef Parthoens and Evelien Van Hamme (VIB-DMBR microscopy core facility) for technical assistance with the Leica confocal microscope and image processing.

11. References

Benavides, S.H.; Monserrat, A.J.; Farina, S. & Porta, E.A. (2002). Sequential histochemical studies of neuronal lipofuscin in the human cerebral cortex from the first to the ninth decade of life. *Archives of Gerontology and Geriatrics*, Vol. 34, No. 3, (May –June 2002), pp. 219-231, ISSN 0167-4943

Beisker, W.; Dolbeare, F. & Gray J.W. (1987) An improved immunocytochemical procedure for high-sensitivity detection of incorporated bromodeoxyuridine. *Cytometry*, Vol. 8, No. 2 (March 1987), pp. 235-239, ISSN 1552-4922

Bratthauer, G.L. (2010) The avidin-biotin complex (ABC) method and other avidin-biotin binding methods. *Methods in Molecular Biology*, Vol. 588, pp. 257-270, ISSN 1064-3745

Brehmer, A.; Blaser, B.; Seitz, G.; Schrödl, F. & Neuhuber, W. (2004). Pattern of lipofuscin pigmentation in nitrergic and non-nitrergic, neurofilament immunoreactive myenteric neuron types of human small intestine. *Histochemistry and Cell Biology*, Vol. 121, No. 1, (January 2004), pp. 13-20, ISSN 0948-6143

Brizzee, K.R.; Ordy, J.M. & Kaack, B. (1974). Early appearance and regional differences in intraneuronal and extraneuronal lipofuscin accumulation with age in the brain of a nonhuman primate (Macaca mulatta). *Journal of Gerontology*, Vol. 29, No. 4, (July 1974), pp. 366-381, ISSN 0022-1422

Brouns, I.; Van Nassauw, L.; Van Genechten, J.; Majewski, M.; Scheuermann, D.W.; Timmermans, J.P. & Adriaensen, D. (2002). Triple immunofluorescence staining with antibodies raised in the same species to study the complex innervation pattern of intrapulmonary chemoreceptors. *Journal of Histochemistry and Cytochemistry*, Vol. 50, No. 4, (April 2002), pp. 575-582, ISSN 0022-1554

Buchwalow, I.B.; Minin, E.A. & Boecker, W. (2005). A multicolor fluorescence immunostaining technique for simultaneous antigen targeting. *Acta Histochemica* Vol 107, No. 2, (July 2005), pp. 143-148, ISSN 0065-1281

Callis, G. (2010). Glutaraldehyde-induced autofluorescence. *Biotechnic and Histochemistry*, Vol. 85, No. 4, (August 2010), pp. 269, ISNN 1052-0295

Clancy, B. & Cauller, L.J. (1998). Reduction of background autofluorescence in brain sections following immersion in sodium borohydride. *Journal of Neuroscience Methods*, Vol. 83, No. 2, (September 1998), pp. 97-102, ISNN 0165-0270

Collins, J.S. & Goldsmith, T.H. (1981) Spectral properties of fluorescence induced by glutaraldehyde fixation. *Journal of Histochemistry and Cytochemistry*, Vol. 29, No. 3, (March 1981), pp. 411-414, ISNN 0022-1554

Dailey, M.; Marrs, G.; Satz, J. & Waite, M. (1999). Exploring biological structure and function with confocal microscopy. *Biological Bulletin*, Vol. 197, No. 2, (October 1999), pp. 115-122, ISSN 0006-3185

De Groot, C.J.; Theeuwes, J.W.; Dijkstra, C.D. & van der Valk, P. (1995). Postmortem delay effects on neuroglial cells and brain macrophages from Lewis rats with acute experimental allergic encephalomyelitis: an immunohistochemical and cytochemical study. *Journal of Neuroimmunology*, Vol. 59, No. 1-2, (June 1995), pp. 123-134, ISSN 0165-5728

Del Castillo, P.; Llorente, A.R. & Stockert, J.C. (1989). Influence of fixation, exciting light and section thickness on the primary fluorescence of samples for microfluorometric analysis. *Basic and Applied Histochemistry*, Vol. 33, No. 3, pp. 251-257, ISSN 0391-7258

Double, K.L.; Dedov, V.N.; Fedorow, H.; Kettle, E.; Halliday, G.M.; Garner, B. & Brunk, U.T. (2008). The comparative biology of neuromelanin and lipofuscin in the human brain. *Cellular and Molecular Life Sciences*, Vol. 65, No. 11, (February 2008), pp. 1669-1682, ISSN 1420-682X

Duhamel, R.C. & Johnson, D.A. (1985) Use of nonfat dry milk to block nonspecific nuclear and membrane staining by avidin conjugates. *Journal of Histochemistry and Cytochemistry*, Vol. 33, No. 7, (July 1985), pp. 711-714, ISSN 0022-1554

Eldred, G.E.; Miller, G.V.; Stark, W. & Feeney-Burns, L. (1982). Lipofuscin: resolution of discrepant fluorescent data. *Science*, Vol. 216, No. 4547, (May 1982), pp. 757-759, ISSN 0036-8075

Fodor, M.; van Leeuwen, F.W. & Swaab, D.F. (2002). Differences in postmortem stability of sex steroid receptor immunoreactivity in rat brain. *The Journal of Histochemistry and Cytochemistry*, Vol. 50, No. 5, (May 2002), pp. 641-650, ISSN 0022-1554

Fox, C.H.; Johnson, F.B.; Whiting, J. & Roller, P.P. (1985). Formaldehyde fixation. *The Journal of Histochemistry and Cytochemistry*, Vol. 33, No. 8, (Augustus 1985), pp. 845-853, ISSN 0022-1554

Frisch, J.; Houchins, J.P.; Grahek, M.; Schoephoerster, J.; Hagen, J.; Sweet, J.; Mendoza, L.; Schwartz, D. & Kalyuzhny, A.E. (2011) Novel multicolor immunofluorescence technique using primary antibodies raised in the same host species. In: *Signal Transduction Immunohistochemistry: Methods and Protocols, Methods in Molecular Biology*, A.E. Kalyuzhny, (Ed.), Vol. 717, No. 4, pp. 233-244 ISSN 1064-3745 e-ISSN 1940-6029 ISBN 978-1-61779-023-2 e-ISBN 978-1-61779-024-9

Haralampus-Grynaviski, N.M.; Lamb, L.E.; Clancy, C.M.R.; Skumatz, C.; Burke, J.M.; Sarna, T. & Simon, J.D. (2003). Spectroscopic and morphological studies of human retinal lipofuscin granules. *Proceedings of the National Academy of Sciences of the United States of America*, Vol.100, No. 6, (March 2003), pp. 3179-3184, ISSN 0027-8424

Harman, D. (1989) Lipofuscin and ceroid formation: the cellular recycling system. *Advances in Experimental Medicine and Biology*, Vol. 266, pp. 3-15, ISSN 0065-2598

Eilbig, H.; Bidmon, H.J.; Oppermann, O.T. & Remmerbach, T. (2004). Influence of post-mortem delay and storage temperature on the immunohistochemical detection of antigens in the CNS of mice. *Experimental and Toxicologic Pathology*, Vol. 56, No. 3, (December 2004), pp. 159-171, ISSN 0940-2993

Hyden, H. & Lindstrom, B. (1950). Microspectrographic studies on the yellow pigment in nerve cells. *Discussions of the Faraday Society*, Vol. 9, (January 1950), pp. 436-441, ISSN 0014-7664

Katz, M.L.; Robinson, W.G.; Herrmann, R.K.; Groome, A.B. & Bieri, J.G. (1984). Lipofuscin accumulation resulting from senescence and vitamin E deficiency: spectral properties and tissue distribution. *Mechanisms of Ageing and Development*, Vol. 25, No. 1-2 , (April-May 1984), pp. 149-159, ISSN 0047-6374

Kaur, R.; Dikshit, K.L. & Raje M. (2002). Optimization of immunogold labeling TEM: an ELISA-based method for evaluation of blocking agents for quantitative detection of antigen. *Journal of Histochemistry and Cytochemistry*, Vol. 50, No. 6, June 2002), pp. 863-873, ISSN ISSN 0022-1554

Kikugawa, K.; Beppu, M.; Sato, A. & Kasai, H. (1997). Separation of multiple yellow fluorescent lipofuscin components in rat kidney and their characterization.

Mechanisms of Ageing and Development, Vol. 97, No. 2, (August 1997), pp. 93-107, ISSN 0047-6374

Mann, D.M.A. & Yates, P.O. (1987). Ageing, nucleic acids and pigments, In: *Recent advances in Neuropathology*, Cavanagh, J.B., pp. 109-137, Livingston, ISBN 9780443032264, Edinburgh.

Martin, P.M. & O'Callaghan, J.P. (1995). A direct comparison of GFAP immunocytochemistry and GFAP concentration in various regions of ethanol-fixed rat and mouse brain. *Journal of Neuroscience Methods*, Vol. 58, No. (1-2), (May 1995), pp. 181-192, ISSN 0165-0270

McKee, A.C. (1999). Brain banking: basic science methods. *Alzheimer Disease and Associated Disorders*, Vol. 13, No. Suppl 1, (April-June 1999), pp. S39-44, ISSN 0893-0341

Ngwenya, L.B.; Peters, A. & Rosene, D.L. (2005). Light and electron microscopic immunohistochemical detection of bromodeoxyuridine-labeled cells in the brain: different fixation and processing protocols. *The Journal of Histochemistry and Cytochemistry*, Vol. 53, No. 7, (July 2005), pp. 821-832, ISSN 0022-1554

Oliveira, V.C.; Carrara, R.C.V.; Simoes, D.L.C.; Saggioro, F.P.; Carlotti, C.G. Jr.; Covas, D.T. & Neder, L. (2010). Sudan Black B treatment reduces autofluorescence and improves resolution of in situ hybridization specific fluorescent signals of brain sections. *Cellular and Molecular Biology*, Vol. 25, No. 8, (August 2010), pp. 1017-1024, ISSN 0006-3088

Peters, A.; Palay, S.L. & Webster, H.D. (1991). *The fine structure of the nervous system* (3rd ed.), Oxfort University Press, ISBN 0195065719, New York.

Porta, E.A. (2002) Pigments in aging: an overview. *Annals of the New York Academy of Sciences*, Vol. 959 , No. 1, (January 2002), pp. 57-65, ISSN 0077-8923

Ravid, R. & Grinberg, L.T. (2008). How to run a brain bank--revisited. *Cell Tissue Bank*, Vol. 9, No. 3, (September 2008), pp. 149-150, ISSN 1389-9333

Riga, D.; Riga, S.; Halalau, F. & Schneider, F. (2006). Brain lipopigment accumulation on normal and pathological aging. *Annals of the New York Academy of Sciences*, Vol. 1067, No. 1, (May 2006), pp. 158-163, ISSN 0077-8923

Romijn, H.J.; van Uum, J.F.M.; Breedijk, I.; Emmering, J.; Radu, I. & Pool, C.W. (1999). Double immunolabeling of neuropeptides in the human hypothalamus as analyzed by confocal laser scanning fluorescence microscopy. *Journal of Histochemistry and Cytochemistry*, Vol. 47, No. 2, (February 1999), pp. 229-236, ISSN 0022-1554

Schnell, S.A.; Staines, W. & Wessendorf, M.W. (1999). Reduction of lipofuscin-like autofluorescence in fluorescently labeled tissue. *Journal of Histochemistry and Cytochemistry*, Vol. 47, No. 6, (June 1999), pp. 719-730, ISSN 0022-1554

Selever, J.; Kong, J.Q. & Arenkiel, B.R. (2011). A rapid approach to high-resolution fluorescence imaging in semi-thick brain slices. *Journal of Visualized Experiments*, Vol. 26, No. 53, (July 2011), pii. 2807, ISSN 1940-087X

Sharma, V.; Nag, T.C.; Wadhwa, S. & Roy T.S. (2009). Stereological investigation and expression of calcium-binding proteins in developing human inferior colliculus. *Journal of Chemical Neuroanatomy*, Vol. 37, No. 2, (March 2009), pp. 78-86, ISSN 0891-0618

Sulzer, D.; Mosharov, E.; Talloczy, Z.; Zucca, F.A.; Simon, J.D. & Zecca, L. (2008). Neuronal pigmented autophagic vacuoles: lipofuscin, neuromelanin, and ceroid as macroautophagic responses during aging and disease. *Journal of Neurochemistry*, Vol. 106, No. 1, (July 2008), pp. 24-36, ISSN 1471-4159

Sun, A.; Liu, M. & Bing, G. (2009). Improving the specificity of immunological detection in aged human brain tissue samples. *Internal Journal of Physiology, Pathophysiology and Pharmacology*, Vol. 2, No. 1, (Dec 2009), pp. 29-35, ISSN 1944-8171

Tagliaferro, P.; Tandler, C.J.; Ramos, A.J.; Saavedra, J.P. & Brusco, A. (1997). Immunofluorescence and glutaraldehyde fixation. A new proceure based on the Schiff-quenching method. *Journal of Neuroscience Methods*, Vol. 77, No. 2, (December 1997), pp.191-197, ISSN 0165-0270

Tardif, E.; Chiry, O.; Probst, A.; Magistretti, P.J. & Clarke, S. 2003. Patterns of calcium binding proteins in the human inferior colliculus: identification of subdivisions and evidence for putative parallel systems. *Neuroscience*, Vol. 116, No. 4, pp. 1111-1121, ISSN 0306-4522

Terman, A. & Brunk, U.T. (1998). Lipofuscin: mechanisms of formation and increase with aging. *Acta Pathologica, Microbiologica et Immunologica Scandinavica*, Vol. 106, No. 2, (February 1998), pp. 265-276, ISSN 09034641

Terman, A. & Brunk, U. (2004). Lipofuscin. *International Journal of Biochemistry and Cell Biology*, Vol. 36, No. 8, (August 2004), pp. 1400-1404, ISSN 1357-2725

Terman, A. & Brunk, U. (2006). Oxidative stress, accumulation of biological "garbage" and aging. *Antioxidants and Redox Signaling*, Vol. 8, No. 1-2, (January-February 2006), pp. 197-204, ISSN 1523-0864

Tokuyasu, K.T. (1973). A technique for ultracryotomy of cell suspensions and tissues. *Journal of Cell Biology*, Vol. 57, No. 2, (May 1973), pp. 551–565, ISNN 0021-9525

Tóth, Z.E. & Mezey, E. (2007). Simultaneous visualization of multiple antigens with tyramide signal amplification using antibodies from the same species. *Journal of Histochemistry and Cytochemistry*, Vol. 55, No. 6, (June 2007), pp. 545-554, ISSN 0022-1554

Waldvogel, H.J.; Bullock, J.Y.; Synek, B.J.; Curtis, M.A.; van Roon-Mom, W.M. & Faull, R.L. (2008). The collection and processing of human brain tissue for research. *Cell Tissue Bank*, Vol. 9, No. 3, (September 2008), pp. 169-179, ISSN 1389-9333

Waldvogel, H.J.; Curtis, M.A.; Baer, K.; Rees, M.I. & Faull, R.L. (2006). Immunohistochemical staining of post-mortem adult human brain sections. *Nature Protocols*, Vol. 1, No. 6, pp. 2719-2732, ISSN 1754-2189

Willingham, M.C. (1983). An alternative fixation-processing method for pre-embedding ultrastructural immunocytochemistry of cytoplasmic antigens: the GBS (glutaraldehyde-borohydride-saponin) procedure. *Journal of Histochemistry and Cytochemistry*, Vol. 31, No. 6, (June 1983), pp.791-798, ISSN 0022-1554

Willingham, M.C. (1999). Chapter 17 Fluorescence labeling of intracellular antigens of attached or suspended tissue-cultured cells. In: *Methods in Molecular Biology™*, Series: *Methods in Molecular Biology*, LC Javois (Ed.), Vol. 115., pp. 121-130, Humana Press Inc, ISBN 0-89603-570-0 II, Totowa, NJ

Zhang, Y.; Zhang, W.; Johnston, A.H.; Newman, T.A.; Pyykkö, I. & Zou, J. (2010). Improving the visualization of fluorescently tagged nanoparticles and fluorophore-labeled molecular probes by treatment with $CuSO_4$ to quench autofluorescence in the rat inner ear. *Hearing Research*, Vol. 269, No. 1-2, (October 2010), pp. 1-11, ISSN 0378-5955

3

Immunoelectron Microscopy: A Reliable Tool for the Analysis of Cellular Processes

Ana L. De Paul, Jorge H. Mukdsi, Juan P. Petiti, Silvina Gutiérrez,
Amado A. Quintar, Cristina A. Maldonado and Alicia I. Torres
Centro de Microscopía Electrónica, Facultad de Ciencias Médicas,
Universidad Nacional de Córdoba,
Argentina

1. Introduction

Electron Microscopy is an indispensable tool to investigate the intricate structures of the cell and organelles, and also to study the cellular biological processes implicated in the responses to changes in the microenvironment. However, several cellular events may be missed if conventional ultrastructural studies are not complemented with details concerning the subcellular localization of a wide range of specific proteins which can become rearranged as part of their own dynamic processes. Thus, immunoelectron microscopy emerges as a technique that links the information gap between biochemistry, molecular biology, and ultrastructural studies, by placing macromolecular functions within a cellular context.

The present chapter is intended to describe the main scope and protocols of the immunogold methods which have been successfully utilized at our research center for the examination and analysis of intracellular and cell surface proteins in mammalian cells. Furthermore, at the ultrastructural level, we demonstrate the role of immunogold labeling in the study of biological processes induced by different stimuli from the environment.

1.1 General considerations concerning immunoelectron microscopy

Immunoelectron microscopy is one of the best methods for detecting and localizing proteins in cells and tissues. This procedure can be used on practically every unicellular and multicellular organism, and often provides unexpected insights into the structure-function associations. The use of primary antibodies conjugated with gold particles allows high-resolution detection and localization of a multiplicity of antigens, both on and within the cells. However, the successful application of immunoelectron microscopy depends on the preservation of the protein antigenicity, the capacity of antibodies to infiltrate throughout the cell, and finally the specificity of recognition between antigen-primary antibodies. In addition, an adequate handling of biological samples is required, which involves fixation, an appropriate selection of an embedding resin and the ready availability of the specific antibodies for the molecules whose ultrastructural location needs to be determined.

Since 1971, when W.P. Faulk and G.M. Taylor published "An immunocolloid method for the electron microscope", colloidal gold has become a very extensively used marker in microscopy. As knowledge of protein location and distribution at the subcellular level plays a pivotal role in cell biology, this tool has been applied to detect a vast range of cellular and extracellular constituents by using *in situ* hybridization, immunogold, lectin-gold, and enzyme-gold labeling. In addition to its use at light microscopy level, colloidal gold remains the label of choice in transmission electron microscopy (TEM) for studying thin sections, freeze-etch, and surface replicas, as well as in scanning electron microscopy.

While conventional electron microscopy provides no information about specific molecules, immunogold labeling can help to connect a visible structure with a specific *in situ* localization and distribution of molecules at a high resolution. In this way, the use of the colloidal gold particles undoubtedly represents a significant event in the improvement of the immunochemistry method.

Many protocols have been developed since the introduction of colloidal gold to immunocytochemistry, with the two most widely used techniques, however, being actually based on transmission electron microscopy and consisting of either immunolabeling after embedding in resin (post-embedding immunogold labeling) or immunostaining prior to this process (pre-embedding immunogold labeling). Here, we also present some approaches from our own work, which reveal how the use of immunoelectron microscopy can offer additional insights into the structure–function relationships.

These detailed methods and notes should facilitate the selection of the best method to use for the antibody and biological material to be studied. It is our hope that this chapter will also facilitate an improved understanding of immunoelectron microscopy and its use in the biological sciences.

1.2 Criteria for selection of the immunoelectron microscopy technique

Before deciding to use the immunoelectron microscopy technique, researchers need to consider some relevant questions in order to select the most appropriate procedure among the methods typically used.

With the aim of guiding researchers, we propose that the following sequence of questions should be answered:

- What is the scope of immunogold labeling?
- What kind of information can be obtained by applying immunocytochemistry?
- What is the subcellular localization of the molecule of interest?
- Does the target molecule to be detected by this methodology possess a single and static location? Or does it exhibit a dynamic behaviour?
- What is the most recommended fixative mixture for the antigen under study?
- How can the antigenicity of the target molecules be preserved?
- Which is the most appropriate type of resin to be used in the immunoelectron microscopy? Does the selected resin depend on the method to be used in immunolabeling?

1.3 General rules of thumb in immunogold electron microscopy

- For immunoelectron microscopy, fixation is one of the most important steps in sample preparation due to the need to preserve as much biochemical reactivity as possible.
- Although there are many types of fixatives, the mixture of formaldehyde and a low concentration of glutaraldehyde is the most generally accepted one for immunoelectron microscopy.
- Immersion fixation is useful for the majority of the biological specimens (e.g. biopsy specimens, cultured cells).
- The relation between the fixative solution and size sample should be at least 40:1.
- It is critical that tissue be cut into pieces no larger than 1–2 mm³ due to the penetration of the fixative and embedding resin into specimens being relatively slow.
- Immunocytochemistry at ultrastructural level can be performed on semithin and ultrathin sections of specimens embedded in epoxy or acrylic resins and also on ultrathin cryo-sections.
- Examination of the semithin sections is important to be able to assess the quality of the fixation and to select the area to be investigated by electron microscopy.
- Multiple labeling with colloidal gold conjugates of different sizes is easily performed by the post-embedding method.
- After fixation and embedding in acrylic resins, specimens without osmication show poor preservation of cellular membranes.

2. Protocols in immunoelectron microscopy

The success of immunolabeling at ultrastructural level depends on various factors, including the initial quantity and quality of antigens, and the preservation of the cell ultrastructure, to be able to finally achieve an accurate localization of the antigen within the cell. Therefore, it is necessary to attain a correct balance between antigen preservation and a good morphology at the ultrastructural level. Although it is difficult to provide a standard protocol because antigens and tissues need individual evaluations to reveal the best experimental conditions, there are many recommendations that researchers have introduced to increase the possibility of arriving at a positive result.

The two following general approaches can be applied to localize cell antigens:

- When the objective is to localize the intracellular antigens, there are at least three different protocols that can be used: *post-embedding*, after embedding in acrylic resins; *cryo-ultramicrotomy* in tissue sections obtained without embedding, and *pre-embedding* combined with membrane permeabilization.
- When the interest is focused on identifying cell surface proteins, the *pre-embedding labeling* protocol is the most convenient as antigens and ultrastructure are well preserved by this method.

However, in each case, the selection of the protocol will frequently depend on the availability of the necessary infrastructure.

Next, we propose the following graphical flowchart that can be used to compare the different approaches of immunoelectron microscopy discussed in the present chapter.

COMPARATIVE DIAGRAM OF
IMMUNOSTAINING METHODS IN ELECTRON MICROSCOPY

INTRACELLULAR ANTIGENS			SUPERFICIAL ANTIGENS
POST-EMBEDDING	CRYO-ULTRAMICROTOMY	PRE-EMBEDDING	PRE-EMBEDDING
Fixation	Weak fixation	Weak fixation	Immunogold labeling
Acrylic resin embedding	Infiltration in sucrose	Ultrasmall Immunogold-silver labeling	Fixation
	Specimen frozen	Epoxy resin embedding	Epoxy resin embedding
Semithin sections/Toluidine Blue staining	Ultrathin cryo-sectioning	Ultrathin sectioning	Semithin sections/Toluidine Blue staining
Ultrathin sectioning			Ultrathin sectioning
Immunogold labeling	Immunogold labeling	Ultrathin section staining	Ultrathin section staining
LIGHT MICROSCOPY OBSERVATION			
Ultrathin section staining			

TRANSMISSION ELECTRON MICROSCOPY OBSERVATION

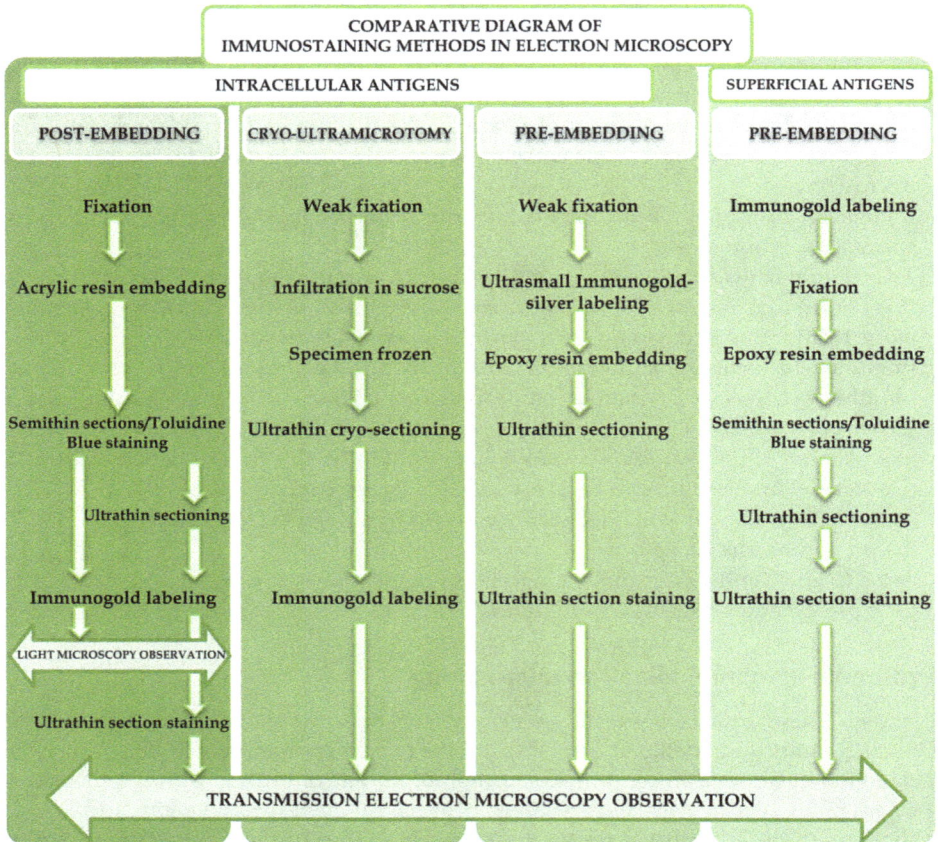

Flowchart 1. Differential procedures of immunogold methods to detect intracellular and superficial antigens.

2.1 Post-embedding technique

2.1.1 Specimen preparation

The post-embedding technique is a good alternative to produce contact between the antibodies and internal antigens exposed at the surface of thin sections obtained from resin-embedded tissues. By using this procedure, specimens are first embedded in resin and sectioned at 60–90 nm. However, post-embedding immunocytochemistry has its own limitations, with the principal drawback being that antibodies cannot penetrate into the resin, and consequently only the antigens that are exposed at thin section surfaces can be labeled. In addition, antigens are affected by fixatives, solvents, resins and heat during polymerization, thus compromising antigenicity. Nevertheless, each of these "threats" to antigenicity can be avoided, at least in part, by implementing various modifications to the conventional protocols used for electron microscopy related to fixation, dehydration, inclusion media, and temperature.

2.1.1.1 Chemical fixation

As a general definition, fixation is an attempt to induce the fast arrest of biological activities and to stabilize the subcellular components with a minimal distortion of the cellular structures. Chemical fixation for TEM allows biological samples to be prepared for subsequent procedures, involving washing with aqueous solvents, dehydration with organic solvents (ethanol or acetone), embedment and polymerization in plastic resins, and subsequent imaging with high-energy electron beams in an electron microscope. The speed of the penetration of the fixatives determines the success of the fixation procedures. Although small fixative molecules (formaldehyde) penetrate more rapidly than larger ones (glutaraldehyde), the latter possess more reactive sites and can thus cross-link and stabilize the cellular components more thoroughly.

The glutaraldehyde concentration should first be reduced as much as necessary. However, this may affect the ultrastructure preservation and make it difficult to locate the antigen in a specific organelle. Therefore, when a new antigen is being studied by immunoanalysis, it is desirable to use a fixative strong enough to have a good ultrastructural preservation, but reduced in glutaraldehyde concentration in comparison with a conventional fixative. For example, this fixative could have 4% of formaldehyde and 1.5% of glutaraldehyde prepared in 0.1 M Cacodylate or Phosphate buffer and used for immersion. However, if the result is poor, the next step should be to lower the glutaraldehyde concentration, with a possible reduction of up to 0.2%. Then, in spite of a poorer preservation of the ultrastructure, it may still be possible to identify the structure where the antigen localizes.

Here, we present several examples of antigen preservation achieved with different fixatives: in Figure 1 the murine mammary tumor virus is well identified after fixation with a mix of formaldehyde and glutaraldehyde at 1.5% and included in LR-White acrylic resin [Peralta Soler, et al.; 1988]; in Figure 2, we localized galectin-1 in Müller cells on chicken retina after fixation in 2% glutaraldehyde by immersion; in Figure 3, a good antigen preservation in the lactotrophs of pituitary gland, as evaluated by labeling on Golgi cisternae as well as on secretory granules, was obtained with a high fixative concentration (3% formaldehyde and 3% glutaraldehyde). This was possible mainly due to the high concentration of antigens stored in the secretory granules of endocrine cells [Maldonado & Aoki; 1986-a].

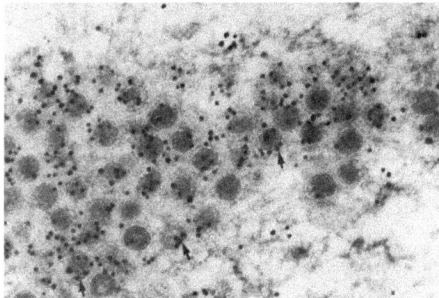

Fig. 1. Identification of intracellular viral antigens. Mammary tumor virus particles in mouse spontaneous mammary carcinoma, are identified with a specific antibody followed by immunogold on thin sections (Reproduced from Peralta Soler et al., 1988; with permission from Micr Electr Biol Cel). Original magnification X 120000.

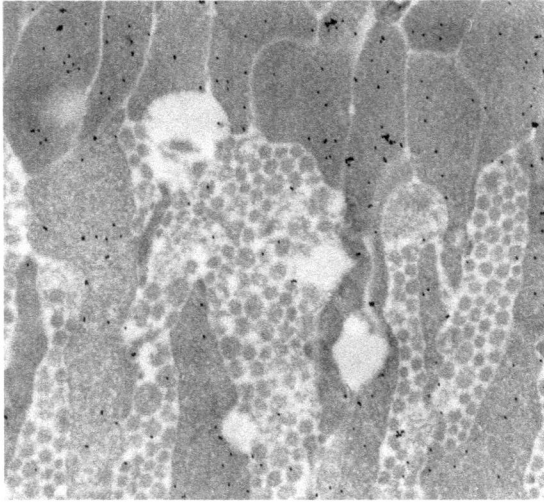

Fig. 2. Localization of galectin-1 in Müller cells of retina. Cytoplasmic projections of Müller cells, at the inner nuclear layer of chicken retina exhibit specific gold labeling. (Reproduced from Maldonado et al., 1999; with permission from Invest Ophthalmol Vis Sci). Original magnification X 25000.

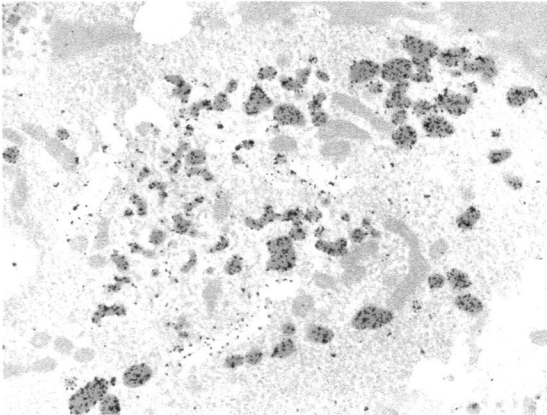

Fig. 3. Preservation of prolactin antigenicity after strong aldehyde fixation. An intense immunogold labeling of prolactin on secretory granules and Golgi complex cisternae can be seen. Original magnification X 20000.

Another way to preserve antigenicity is to omit osmium fixation, as this is a very strong fixative that cross-links proteins and polypeptide chains and consequently diminishes their antigenicity. In fact, only when there is a concentrated mass of antigens can osmium be used. Even then, before starting the immunocytochemical protocol, it requires antigens to be unmasked by means of incubation with a strong oxidant in order to reoxidize and solubilize the reduced osmium molecules.

In Figure 4, a saturated aqueous solution of sodium metaperiodate [Bedeyan & Zollinger; 1983] was applied to thin sections of osmicated pituitary gland to identify the prolactin hormone stored in secretory granules [Maldonado & Aoki; 1986-b], with the absence of immunostaining at the Golgi level being due to a decrease in the antigenicity.

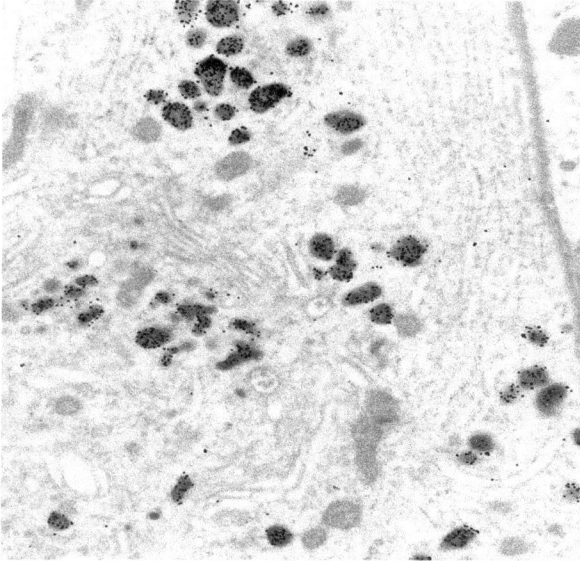

Fig. 4. Immunogold labeling on osmium fixed tissue. In contrast with Fig. 3, prolactin immunoreactivity was only preserved at secretory granules when osmium postfixation was carried out after aldehyde fixation. Immunolabeling was performed after a previous etching with sodium metaperiodate (Reproduced from Maldonado & Aoki, 1986; with permission from Histochem J). Original magnification X 20000.

In Figure 5, the identification of secretory granules positive for ANP (atrial natriuretic factor) was performed with aldehyde fixation (1.5% formaldehyde and glutaraldehyde), which were then postfixed with 1% osmium tetroxide, and removed with 10% hydrogen peroxide prepared in water [Maldonado, et al.; 1986]. The effects of not using osmium tetroxide can be partially compensated for by the inclusion of 0.2% of picric acid in the glutaraldehyde solution.

2.1.1.2 Inclusion media

Epoxi resins used for conventional electron microscopy are not adequate for immunolabeling because they are hydrophobic resins that need extreme dehydration and high temperatures to be able to polymerize. For this reason, during the 1980´s, experts developed water-miscible (acrylic) embedding media, which are more permeable to aqueous solutions and therefore more suitable for immunostaining at the ultrastructural level. Nowadays, various resins are available, which have been manufactured by companies specializing in electron microscopy products, with Lowicryl and LR-White being the most common. These resins have proved to be quite useful in retaining biochemical reactivity

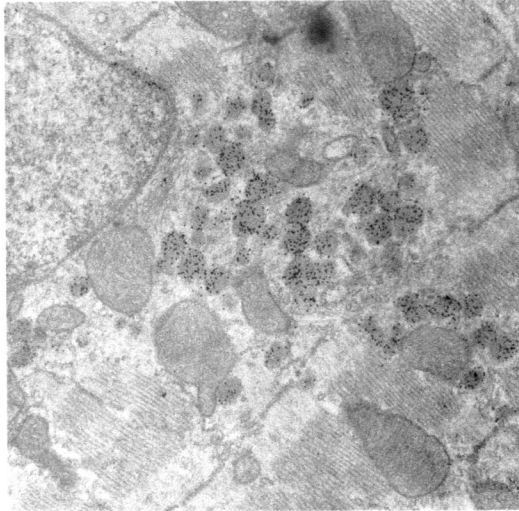

Fig. 5. Immunogold labeling on osmium fixed tissue. Immunogold identification of ANP in secretory granules of myocardiocytes in thin sections of rat atrium fixed with aldehyde osmium and included in LR White. Immunolabeling was performed after a previous etching with 10% H_2O_2 (Reproduced from Maldonado, et al., 1986; with permission from Anat Embryol (Berl)). Original magnification X 25000.

within samples, although the final images are structurally more poorly defined than those produced using the epoxy-embedded technique.

Lowicryl has the advantage that it polymerizes at low temperatures, which contributes significantly to antigen preservation. In Figure 6, a thin section of pituitary gland that was included in Lowicryl exhibited a strong staining for prolactin, even at the Golgi cisternae of the lactotrophs where the hormone concentration was lower than in secretory granules [Maldonado & Aoki; 1986-a]. This resin is recommended when there is scarce antigen concentration; but it has the disadvantages that a special chamber is necessary to polymerize at -30° under U.V., and it is also very toxic.

LR-White polymerize at 50°C, which helps to provide a good antigenicity. Furthermore, it does not need additional equipment and is much less toxic than Lowicryl. Other advantages are that LR-White tolerates partial dehydration (accepting tissue from 70% ethanol), it is beam-stable and easy to be sectioned, and osmium fixation does not interfere with polymerization as in the case of Lowicryl.

All these resins exhibit less stability under the electron beam than epoxides, and also post-stain more quickly than most of them (since they are more hydrophilic than epoxides and react more quickly with the typical aqueous post-stain). However, they will not polymerize properly if they are contaminated with acetone or have been exposed to air. For this reason, dehydration with increasing concentrations of ethanol should be followed by resin infiltration, and polymerization should be carried out in gelatin capsules (anaerobic conditions). In addition, dehydration is not as critical as with epoxide resins, since common

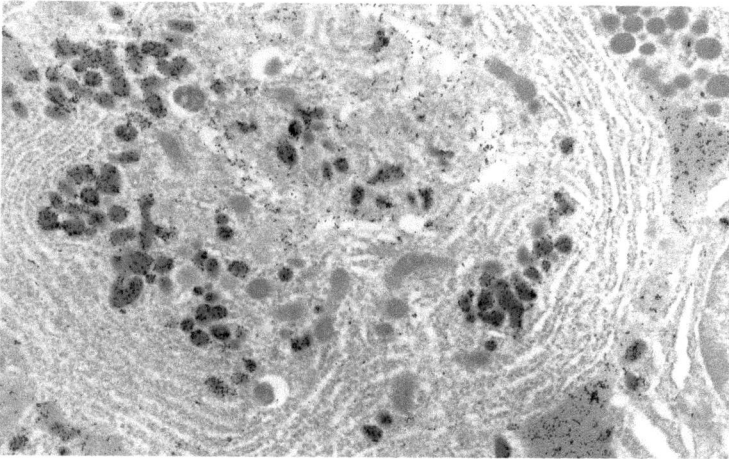

Fig. 6. High preservation of prolactin antigenicity in Lowicryl included pituitary gland. This labeling was obtained in pituitary gland fixed in 3% formaldehyde and 3% glutaraldehyde and included in Lowicryl at -30°C with U.V. for 24 h. Original magnification X 17000.

acrylic resins can tolerate small quantities of water remaining in the sample during polymerization.

Alternatively, antigenic alterations derived from dehydration and inclusion resins can be completely avoided by using cryo-ultramicrotomy. In this technique, immunolabeling is performed on ultrathin thawed cryo-sections according to the method of Tokuyasu [Tokuyasu; 1973], thereby providing a very sensitive high-resolution localization. Its main advantages are that the antigens remain in a hydrated environment prior to immunolabeling, and that the antigen accessibility is improved compared with resin section labeling. In the case of biological samples, they need to be chemically fixed (usually with aldehydes), cryo-protected (and partly dehydrated) with high concentrations of sucrose (70-80%), frozen in liquid nitrogen and sectioned at approximately -115°C. Dry-frozen ultrathin sections can then be picked up with a drop of methyl cellulose and/or sucrose and transferred onto an electron microscopic support grid for a subsequent immunogold procedure. There are many reviews concerning this technique such as that of Peters and Pierson [Peters & Pearson; 2008]. However, the main limitation of this technique is that it requires special laboratory infrastructure and equipment and trained personal. Therefore, it is only possible to use in specialized laboratories.

2.1.1.3 Immunoreactives

Gold particles have definitively replaced enzymatic markers such as horseradish peroxidase, with a disadvantage of the enzyme-based detection systems being that the reaction product formed may diffuse away from the reaction site and precipitate over a greater area. Therefore, particulate labeling with gold particles is now preferred since it provides a higher subcellular resolution. Furthermore, gold complexes have various advantages that have made them into the most used markers: the high electron density, the possibility of being prepared with different sizes, and the fact that they are multiple markers with gold particles that can be easily bound to different proteins besides those of antibodies. Regarding the colloidal gold particle diameters usually applied, the most frequent are in a range from 5 to 25 nm when the antigens are exposed on the surface of ultrathin sections [Hagiwara, et al.; 2000]. However, these are of little use for detecting intracellular molecules with the pre-embedding method [Ferguson, et al.; 1998]. More recently, the development of ultrasmall gold (<1.0 nm) has widened significantly the scope of gold complex conjugates, thus providing a marker system with a greater labeling sensitivity. This is due to these particles being less prone to steric hindrance and able to penetrate better, even without pre-treatment with detergent [Hainfeld & Furuya; 1992].

In addition, the combination of ultrasmall gold conjugates with silver enhancement has significantly improved the ultrastructural detection of intracellular antigens by using pre-embedding procedures. This has been successfully applied by Van Lookeren Campagne to analyze distribution of the growth-associated protein B-50 in hippocampal neurons [Van Lookeren Campagne, et al.; 1992].

2.1.1.4 Blocking solutions

A critical aspect to be considered when performing immunoreactions at electron microscopy level is to control the different sites that can contribute to the background in the reaction. Background is the result of "non-specific" reactions, which are due to the general physical chemical properties of both the specimen and the primary antibodies, and also result from secondary antibody/marker conjugates. For this purpose, a blocking solution should be applied immediately before incubation with the primary antibody, with the ideal blocking solution being able to bind to all potential sites of non-specific interactions, and to eliminate background or "noise" altogether without altering or obscuring the epitope for antibody binding, i.e. the "signal". Although there is no single recipe for making a blocking solution, a general recommended rule, when investigating a new target antigen or using a new antibody, is to test several different blockers in order to determine the highest signal/noise ratio of the assay.

We have long used 1% Bovine Serum Albumin or 1% normal serum of the same species as the secondary antibody. However, it is not infrequent to find non-specific markers (e.g. on the heterochromatin), which cannot be controlled by this blocking solution, particularly when IgG-gold complexes are applied. In these cases, a more robust blocker solution should be sought.

An appropriate blocking has to be able interact with both hydrophobic and hydrophilic properties. The following three proteins are frequently used together to obtain an adequate blocking solution:

-   Bovine Serum Albumin (BSA).
-   Normal serum, obtained from the same species as the secondary antibody (but which must not be used when protein A is the reagent).
-   Cold fish skin gelatin at 0.1%.

The two first components can be used in concentrations varying from 1-5%, and should be prepared in PBS containing 0.05% sodium azide to avoid contaminations. Sufficient washing after the blocking step is necessary in order to remove any excess protein that may prevent detection of the target antigen.

2.1.1.5 Control of specificity of immunogold staining

A critical point for immunoreactions, independently of whether they occur, at light or electron microscopy level or whatever marker is being used, is to validate the results obtained by including adequate controls. These can be designed to detect any unspecific interactions related to the different reactives used during the protocols.

2.1.1.5.1 Controlling primary antibody

In order to determine the part of the immunolabeling that is due to unspecific binding of the IgG molecules of the primary antibody, it is necessary to replace it with the same dilution of the pre-immune serum or a serum of the same specie as the primary antibody, and also to maintain the same incubation time as that used for the specific antibody. This will then reveal unspecific interactions due to the presence of hydrophobic and hydrophilic regions in the biological structures. Nevertheless, another source of unspecific label arises due to epitopes of other molecules being able to interact or cross-react with the primary antibody. This can be demonstrated by absorbing the primary antibody with the protein or peptide used to immunize, and then incubating for 24 h at 4°C in order to promote interaction with the antibody. For example, primary antibody could be combined with a fivefold excess of blocking antigen peptide [Mukdsi, et al.; 2006]. Then, after centrifugation to precipitate the antigen-antibody complex, the supernatant can be used to incubate the grid, followed by the gold complexes.

2.1.1.5.2 Controlling secondary antibody

To determine whether the gold complex is contributing to the labeling through unspecific bindings to the tissue, after applying the blocking solution the grids need to be incubated with gold complex but omitting the primary antibody or the pre-immune serum.

The introduction of adequate controls, together with the introduction of blocking solution, can render a confident label that will certainly contribute to the analysis of the biological processes. Finally, when a new antigen is being studied, it is recommended that positive controls are introduced.

2.1.2 Post-embedding protocol

2.1.2.1 Fixation

- Immerse samples in a mix of 1.5% (v/v) glutaraldehyde and 4% (w/v) formaldehyde in 0.1 M Cacodylate buffer, pH 7.3, at room temperature for 5-6 h [1]. Osmium post-fixation must be omitted.
- Remove fixative and wash three times in wash buffer sodium phosphate buffer (0.1M, PBS), pH 7.4.

2.1.2.2 Dehydration and embedding

- Dehydrate tissue pieces or cellular pellets in a series of increasing concentrations of ethanol: 50%, 70% and 90% (15 min each one) [2].
- Remove the 90% ethanol and infiltrate with a mixture of LR-White and 90% ethanol (1:1). Shake gently in a rotor for 2 h.
- Remove the mixture and replace by 100% embedding LR-White and leave overnight at 4°C. Ensure that the tissue pieces remain completely covered in embedding resin.

- Transfer the samples to gelatin capsules filled with LR-White, and cap and polymerize in an oven at 55°C for 24 h.

2.1.2.3 Sectioning

- Trim the LR-White blocks to form a pyramid, and obtain semithin sections (150-200 nm) which can be examined under a light microscope after staining with the toluidine blue solution [3].
- Trim the blocks for immunoelectron microscopy, forming a small pyramid with the region containing the tissue/cells selected from semithin sections.
- Obtain ultrathin sections (60-90 nm) with a diamond knife and mount the sections on a nickel grid (250 mesh). Store these in a grid case at room temperature [4].
- To inactivate the residual aldehyde groups, incubate grids on drops of 0.05 M Glycine in PBS buffer for 10-20 min.

2.1.2.4 Immunostaining procedure

- Transfer the grids to drops of 1% PBS-BSA (blocking buffer) placed on a piece of Parafilm, and incubate in a wet chamber in a Petri dish for 30 min at room temperature in order to block unspecific sites.
- Incubate the grids in 50 µl of specific primary antibody diluted in blocking buffer, followed by an overnight incubation in a wet chamber, at 4°C [5].
- Wash the grids with PBS under a jet using a wash bottle for 2 min.
- Incubate the grids in a drop of 50 µl protein A/colloidal gold complex (16 nm) [Maldonado & Aoki, 1986-b] or colloidal gold-IgG complex (available in 6, 10, or 15 nm) diluted in blocking buffer (1:20), for 30 min at 37°C.
- Wash the grids with PBS using a wash bottle and then with distilled water.

2.1.2.5 Electron microscopy contrast staining

- Incubate the grids on a drop of aqueous uranyl acetate saturate solution for 1 min, and then wash with distilled water.
- Use lead citrate if more contrast is needed [6].
- Examine the immunolabeled ultrathin sections in a TEM.

2.1.2.6 Control of specificity of the immune gold staining (see 2.1.1.4.1 section)

Footnotes

[1] Glutaraldehyde and the other chemicals used for fixation and staining are extremely hazardous, so solutions should be prepared and used in a hood. Gloves and protective eyewear should be worn whenever handling chemicals.

- There are no standard fixation protocols, and optimal conditions must be previously established for each antigen with the essential goal being their immobilization.
- Ensure that the individual tissue pieces are fully in contact with the fixative.
- If the tissue requires fixation by perfusion, a mix of 1.5% formaldehyde-glutaraldehyde in 0.1 M Cacodylate buffer, pH 7.4, is useful [Pozzo Miller & Aoki; 1991].
- Osmolarity of the fixation vehicle is one of the most essential factors for the preservation of the ultrastructure.
- The use of distilled formaldehyde, rather than formalin, is recommended.

- Some protocols indicate incubating the specimen in 0.1 M glycine solution for 20 min to quench the free aldehyde groups.

(2) N, N-Dimethylformamide (DMF) and methanol are other possible dehydration media.

(3) Thick sections can initially be cut and a light microscope used to identify the region of the block containing the sample. To carry this out blue toluidine staining applied for 30 s is useful.

(4) The use of nickel grids is recommended, especially if silver enhancement procedures are going to be used.

(5) The optimal antibody dilution and incubation conditions must be determined empirically for each antibody.

- When excessive background labeling occurs, increasing the amount of BSA in the blocking solution, lowering the antibody concentration, and/or shortening the incubation time may decrease the non-specific labeling.
- Primary antibody dilution for immunoelectron microscopy is usually 10 times more concentrated than that for immunostaining at light microscopy level.

(6) Other protocols recommend counterstaining the grids with 1% OsO_4 solution, 2% uranyl acetate for 5 min, followed by Reynolds' lead citrate solution for 1 min, and after each staining washing with distilled water.

2.1.3 Double labeling protocol

Post-embedding protocols allow the simultaneous localization of two antigens in the grid with two gold complexes of different gold particle sizes. Each antigen can usually be localized by two independent labels on each side of the grid, especially if the antibodies available were prepared in the same species. Otherwise, antibodies can be mixed and the reaction performed in a single step on one side of the grid. Figure 7 shows a double immunogold with a rabbit anti-prolactin antibody and a monkey anti-growth hormone in a ultrathin section of pituitary gland, with both being revealed using complexes of protein A-gold particles of 5 and 15 nm respectively. Technically, one face of the grid was labeled for PRL with the small gold particles, and after a brief drying, the other face was stained for GH with the particle complex [Pasolli, et al.; 1994]. In Figure 7, both hormones colocalized in the same secretory granule, while in Figure 8, neighbouring cells in primary pituitary cell cultures were identified as lactotrophs and somatotrophs in the same section using an identical protocol [Orgnero de Gaisan, et al.; 1997].

2.1.4 Immunogold applied at light microscopy level

Gold particles applied in immunolabeling have the advantage that they can also be used at light microscope level on semithin sections of the same inclusion before they are prepared. For this purpose, 1 nm gold particles (ultrasmall) offer optimal advantages mainly because their small size allows a great concentration of gold particles in a small area. However, this is not sufficient to be able to visualize the specific label in a light microscope, with it being essential to apply a silver solution to enhance the gold particles. This silver enhancement reaction is based on the gold particle catalyzed reduction of Ag^+ to metallic silver by using

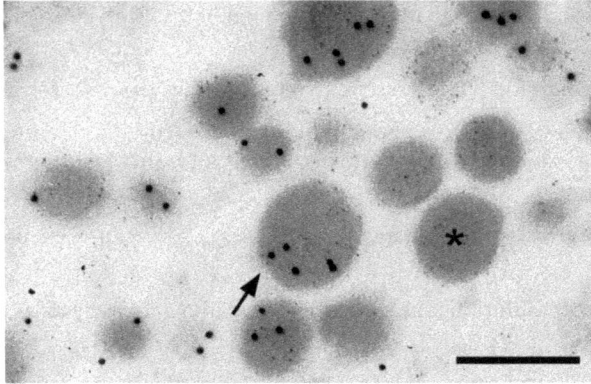

Fig. 7. Double labeling applied to identify antigen co-localization. Co-localization of growth hormone (15 nm-gold particles) and prolactin (5 nm-gold particles) in same secretory granule (arrow), and others containing only prolactin hormone are also seen (asterisk). Bar=0.25 μm. (Reproduced from Pasolli et al., 1994; with permission from Histochemistry).

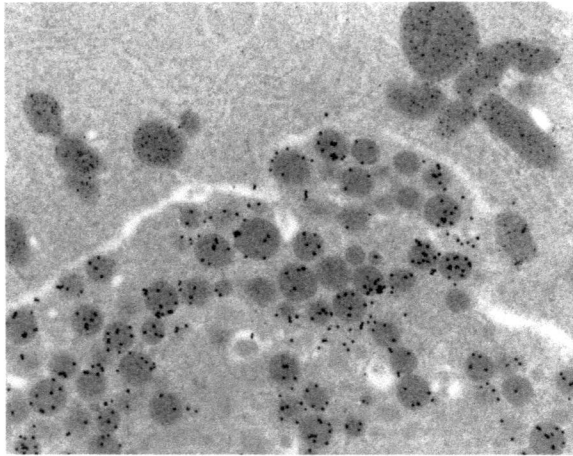

Fig. 8. Double labeling to recognize different cell types in the same section. Micrograph from normal pituitary gland in culture showing a somatotroph cell containing growth hormone (15 nm-gold particles) and a lactotroph cell expressing prolactin (5 nm-gold particles) (Reproduced from Orgnero de Gaisan et al., 1997; with permission from Ann Anat). Original magnification X 25000.

photographic developing compounds as the electron source, a system first developed by Danscher in 1981 [Danscher; 1981].

In Figure 9, silver gold enhancer was used to visualize growth hormone in sections of pituitary gland fixed in 4% formaldehyde and 2% glutaraldehyde and then included in LR-White. In this procedure, the labeling fills up the cytoplasm leaving the nuclei free and is ideal to pre-evaluate immunoreaction in a panoramic view before observation at ultrastructural level. It is also adequate for obtaining light microscopy labeling with a better resolution than in paraffin sections, particularly when used in combination at ultrastructural level. Furthermore, it is ideal for counting positive cells at great resolution [Bonaterra, et al.;1998].

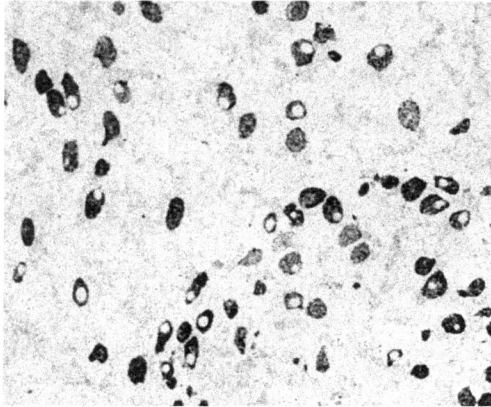

Fig. 9. Immunogold staining at light microscopy level. Growth hormone secretory cells were identified in semithin sections of pituitary gland. After gold labeled secondary antibody, a silver enhancement was performed to allow visualization by light microscopy. Original magnification X 200.

In Figure 10A, gold labeling combined with silver enhancement allowed to identify Galectin 1 (Gal-1), an immunomodulatory protein, in cytoplasmic projections of Müller cells surrounding the neuronal bodies at the inner nuclear layer in semithin sections of chicken retina. This type of staining helped to interpret the labeling obtained at the ultrastructural level (Fig. 10B) [Maldonado, et al.; 1999].

2.2 Pre-embedding technique

The pre-embedding method is applied to perform immunogold labeling before fixation (e.g. intact living cells) or after a weak fixation of biological samples (e.g. brain tissue), thus resulting in a greater preservation of the antigenicity of the molecules [Gutiérrez, et al.; 2008; Yi, et al.; 2001].

In this section, we address the protocols and basic applications of the pre-embedding immunogold labeling for transmission electron microscopy.

One of the main advantages of this approach is that specimens are not exposed to harmful or possibly damaging chemicals that can lead to the blocking or loss of target proteins. In addition, after performing immunolabeling, the samples are processed for conventional electron microscopy and included in epoxy resins that allow an improved preservation of the cellular ultrastructure.

In our laboratory, pre-embedding immunogold labeling for the localization of antigens on cell surfaces in culture cells has been achieved. This technique enables the immuno-

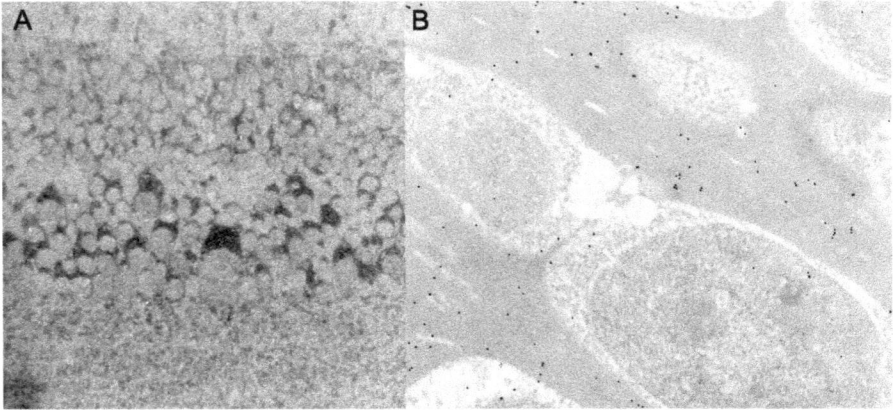

Fig. 10. Immunolocalization of Galectin 1 in chicken retina. A: Silver enhancement to localize
Gal-1 in chicken retina at light level. Gold labeling combined with silver enhancement
allowed identification of this protein on cytoplasmic projections of Müller cells surrounding
neuronal bodies at the inner nuclear layer of chicken retina. Original magnification X 200.
B: Gold labeling of Gal-1 positive structure at ultrastructural level. To verify results at light
microscopy level, cytoplasmic projections of Müller cells, identified by their high electron
density, exhibit specific gold stain for Gal-1 (Reproduced from Maldonado et al., 1999; with
permission from Invest Ophthalmol Vis Sci). Original magnification X 25000.

detection of antigens on the surface of isolated cells, viruses or bacteria, with gold particles
being localized not only on the cell surface but also on membrane extensions (pili, flagella).

Pre-embedding immunogold labeling has been useful to label virus particles in the
morphogenesis stages that imply interaction with the plasma membrane. In this particular
situation, viral replication amplifies the quantity of antigens exposed in the plasmalemma,
and a brief fixation with 1% glutaraldehyde for 10 min to preserve the cell ultrastructure
during overnight primary antibody incubation can be tolerated. In Figure 11, gold particles
decorate some virus particles being released from mouse embryonic cells infected with
Togavirus [Maldonado & Aoki, 1983].

On the other hand, when the intention is to detect intracellular antigens before resin
inclusion, the main problem is to internalize the primary and secondary antibodies inside
the cell. However, this problem has been overcome since the development of ultrasmall gold
particles. Using these in combination with posterior silver enhancement facilitates the
localization of these nanoparticles at electron microscopy level.

If the biological sample is a tissue block, it is recommended to obtain slices in a vibratome
(50 μm) after a weak fixation with 4% formaldehyde and 0.05% glutaraldehyde. Then, slices
may be treated with 0.05% Triton-X-100 in PBS for 30 min in order to permeabilize the
plasma membrane and allow the access of reactives.

This protocol has been applied successfully in the central nervous system by Sesack et al.
[Sesack, et al.; 2006]. In the case of cultured nerve cells, detergents can be replaced by
incubation with 0.05% NaBH$_4$ and 0.1%glycine in phosphate buffer overnight at 4°C to

permeabilize cell membranes, after applying the same weak fixation referred to above for tissue blocks [Van Lookeren Campagne, et al.; 1992].

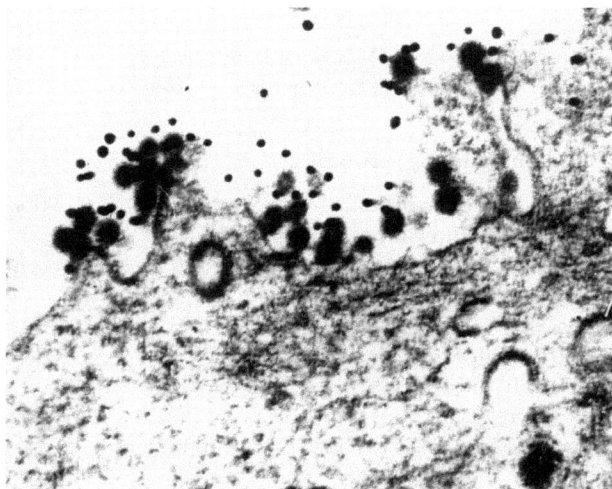

Fig. 11. Identification of superficial viral antigens. Togavirus particles budding from the plasma membrane of infected fibroblasts (Reproduced from Maldonado & Aoki, 1983; with permission from Micr Electr Biol Cel). Original magnification X 110000.

2.2.1 Pre-embedding protocol

In this section, we describe the pre-embedding immunogold labeling performed on intact living pituitary cells for the localization of antigens on cell surfaces.

2.2.1.1 Immunostaining procedure

-   Remove the culture medium from a 35 mm dish of 70-80% confluent cells and rinse gently with Hanks' Balanced Salt Solution, pH 7.0 (HBSS) [1].
-   Block unspecific antigens by employing 1% PBS-BSA (blocking buffer), pH 7.4, for 15 min at 37°C.
-   Incubate the cell monolayers with specific primary antibodies diluted in HBSS, pH 7.0 for 1 h at 37°C [2].
-   Wash 3 times with HBSS.
-   Block unspecific sites by incubating with blocking buffer for 15 min at 37°C.
-   Incubate the cell monolayers with an appropriate secondary antibody conjugated to gold particles (16 nm) diluted in blocking buffer (1:20), for 60 min at 37°C [3].
-   Wash 3 times with HBSS.
-   Lift the cells by applying a soft scraping in order to minimize shear [4].
-   Centrifuge the cell suspension at 1000 rpm for 5 min to pellet.

2.2.1.2 Fixation

-   Fix the cellular pellet with a mixture of 4% formaldehyde, 1.5% glutaraldehyde in 0.1 M Cacodylate buffer, pH 7.4 plus 7% sucrose for 2 h at room temperature.

- Wash in 0.1 M Cacodylate buffer, pH 7.4 containing 7% sucrose three times, 5 min each wash.
- Post-fix with 1% osmium tetroxide (OsO₄) in Cacodylate buffer for 1 h, in a rotor at room temperature [5].
- Rinse the cellular pellet twice in 0.1 M acetate buffer, pH 5.2.
- Stain in 1% uranyl acetate in 0.1 M acetate buffer, pH 5.2, for 20 min (bloc staining) [6].

2.2.1.3 Dehydration and Embedding Sectioning

- Dehydrate with a graded series of cold (4°C) acetones: 50%, 70%, 90%, 10 min each one; followed by acetone 100% (three times, 15 min each step).
- Perform pre-inclusion in pure Araldite/acetone 100% (1:1), overnight (covered) in rotor.
- Perform pre-inclusion in fresh 100% Araldite, 60 min in rotor.
- Embed in fresh 100% Araldite using flat embedding molds (put typed or pencil-written label in order to identify the sample). Place in 60°C oven for 24-48 h.
- Obtain ultrathin sections (60-90 nm) with a diamond knife and mount the sections on a nickel grid (250 mesh). Store these in a grid case at room temperature.

2.2.1.4 Electron microscopy contrast staining

- Stain ultrathin sections with alcoholic uranyl acetate/lead citrate [7].
- Examine the immunolabeled ultrathin sections in a TEM (Fig. 12).

Footnotes

[1] An important consideration is to start from a high cellular density due to the fact that during the procedure it is possible to lose a small number of cells, mainly during the washing step. A cell confluence of 70-80% is recommended.

[2] Primary antibody incubation is carried out in the plate wells. A piece of Parafilm set on the cell monolayer ensures full access to antibodies, and gentle shaking on a rocking platform is recommended.

[3] Appropriate dilutions should be tested and prepared in the same HBSS buffer. Excessive clustering of gold particles should be avoided, and it is recommended to test the antibody dilutions.

[4] Although trypsin is widely used for cell dissociation, it is not recommended in this procedure because it can affect the immunostaining.

[5] OsO₄ acts as a secondary fixative by reacting with lipids and oxidizing the unsaturated bonds of fatty acids. The OsO₄ is reduced to a black metallic osmium which is electron dense and adds contrast to the biological tissues.

[6] The incubation of specimens with uranyl acetate before dehydration, also called bloc staining, improves the contrast of the different structures of biological samples.

[7] This staining is used to increase the electron density of the membranes and enhance the contrast.

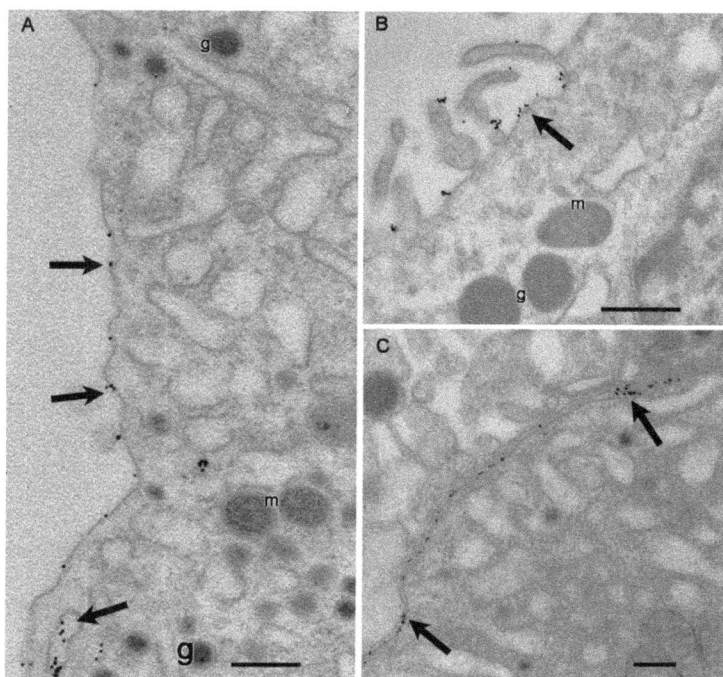

Fig. 12. Immuno detection of estrogen receptor α in lactotroph cells by pre-embedding method. Colloidal gold-particles can be observed adhered to the plasmatic membrane and attached to the cytoplasmic vesicles membranes (arrows). m: mitochondria; g: granules. Bar=0.5 μm (Reproduced from Gutiérrez et al., 2008; with permission from Steroids).

In table 1, we summarized the main differences among different protocols for immunolectron microscopy related to the antigen cell localization explained above.

3. Contribution of immunogold localization of protein in revealing morpho-functional aspects of dynamic cellular processes

Immunoelectron microscopy has been an important tool in cell biology for many decades, supplying powerful, highly valuable visual evidence to help reveal the specific localization of cell proteins, which consequently allows different cellular types to be identified. In addition, the knowledge concerning the subcellular immunogold localization of a wide range of proteins has given novel information on structure-function relationships and led to breakthroughs in the understanding of the cell biology in various physiological and pathological conditions.

In this section, we report on our studies on the cell biology of different tissues, wherein the identification of key molecules by immunoelectron microscopy has provided relevant information from the static localization of proteins to their subcellular translocation, thus revealing new insights into dynamic cell processes such as metabolic activity, differentiation, transdifferentation, proliferation and death.

INTRACELLULAR ANTIGENS			SUPERFICIAL ANTIGENS
Post-embedding	Pre-embedding		Pre-embedding
CELL SUSPENSIONS OR TISSUES	CULTURED CELLS	TISSUES	CULTURED CELLS
4% formaldehyde 1.5% glutaraldehyde	4% formaldehyde 0.05% glutaraldehyde	4% formaldehyde 0.05-0.1% glutaraldehyde	No fixation is required
Inclusion in **LR White acrylic resin**, and mounting of thin sections on **nickel grids**			
Grid incubation with 0,05M **glycine** in PBS buffer for 10-20 min to inactivate residual aldehyde	0,1% **NaBH$_4$** in PBS for 10 min to inactivate residual aldehyde	0,1% **NaBH$_4$** in PBS for 10 min to inactivate residual aldehyde	No required
Permeabilization is not required	Permeabilization with 0.05% Triton-X-100 in PBS for 30 min	Permeabilization of **vibratome slices** with 0.05% Triton-X-100 in PBS for 30 min	Permeabilization is not required
Blocking solution	Blocking solution	Blocking solution	Blocking solution
Incubation with primary antibody	Incubation with primary antibody	Incubation with primary antibody	Incubation with primary antibody
Gold complex of 16 nm particles, or combination of 5 and 15 nm particles for double labeling	Incubation with Ultra-small gold complex	Incubation with Ultra-small gold complex	Incubation with gold complex of 5, 10 or 15 nm particles
	Fixation with 2% glutaraldehyde	Fixation with 2% glutaraldehyde	Fixation with 1,5% glutaraldehyde
Silver enhancement if small or ultra small gold particles were applied	Silver enhancement	Silver enhancement	No required
	Inclusion in epoxy resins. Mounting of thin sections on copper grids	Inclusion in epoxy resins. Mounting of thin sections on copper grids	Inclusion in epoxy resins. Mounting of thin sections on copper grids
Counterstaining with **aqueous uranyl** and lead citrate (optional)	Counterstaining with alcoholic uranyl acetate and lead citrate (optional)	Counterstaining with alcoholic uranyl acetate and lead citrate (optional)	Counterstaining with alcoholic uranyl acetate and lead citrate

Table 1. Comparative table of protocols that can be applied to label superficial or intracellular antigens.

3.1 Immunocytochemical studies of cell populations are indispensible for evaluating their functional status

The anterior pituitary gland consists of several endocrine cell types which can be identified by ultrastructural features based on the size, shape, electron density, and the distribution of their secretory granules. However, under different physiological, pathological and experimental conditions, the pituitary cells display changes in the cytoplasmic organelles and in the secretory granule profiles, which make it difficult identify the various cell populations. In these circumstances, immunocytochemistry using gold particles is indispensable for the specific recognition of pituitary cell types.

We performed several studies aimed at identifying the pituitary cellular hormonal content, by applying specific antibodies against prolactin (PRL), and growth (GH), luteinizing (LHβ), adrenocorticotrophic (ACTH) and thyrotrophic (TSH) hormones.

The typical lactotrophs, somatotrophs and gonadotrophs are easily recognized because they retain the main features which have been classically described by conventional electron microscopy studies (Fig. 13). However, other undefined groups of cells displaying small round, oval or sharply pointed secretory granules can only be identified by immunoelectron microscopy, thereby highlighting the fact that this technique is essential for identifying the hormonal content of secretory granules [Orgnero, et al.; 1997].

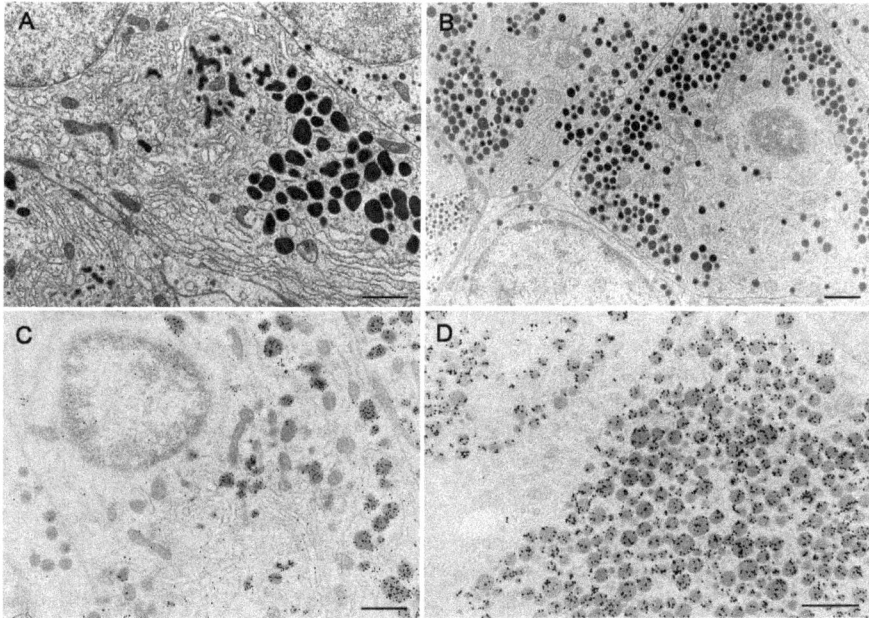

Fig. 13. Typical lactotroph and somatotroph cells identified specifically by immunoelectron microscopy. A: Typical lactotroph with immature secretory granules associated with the trans face of Golgi stacks with numerous polymorphic mature secretory granules (500-900 nm) being stored in the cytoplasm. B: Somatotroph cell with abundant round secretory granules ranging from 200 to 350 nm in diameter. C: Immunocytochemical localization of PRL in a typical lactotroph. Irregular shaped secretory granules surrounded by Golgy saccules are immunolabeled with colloidal gold particles. D: Somatotroph containing numerous spherical secretory granules immunolabeled with growth hormone antiserum. Bar=1 µm (Reproduced from Bonaterra et al., 1998; with permission from Exp Clin Endocrinol Diabetes).

In response to different stimuli, the pituitary cell populations acquire significant ultrastructural changes related to their metabolic activity. The lactotrophs or PRL-producing cells have been extensively studied in our laboratory, and we have identified different morphological types of lactotrophs (typical and atypical) by using the specific PRL immunogold label in secretory granules that display different shapes and sizes, distinguishable only by the application of the immunoelectron microscopy technique (Fig. 14). The proportions of each lactotroph subtype fluctuate in male and female rats, with these morphological variations appearing to be associated with changes in their secretory activities [Maldonado, et al.; 1994]. In another report, we demonstrated that the presence of morphological subtypes of lactotroph cells in rat pituitary cell cultures produced different secretory responses to neuropeptides, with the type I PRL cell population showing the highest response to angiotensin II and TRH action [De Paul, et al.; 1997].

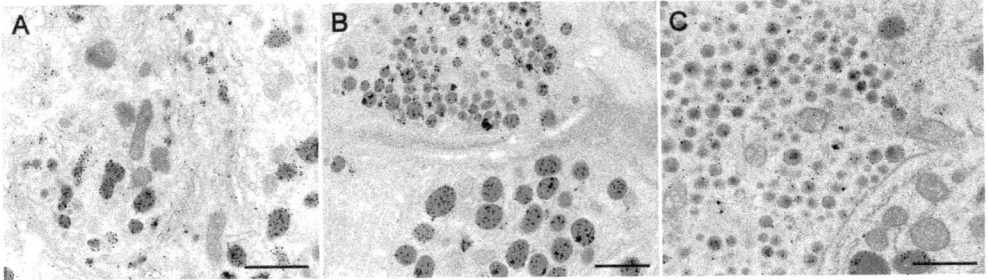

Fig. 14. Immunogold labeled lactotroph subtypes. A: Typical lactotrophs containing heavily immature and mature irregular-shaped electron-dense secretory PRL positive granules (diameter 500-900 nm). B: Atypical lactotroph characterized by the presence of spherical electron-dense granules ranging between 100-250 nm diameters atn the upper side of the picture and other typical subtypes, with bigger mature secretory granules, seen at the bottom of figure. C: Atypical lactotroph containing weakly labeled spherical granules, exhibiting an electron dense core eccentrically placed. A typical lactotroph on the right of the figure is also seen. Bar=1 μm (Reproduced from Maldonado & Aoki, 1994; with permission from Biocell).

Classical studies on the pituitary gland performed at electron microscopy level have shown that PRL and GH are synthesized by the distinct and specific cell types, lactotrophs and somatotrophs, respectively. In addition, the application of newer techniques, with a greater resolution and sensitivity has permitted the detection of a cell that produces both PRL and GH (designated as mammosomatotroph, MS), which was able to participate as an intermediate cell in prospective functional interconversions between somatotroph and lactotroph cells in the adult pituitary gland [Beresford; 1990]. In a previous work, we reported that the colocalization of PRL and GH by double immunogold labeling in the same cell was rarely observed in the pituitary of adult rats. Moreover, oestrogen treatment, which is implicated in the transdifferentiation of GH cells into PRL cells, had no effects on the MS population (Fig. 15). Thus the data obtained do not support the suggested role for MS, as transitional cells in the presumptive interconversion of PRL and GH producing cells [Pasolli, et al.; 1994].

In our laboratory, it has been demonstrated that the various endocrine cells constituting the pituitary gland do not occur at fixed proportions and undergo extensive changes depending on the physiological and experimental conditions. Each type of cell proliferates in response to sustained stimuli induced by trophic factors and hypothalamic hormones, with this situation being reversed by a degeneration of surplus cells after the interruption of a specific

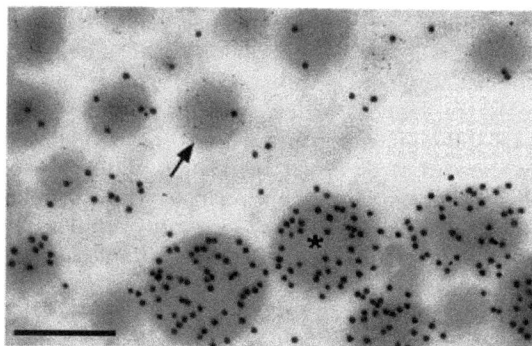

Fig. 15. Mammosomatotroph cell double immunogold labeled in pituitary rat. Co-localization of growth hormone (20 nm-gold particles) and prolactin (5 nm-gold particles) in a mammosomatotroph cell (arrow). A portion of a somatotroph labeled only for growth hormone (asterisk) is at the bottom of the figure. Bar=0.25 μm. (Reproduced from Pasolli et al., 1994; with permission from Histochemistry).

stimulus [Haggi, et al.; 1986; Pasolli, et al.; 1992; Torres, et al.; 1995; Bonaterra, et al.; 1998]. Even though the involuted pituitary cells can be recognized by just using conventional electron microscopy, specific immunogold labeling of their hormonal secretory content allows the cell type involved to be identified with complete confidence (Fig. 16). Also, the presence of an increased number of dead cells activated the phagocytosis of the cell remnants and debris through the stellate cells immunolabeled to S-100 protein [Orgnero, et al.; 1993].

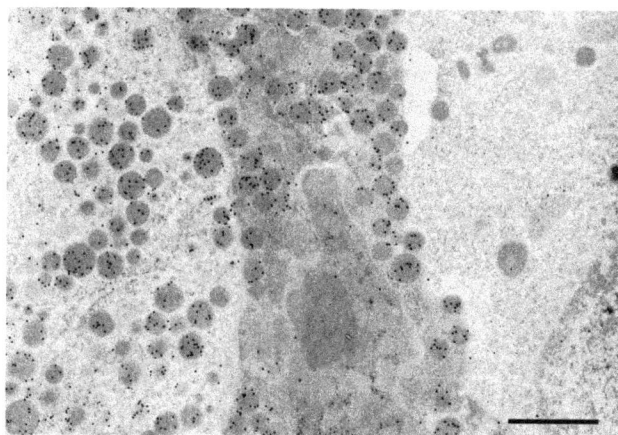

Fig. 16. Immunoelectron microscopy of somatotroph cells in different functional states. Two immunogold labeled somatotrophs in a section from pituitary male rats. A degenerated somatotroph can be seen in the centre and an active growth hormone positive cell on the left, both containing characteristic spherical secretory granules immunolabeled with specific antiserum. On the right, an unlabeled cell serves as the negative control. Bar=1 μm (Reproduced from Torres et al., 1995; with permission from Histochem J).

The size of the different pituitary cell populations was assessed by morphometric analyses of the specific immunostaining of each endocrine cell type, using light microscopy immunocytochemistry by applying immunogold complexes followed by silver enhancement. A comparative analysis of the variations in each pituitary cell population from rats submitted to different experimental and physiological conditions, revealed marked fluctuations in the size of their cell population, which was closely related to the levels of secretory activity. The response of endocrine cells exposed to increased hormonal demands was expressed not only by a release of secretory granules but also by hypertrophy and hyperplasia occurring in proportion to the levels of stimulation. In contrast, cessation of the stimulus activated regressive processes that restored the various pituitary cell populations to their initial values.

Summing up, the evaluation of the pituitary cell types by immunocytochemistry correlated with their ultrastructural changes and the serum hormone levels, which permit the metabolic cell activities and the homeostatic regulation of pituitary gland to be inferred.

3.2 Subcellular translocation of kinases evidenced by immunoelectron microscopy is indicative of biological cell activities

In order to study the mechanisms involved in modulating the proliferation and cell death of normal and tumoral pituitary cells, we investigated the intracellular translocation of protein mediators which are activated in response to different stimuli. It is well known that variations in the intracellular localization of proteins are related to their different biological functions, thus revealing dynamic processes in the cells. Immunoelectron microscopy is a key technique that can be applied to identify the specific distribution of these proteins, thereby contributing to the understanding of their functions. It has been widely demonstrated that the same protein can display opposite functions, depending on its subcellular localization, with a common example of this being the family of protein kinases C (PKC) [Akita; 2002]. These are normally inactive in the cytosolic, but once activated translocate to different cellular compartments.

In our laboratory, we investigated the regulation of cell proliferation by specific PKC isozyme translocation in normal and tumoral pituitary cells. This dynamic process was studied by using an ultrastructural immunocytochemical technique with post-embedding immunogold labeling, following protocols described previously [Petiti, et al.; 2009]. Immunoelectron microscopy demonstrated that in unstimulated normal lactotroph cells, PKCα and PKCε were evenly distributed throughout the cytosol and nucleus, but were not associated with any specific organelles. Interestingly, after a mitogenic treatment, the PRL cell population showed an association of gold particles, thus indicating the presence of PKCα at the plasma membrane, while PKCε was mainly localized to the rough endoplasmic reticulum (RER) and Golgi networks [Petiti, et al.; 2008].

In tumoral pituitary GH3B6 cells, the proliferative stimulus induced the specific translocation of both PKCs to the plasma and nuclear membrane. These results indicate the existence of a close correlation between the subcellular localization of PKC isozymes with different biological functions, revealing that the localization in the plasma and nuclear membranes is associated with cell proliferation and that their immunolabeling in Golgi networks might be linked with vesicular trafficking (Fig. 17A-C) [Petiti, et al.; 2009]. Our investigation illustrates that the specific subcellular targeting of PKCα and PKCε is indicative of the role of these enzymes in the regulation of the activity of normal and tumoral lactotroph cells.

Bearing in mind that different PKC isozymes are implicated in modulating almost all aspects of pituitary tumorigenesis, it is important to explore the participation of these kinases in cell proliferation, survival and cell death. In contrast with the role of PKCα and PKCε in promoting pituitary cell growth, it has been demonstrated in GH3B6 pituitary adenoma cells that PKCδ activation plays an crucial role in programmed cell death [Leverrier, et al.; 2002]. Also, in a recent study, we identified the non-apoptotic mechanism parapoptosis as being the predominant cell death type involved in the regression of pituitary tumors after bromocriptine treatment.

In tumoral pituitary cells, we determined the fine localization of PKCδ by means of immuno-electron microscopy and observed immunolabeling in the cytosolic matrix, which was associated with RER and some isolated mitochondria in intact male rats. Furthermore, bromocriptine treatment was able to enhance the immunogold labeling to PKCδ in the nuclear compartment of pituitary cells at different involutive and dying stages (Fig. 17D). These results are indicative of PKCδ intracellular translocation, inferring that a dynamic process occurs in pituitary cell death [Palmeri, et al.; 2009].

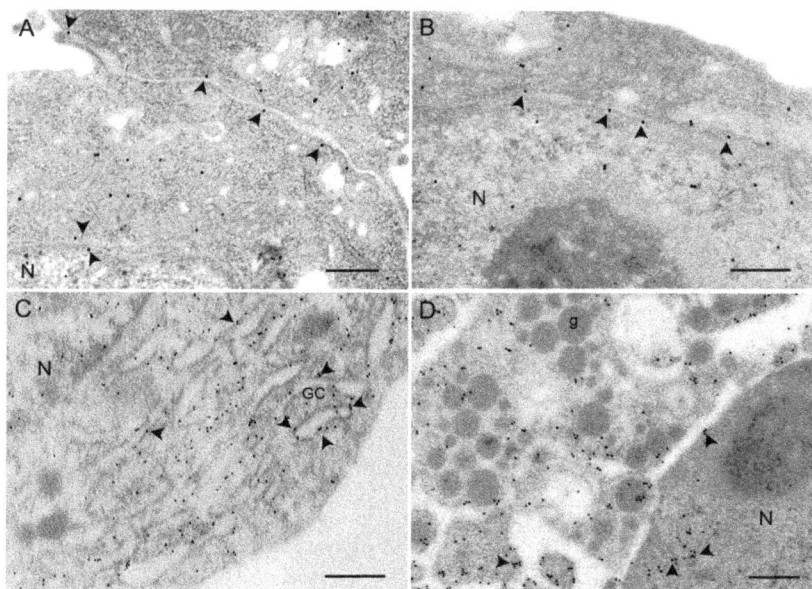

Fig. 17. Immuno-electron labeling for PKC isozymes in pituitary cells. Immunogold particles for PKCα attached to the plasma membrane (A) and PKCε associated with nuclear envelope (B) (arrowhead) after mitogenic stimulation in pituitary GH3B6 tumoral cells. PKCε was mainly localized to the Golgi complex (GC) (arrowhead) in proliferating lactotroph cells (C). Bc treatment enhanced PKCδ labeling in the nuclear compartment associated with euchromatin and nuclear envelope (arrowhead). N: nucleus; g: granules. Bar=1 μm. (Reproduced from Petiti et al., with permission from Mol Cell Endocrinol 2008 and J Mol Histol 2009, and Palmeri et al., with permission from Toxicol Appl Pharmacol 2009).

3.3 Immunoelectron microscopy emerges as an indispensable tool to understand the differentiation status of the cell

Diverse changes in the cell microenvironment, such as those generated by inflammatory stimuli, can induce significant alterations in both the epithelial and stromal cells, thus leading to modifications in cell biology and function. However, immunogold labeling at the ultrastructural level is necessary in order to reveal and characterize these processes, by evaluating cytoskeleton components, secretory products, growth factor receptors, etc.

In lung epithelium, we have reported changes induced by allergy inflammation in bronquiolar Clara cells (which are strongly involved in several key homeostatic mechanisms). The apical cytoplasm of these polarized secretory cells is filled with polymorphic mitochondria and scarce secretory granules under the plasma membrane (Fig. 18A). Nevertheless, even under normal conditions, the immunogold technique was able to differentiate between these two organelles: mitochondria stained positive to CYP1E2 (P450 cytochrome), while the secretory granules were easily identified by their immunoreactivity for CC16, the main secretory protein of Clara cells (Fig. 18B). After applying a short allergic stimulus, these cells hypertrophied and filled up with big secretory-like granules of moderate electron density (Fig. 18C). Then, immunogold helped us to identify the mitochondria as scarce CYP1E2-positive structures intermingled among the granules and to observe that the numerous secretory granules strongly gold-labeled for CC16 (Fig. 18D). Following chronic allergic exposition, the cells continued to be hypertrophied, but their secretory granules were bigger and fused with very low electron-densities (Fig. 18E), with immunogold indicating that they were not storage sites for CC16 (Fig. 18F). In this way, immunoelectron microscopy confirmed that a short allergy stimulus induced Clara cells to respond with hypersecreting CC16 as a protective mechanism against inflammation, while a chronic inflammatory microenvironment caused their transdifferentiation to mucous secreting cells that lost their CC16 storage in secretory granules [Roth, et al.; 2007].

Stromal cells are other important targets of microenvironmental modifications. In particular, fibroblasts and smooth muscle cells have proved to be very sensitive to cytokines and growth factors, with the transforming growth factor beta being a well-known stromal remodeling agent. A clear example of this remodeling was observed in prostatic smooth muscle cells after inflammation; under normal conditions. These cells are fully differentiated to a contractile phenotype containing mainly cytoskeleton components, as can be shown by immunolabeling smooth muscle α-actin (ACTA2), and scarce synthesis organelles (Fig. 19A and B). In experimental prostatitis, we observed phenotypic changes in these cells, including an increase in the cisternae of rough endoplasmic reticulum and mitochondria and the loss of cytoskeleton fibers (Fig. 19C), with the contribution of immunogold staining in this process having been focused on confirming the muscular nature of these cells and corroborating the reduction of the ACTA2-positive cytoskeleton compartment within the cells (Fig. 19D) [Quintar, et al.; 2010].

Another important example of remodeling in response to microenvironmental changes is the transformation of fibroblasts (constant components of the stroma) into myofibroblasts by myodifferentiation, a process that could be observed by immunolabeling smooth muscle α-actin at electron microscopy level. In a prostatitis model, we detected cells with morphological features of myofibroblasts, including a well-developed rough endoplasmic reticulum and signs of nuclear activation (Fig. 19E). As these cells are associated with reactive stroma in cancer, it was important to verify that they were present in the stroma in the prostatitis model, which was carried out by applying immunogold labeling to ACTA2 (Fig. 19F).

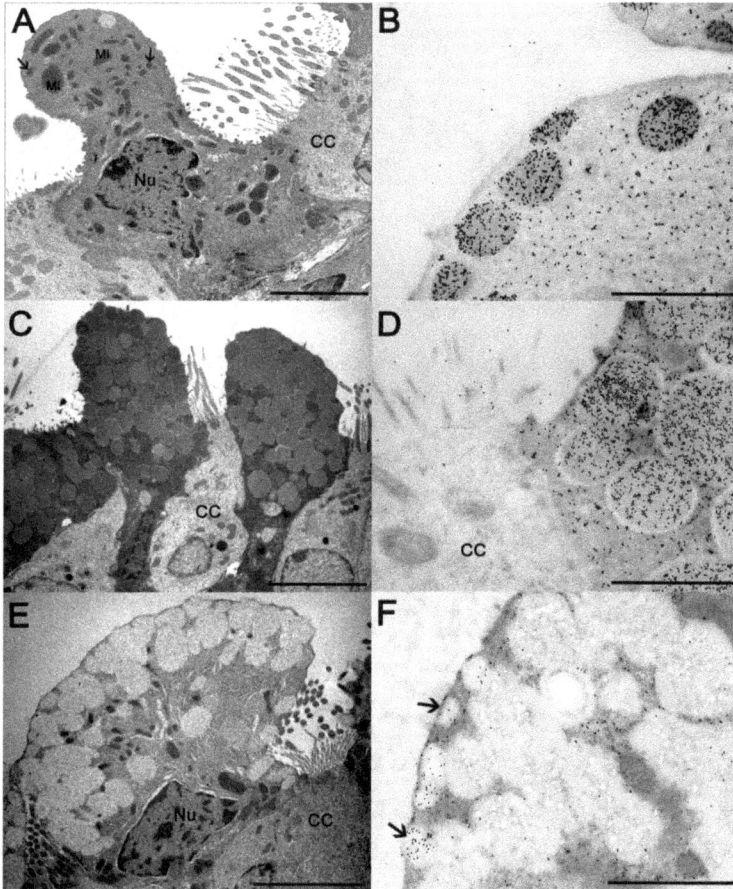

Fig. 18. Immunogold for CC16 to analyze effects of allergic inflammation on bronchiolar Clara cell secretions at ultrastructural level. A: Normal bronquiolar Clara cells with numerous mitochondria (Mi) and few small secretory granules with round profiles and moderate electron density. B: By immunogold labeling, the protein CC16 was shown to be intense at the secretory granules, while it was moderate in the cytoplasm. C: After acute OVA challenge, Clara cells exhibit numerous large secretory granules filling the whole apical cytoplasm. D: Large and electron- lucent secretory granules, appearing after acute OVA challenge, are heavily decorated for CC16 protein. Also, free cytoplasmic labeling appears around the granules. E: After chronic OVA, mucous transformed Clara cells show characteristic large fused electron-lucent granules. F: Mucous transformed Clara cells exhibit scarce CC16 label restricted to a few positive small granules under the plasma membrane with the cytoplasm portions intermingled. CC: ciliated cell; Nu: nucleus; OVA: ovalbumin. Bar=1 μm (Reproduced from Roth et al., 2007; with permission from Histochem Cell Biol).

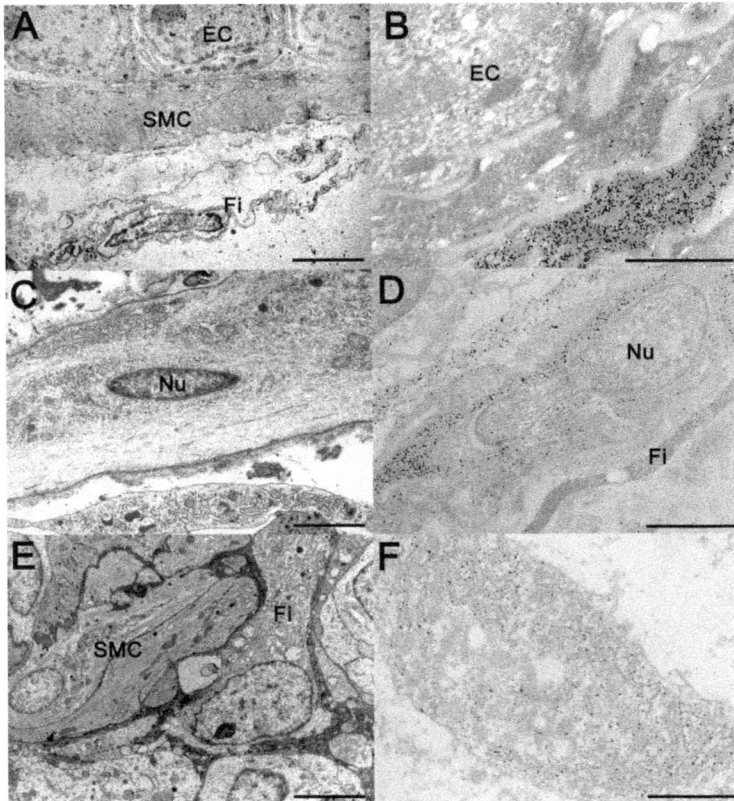

Fig. 19. Immunogold for ACTA2 to analyze the effects of prostatitis on cytoskeleton components of prostatic stromal cells at ultrastructural level. A: Periacinar smooth muscle cell (SMC) in normal prostate gland. B: ACTA2 immunogold labeling indicates a homogenously distributed actin cytoskeleton in periacinar smooth muscle cells on normal prostate gland. C: Smooth muscle cell in prostatitis. Note the large development of cytoplasmic organelles accompanied by loss of contractile filaments. D: ACTA2 immunolabeling on smooth muscle cell indicates that actin filaments are restricted to peripheral zones of these periacinar cells in prostatitis. E: In prostatitis, a periacinar stimulated fibroblast exhibits well developed rough endoplasmic reticulum; smooth muscle cells exhibit signs of dedifferentiation as described in figure C. F: ACTA2 immunolabeling exposes the myodifferentiation process occurring in stimulated periacinar fibroblast in prostatitis. Nu: nucleus; Fi: fibroblast; EC: epithelial cell. Bar=1 μm. (Reproduced from Quintar et al., 2010; with permission from Prostate).

4. Conclusion

This review was written with the main purpose of emphasizing the great potentiality of immunolabeling as an indispensable tool to combine the ultrastructure with the specific identification of the molecules involved in biological processes, thus contributing to the understanding of the structure–function relationships. In order to encourage scientists from different fields of cell biology, the challenge of immunocytochemistry at ultrastructural level has been treated using a wide approach, thereby providing the knowledge and criteria necessary to implement various techniques and to suggest solution the problems inherent to this methodology. Studies performed by our research group have been described, with the identification of key molecules by immunoelectron microscopy permitting dynamic cell processes in different tissues to be inferred.

5. Acknowledgment

We are especially grateful to Prof. Dr. Agustín Aoki, whose tireless dedication to the field of transmission electron microscopy has enabled us to relate, in this chapter, our experience in the immunoelectron microscopy field. Drs. Elsa Orgnero-Gaisán, Alejandro Peralta Soler, Amalia Pasolli, Ernesto Haggi, Mónica Bonaterra, Félix Roth and Claudia Palmeri are gratefully acknowledged for their generous contribution to the immunoelectron microscopy images. Also, the authors wish to thank Roald Pittau, Mrs Mercedes Guevara, Mrs Elena Pereyra and Mrs. Lucía Artino for their excellent technical assistance and native English speaker Dr. Paul Hobson for revising the English of this chapter. This work was supported by grants from the Consejo Nacional de Investigaciones Científicas y Técnicas (CONICET) and Fondo para la Investigación Científica y Tecnológica (FONCyT-ANPCyT).

6. References

Akita, Y. (2002). Protein kinase C-epsilon (PKC-epsilon): its unique structure and function. *J Biochem*, Vol. 132, No. 6, (Dec 2002), pp.847-852, ISSN 0021-924X

Bendayan, M. & Zollinger, M. (1983). Ultrastructural localization of antigenic sites on osmium-fixed tissues applying the protein A-gold technique. *J Histochem Cytochem*, Vol. 31, No. 1, (Jan 1983), pp. 101-109, ISSN 0022-1554

Beresford, W. (1990). Direct transdifferentiation: can cells change their phenotype without dividing?. *Cell Differ Dev*, Vol. 29, No. 2, (Feb 1990), pp. 81-93, ISSN 0922-3371

Bonaterra, M.; De Paul, A.; Aoki, A.; Torres, A. (1998) Residual effects of thyroid hormone on secretory activity of somatotroph population. *Exp Clin Endocrinol Diabetes*,Vol. 106, No. 6, pp. 494-9. ISSN 0947-7349

Danscher, G. (1981). Localization of gold in biological tissue. A photochemical method for light and electronmicroscopy. *Histochemistry*, Vol. 71, No. 1, pp. 81-88, ISSN 0301-5564

De Paul, A.; Pons, P.; Aoki, A. & Torres, A. (1997). Different behavior of lactotroph cell subpopulations in response to angiotensin II and thyrotrophin-releasing hormone. *Cell Mol Neurobiol*, Vol. 17, No. 2, (Apr 1997), pp. 245-258, ISSN 0272-4340

Faulk, W. & Taylor, G. (1971). An immunocolloid method for the electron microscope. *Immunochemistry*. Vol. 8, No. 11, (Nov 1971), pp. 1081-1083, ISSN: 0019-2791

Ferguson, D. J.; Hughes, D. A. & Beesley, J. E. (1998). Immunogold probes in electron microscopy. *Methods Mol Biol*, Vol. 80, pp. 297-311, ISSN:1064-3745

Haggi, E.; Torres, A.; Maldonado, C. & Aoki A. (1986). Regression of redundant lactotrophs in rat pituitary gland after cessation of lactation. *J Endocrinol*, Vol. 111, No. 3, (Dec 1986), pp. 367-373, ISSN 0022-0795

Hagiwara, H.; Aoki, T.; Kano, T. & Ohwada, N. (2000). Immunocytochemistry of the striated rootlets associated with solitary cilia in human oviductal secretory cells. Histochem Cell Biol, Vol. 114, No. 3, pp. 205-212, ISSN:0948-6143

Hainfeld, J. & Furuya, F. (1992). A 1.4-nm gold cluster covalently attached to antibodies improves immunolabeling. *J Histochem Cytochem*, Vol. 40, No. 2, (Feb 1992), pp. 177-184, ISSN 0022-1554

Leverrier, S.; Vallentin, A. & Joubert, D. (2002). Positive feedback of protein kinase C proteolytic activation during apoptosis. *Biochem J*, Vol. 15, No. 368, (Dec 2002), pp. 905-913, ISSN 0264-6021

Maldonado, C. & Aoki, A. (1983). Immuno-gold technique for identification of virus antigens. *Microsc Electron Biol Celular*, Vol. 7, No. 2, pp. 23-26, ISSN 0326-3142

Maldonado, C. & Aoki, A. (1986.-a). Influence of embedding media in prolactin labelling with immunogold techniques. *Histochem J*, Vol. 18, No. 8, (Agost 1986), pp. 429-433, ISSN 0018-2214

Maldonado, C. & Aoki, A. (1986-b). Improvement of prolactin immunolabelling in osmium-fixed acrylic-embedded pituitary gland. *Bas Appl Histochem*, Vol. 30, No. 3, pp. 301-305, ISSN 0391-7258

Maldonado, C.; Saggau, W. & Forssmann, W. (1986). Cardiodilatin-immunoreactivity in specific atrial granules of human heart revealed by the immunogold stain. *Anat Embryol (Berl)*, Vol. 173, No. 3, pp. 295-298, ISSN 0340-2061

Maldonado, C. & Aoki, A. (1994). Occurrence of atypical lactotrophs associated with levels of prolactin secretory activity. *Biocell*, Vol. 18, No. 3, pp. 88-95, ISSN 0327-9545

Maldonado, C.; Castagna, L.; Rabinovich, G. & Landa C. (1999). Immunocytochemical study of the distribution of a 16-kDa galectin in the chicken retina. *Invest Ophthalmol Vis Sci*, Volm. 40, No. 12, (Nov 1999), pp.2971-2977, ISSN 0146-0404

Mukdsi, J.H.; De Paul, A. L.; Petiti, J.P.; Gutiérrez, S.; Aoki, A. & Torres, A.I. (2006). Pattern of FGF-2 isoform expression correlated with its biological action in experimental prolactinomas. *Acta Neuropathol*, Vol. 112, No. 4, (Oct 2006), pp. 491-501, ISSN 0001-6322

Orgnero de Gaisán, E.; Maldonado, C. & Aoki A. (1993). Fate of degenerating lactotrophs in rat pituitary gland after interruption of lactation: a histochemical and immunocytochemical study. *Histochem J*, Vol. 25, No. 2, (Feb 1993), pp. 150-165, ISSN 0018-2214

Orgnero de Gaisán, E.; Maldonado, C.; Diaz Gavier, M. & Aoki, A. (1997). Diversity of pituitary cells in primary cell culture. An immunocytochemical study. *Ann Anat*, Vol. 179, No. 5, (Oct 1997): pp. 453-460, ISSN 0940-9602

Palmeri, C.; Petiti, J.; Sosa, Ldel V.; Gutiérrez, S.; De Paul, A.; Mukdsi, J. & Torres, A. (2009). Bromocriptine induces parapoptosis as the main type of cell death responsible for experimental pituitary tumor shrinkage. *Toxicol Appl Pharmacol*, Vol. 240, No. 1, (Oct 2009), pp. 55-65, ISSN 0041-008X

Pasolli, H.; Torres, A. & Aoki, A. (1992). Influence of lactotroph cell density on prolactin secretion in rats. *J Endocrinol*, Vol. 134, No. 2, (Aug 1992), pp. 241-246, ISSN 0022-0795

Pasolli, H.; Torres A.; Aoki, A. (1994). The mammosomatotroph: a transitional cell between growth hormone and prolactin producing cells? An immunocytochemical study. *Histochemistry*, Vol. 102, No. 4, (Oct 1994), pp. 287-296. ISSN 0301-5564

Peralta Soler, A.; Maldonado, C. & Aoki, A.(1988). Differential detection of murine mammary tumor virus constituents by immuno-electron microscopy. *Microsc Electron Biol Celular*, Vol. 12, No. 1, (Jun 1988), pp. 89-99, ISSN 0326-3142

Peters, P. & Pierson, J. (2008). Immunogold labeling of thawed cryosections. *Methods Cell Biol*, Vol. 88, Chapter 8, pp. 131-149, ISSN 0091-679X

Petiti, J.; De Paul, A.; Gutiérrez, S.; Palmeri, C.; Mukdsi, J. & Torres, A. (2008). Activation of PKC epsilon induces lactotroph proliferation through ERK1/2 in response to phorbol ester. *Mol Cell Endocrinol*, Vol. 16, No. 289 (1-2), (Jul 2008), pp. 77-84, ISSN 0303-7207

Petiti, J.; Gutiérrez, S.; Mukdsi, J.; De Paul, A. & Torres, A. (2009). Specific subcellular targeting of PKCalpha and PKCepsilon in normal and tumoral lactotroph cells by PMA-mitogenic stimulus. *J Mol Histol*, Vol. 40, No. 5-6, (Oct 2009), pp. 417-425, ISSN 1567-2379

Pozzo Miller, L. D.; Aoki, A. (1991). Stereological analysis of the hypothalamic ventromedial nucleus. II. Hormone-induced changes in the synaptogenic pattern. *Brain Res Dev Brain Res*. Vol. 61, No. 2, (Aug 1991), pp. 189-196, ISSN 0165-3806

Quintar A. A.; Doll A.; Leimgruber C.; Palmeri C.M.; Roth F. D.; Maccioni M. & Maldonado, C.A. (2010). Acute inflammation promotes early cellular stimulation of the epithelial and stromal compartments of the rat prostate. *The Prostate*, Vol. 70, No 11, pp. 1153-1165, ISSN 1097-0045

Roth, F.D.; Quintar, A. A.; Uribe Echevarría, E. M.; Torres, A. I.; Aoki, A. & Maldonado C. A. (2007). Budesonide effects on Clara cell under normal and allergic inflammatory condition. *Histochem Cell Biol*, Vol. 127, No 1, pp. 55-68, ISSN 0948-6143

Sesack, S.; Miner, L. & Omelchenko N. (2006). Preembedding Immunoelectron Microscopy: Applications for Studies of the Nervous System, In: *Neuroanatomical tract-tracing 3: molecules, neurons, and systems*, F. Laszlo Zaborszky (Editor), Lanciego J (Editor), pp. 6-71, Publisher Springer, ISBN 0387289410. New York.

Tokuyasu, K. (1973). A technique for ultracryotomy of cell suspensions and tissues. *J Cell Biol*, Vol. 57, No. 2, (May 1973), pp. 551-565, ISSN 0021-9525

Torres, A.; Pasolli, H.; Maldonado, C. & Aoki A. (1995). Changes in thyrotroph and somatotroph cell populations induced by stimulation and inhibition of their secretory activity. *Histochem J*, Vol. 27, No. 5, (May 1995), pp. 370-379, ISSN 0018-2214

Van Lookeren Campagne, M.; Dotti, C.; Jap Tjoen San, E.; Verkleij, A.; Gispen, W. & Oestreicher, A. (1992). B-50/GAP43 localization in polarized hippocampal neurons in vitro: an ultrastructural quantitative study. *Neuroscience*, Vol. 50, No. 1 (Sep 1992), pp.35-52, ISSN 0306-4522

Yi, H.; Leunissen, J.; Shi, G.; Gutekunst, C. & Hersch, S. (2001). A novel procedure for pre-
embedding double immunogold-silver labeling at the ultrastructural level. *J Histochem Cytochem*, Vol 49, No. 3 (Mar 2001), pp.279-284, ISSN 0022-1554.

4

Immunohistochemical Correlation of Novel Biomarkers with Neurodegeneration in Rat Models of Brain Injury

Shyam Gajavelli, Amade Bregy, Markus Spurlock, Daniel Diaz,
Stephen Burks, Christine Bomberger, Carlos J. Bidot, Shoji Yokobori,
Julio Diaz, Jose Sanchez-Chavez and Ross Bullock
Lois Pope LIFE Center, University of Miami
Miller School of Medicine, Miami, FL
USA

1. Introduction

Immunohistochemistry is an important technique used to visualize specific changes in tissues as part of both normal development and pathological conditions. Immunohistochemistry (IHC) combines antibody specificity with high resolution imaging techniques which, together, can provide reliable visual evidence of presumed physiological processes. The technique is amenable to rigorous experimental design. In controlled settings, with the appropriate constraints, and image acquisition settings, data obtained becomes reliable, quantifiable and reproducible. Through the numerous past and current explorations of IHC, novel, antibody-based, 'point of care' diagnostics as well as antibody-based therapeutics will and are becoming a valuable tool for clinicians and researchers. A field that really craves such new diagnostic and therapeutic technologies is that of traumatic brain injury.

Brain injury, especially traumatic brain injury, or TBI, initiates a complex series of neurochemical signalling events. These pathological changes are mediated, at lease in part, by glutamate excitotoxity, inflammation and increased blood-brain barrier (BBB) permeability leading to numerous sequale which include neuronal hyperactivity, increased cellular vulnerability, edematous states, cellular dysfunction and consequent apoptotic and necrotic cell death. Many of these changes produce subtle and slight global manifestations and hence are invisible to current diagnostic imaging techniques such as magnetic resonance imaging or computer aided tomography scans. Recently though, the decades-long efforts of a few pioneering groups has established novel protein biomarkers for TBI (Mondello, Muller et al. 2011). Unlike many traditional biomarkers, these novel protein biomarkers can be investigated using IHC. As their discovery remains recent, the relationship between their presence in the extracellular environment and their intracellular source has not been explored.

Our group investigated one such biomarker, namely ubiquitin carboxy-terminal hydrolase-1 (UCH-L1), and found that it could provide powerful information regarding the degenerative state of neurons. Using immunohistochemistry of serial sectioned brain slices we saw that brain areas containing degenerate neurons were devoid of immuno-signal for UCH-L1. In contrast robust UCH-L1 staining could be seen where degenerate neurons were absent. Similar IHC results were observed in surgically resected human brain tissue from patients with brain injury, although independent verification is needed. The levels seen using UCH-L1 immunohistochemistry were inversely proportional to the extracellular concentrations of this protein. In the next few sections of this chapter we will present two rodent brain injury models, IHC based cell-specific markers, chemical stains associated with neurodegeneration, the role of IHC in assessing the relationship between novel brain injury biomarker UCH-L1 and the polyanionic fluorescein derivitave, Fluorojade B (FJB), which sensitivity and specifically binds to degenerating neurons. Taken together our data suggests that UCH-L1 is lost prior to the onset of FJB target *in vivo* following brain injury

2. Rodent models of brain injury

The establishment and use of animal models of traumatic brain injury (TBI) remains vital to understanding the pathophysiology of this highly complex condition. The purpose of these experimental models is to replicate certain pathological components or phases of clinical trauma in experimental animals aiming to address pathology and/or treatment (Cernak 2005). Such models share the ultimate goals of reproducing patterns of tissue damage observed in humans. Our laboratory has extensive experience with two TBI models (Dietrich, Alonso et al. 1994; Zhou, Sun et al. 2008). In this study we explore the use of those models to assess the relationship between neurodegeneration and immunohistochemical analyses of novel biomarkers.

2.1 Acute Subdural Hematoma (ASDH)

Adult Sprague Dawley (350-400 g) rats are anesthetized with a mixture of 3% isoflurane 70% N_2O and 30% oxygen. Following determination of adequate anesthesia by monitoring toe-pinch (foot reflex-flexion to pain) and the corneal reflexes, the tail artery is catheterized to aid monitoring of vitals such as blood pressure, blood gases and pH. Detailed descriptions of the procedures have been extensively published (Daugherty, Levasseur et al. 2004; Kwon, Sun et al. 2005; Zhou, Sun et al. 2008). Briefly, the animal is intubated using a shielded 14GA i.v. catheter (BD Insyte Autoguard, 14GA 175IN, 2.1x45mm). The animal is paralyzed using Pancuronium Bromide (0.35 mg/kg every ½ hour) and connected to the ventilator. Following reduction of isoflurane level to 0.5-1.0%, the first blood gas analysis is performed to control proper ventilation of the animal. The partial oxygen tension (pO2) of ~100-150mmHg and a partial carbon dioxide tension (pCO2) of 35-45mmHg is attained before continuing with the surgery. A midline scalp incision is made to expose the sagittal and coronal sutures. A parasagittal craniotomy (4.8 mm) using a trephine (Roboz) is performed at 3.8 mm posterior to bregma and 2.5 mm lateral to the midline creating a burr hole 3 mm in diameter. Using an operating microscope the dura is incised and a blunt, pre-bent J-shaped 23-guage needle filled with ~0.4ml non-heparinized, autologous arterial blood is carefully inserted into the subdural space with the curved tip pulled up against the overlying dura; rapid-curing cyanoacrylate glue is used to seal the burr hole and secure the needle. To create the subdural hematoma 0.35 ml of the blood is slowly injected into the subdural space over 6-7 minutes. The needle is crimped and animal is treated according to a previously assigned therapeutic intervention. As shown in the Fig.1 the hematoma is subdural and is wide spread.

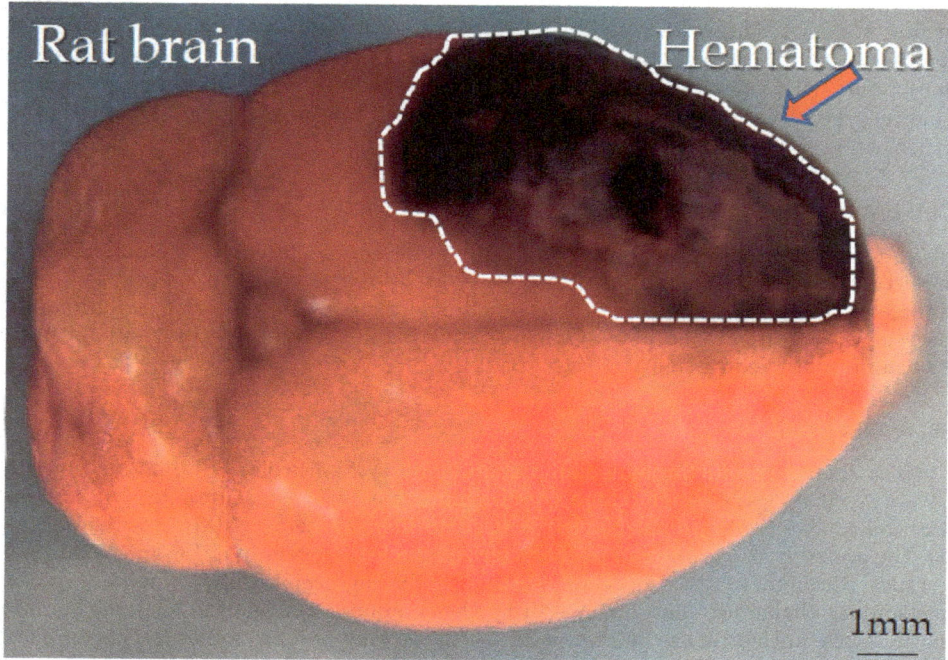

Fig. 1. Acute subdural hematoma model (ASDH) An extensive subdural hematoma (white dashed outline) can be seen on the left frontal and temporal lobe of this rat brain resulting in damage of the cortex in this area (dark area).

2.2 TBI

Adult Sprague-Dawley rats are anesthetized and a 4.8mm burr hole is made as described above in ASDH paragraph. A sterile plastic injury tube is next placed over the exposed dura and glued to the skull using acrylic adhesive. Dental acrylic is poured around the injury tube to obtain a perfect seal. After the acrylic has hardened, the scalp is closed using staples and the animals are returned to their home cage. About 18 hours after the previous preparation the rats are anesthetized, intubated, connected to a respirator and ventilated with 0.5-1% isoflurane and a mixture of 70% nitrous oxide and 30% oxygen. The animal is paralyzed with Pancuronium for mechanical ventilation to maintain arterial blood gases within normal limits. The fluid percussion (F-P) device (Custom Design and Fabrication, VCU Medical Center, University of Virginia, VA, USA) consists of a Plexiglas cylindrical reservoir bounded at one end by a rubber-covered Plexiglas piston with the opposite end fitted with transducer housing and a central injury screw adapted for the rat's skull. The entire system is filled with isotonic saline. The (aseptic) metal injury screw is next firmly connected to the plastic injury tube of the intubated and anesthetized rat. The injury is induced by the descent of a metal pendulum striking the piston, thereby injecting a small volume of saline epidurally into the closed cranial cavity and producing a brief displacement (22 msec) of neural tissue. The amplitude of the resulting pressure pulse is measured in atmospheres by a pressure transducer and recorded on a PowerLab chart recording system (ADinstruments, CO, USA). Brain and body temperature are recorded with a thermistor into the left temporalis muscle and a rectal probe. Animals are maintained normothermic with a combination of heating lamps and a negative feedback control during the entire procedure. In this study, a moderate (1.8-2.2 atmospheres) injury was used (Dietrich, Alonso et al. 1994; Bregy 2010).

3. Cell type specific markers

Antibodies against proteins found predominantly in the central nervous system (CNS) have been successfully used to identify the CNS cells. Molecular markers for specific cell types are intricately associated with structure or function of the cells; they are also evolutionarily conserved, lending them credibility. However, changes in cell type specific markers alone are not sufficient to deduce injury effects on cells.

3.1 Neurons: NeuN

Antibodies against specific proteins found abundantly in a particular cell type can be used to identify cell types. A single clone (A60) against a vertebrate neuron-specific nuclear protein called NeuN (Neuronal Nuclei) is used to identify neurons (Mullen, Buck et al. 1992). The antibody reacts with an uncharacterized nuclear protein found in most neuronal cell types throughout the nervous system including cerebellum, cerebral cortex, hippocampus, thalamus, and spinal cord. The immunohistochemical staining occurs primarily in the nucleus of the neurons with lighter staining in the cytoplasm. The antibody cross-reacts with nervous tissue in a wide range of animals from salamanders, chicks, and rats to humans. The figure shows a confocal image tile of NeuN labeled neurons (green) in gray matter of the rat spinal cord (Fig.2). As the release of NeuN antigen into extracellular media is not characterized, the utility of this protein as a serum or cerebrospinal fluid biomarker remains unexplored.

3.2 Glia: GFAP

Glial fibrillary acidic protein (GFAP) has been identified as a brain specific protein in brains of patients with multiple sclerosis (MS), in which GFAP was purified from a large MS plaque consisting primarily of fibrous astrocytes and demyelinated axons (Eng, Ghirnikar et al. 2000). GFAP mediates astrocyte functions known to be important during regeneration, synaptic plasticity and reactive gliosis (Eng, Ghirnikar et al. 2000) (Middeldorp and Hol 2011). However, GFAP is not only expressed in astrocytes but also in the neural crest-derived non-myelinating Schwann cells (SC) and in spinal cord following injury (Gajavelli, Wood et al. 2004). Several antibodies against GFAP are widely used in research and clinically for immunohistochemical diagnosis of tumors. GFAP positive cells appear green with cytoplasmic GFAP intermediate filaments revealing underlying cellular morphology of astrocytes in an immunostained rat spinal cord section (Fig. 2). Cautious interpretation of anti-GFAP immunocytochemical results is recommended as positive staining may identify an astrocyte in the CNS but a negative result may be false. (Eng, Ghirnikar et al. 2000). Currently, GFAP is used as a biomarker for gliosis (Mondello, Muller et al. 2011).

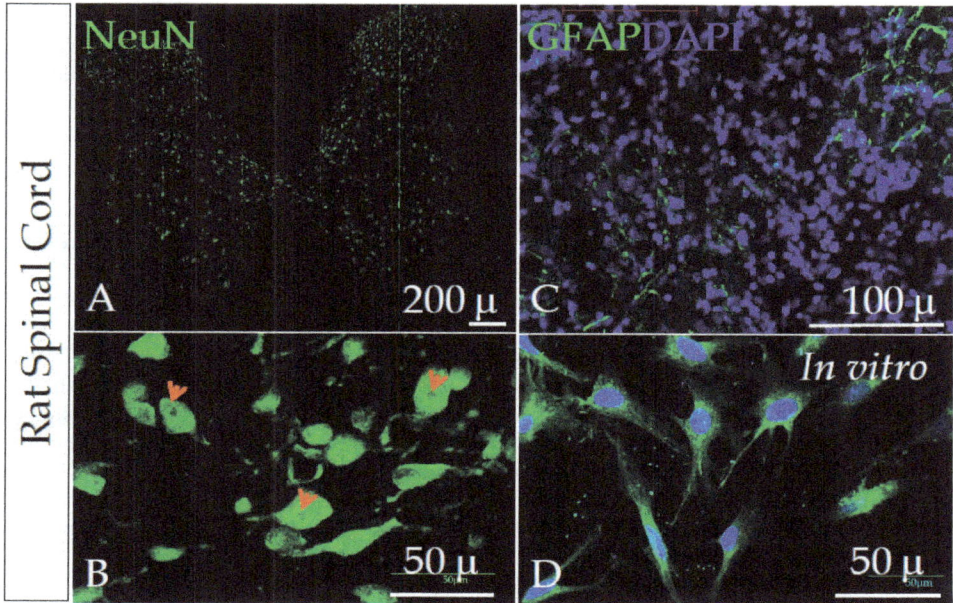

Fig. 2. Coronal sections of rat spinal cord were immunostained for NeuN (A) and GFAP (C). NeuN predominantly stains the nucleus and labels the entire gray matter. At higher magnification a few cytoplasmic processes but not nucleolus (arrowheads in B) stain with NeuN. GFAP+ astrocytes appear throughout the spinal cord (C). At a higher magnification GFAP-immunocytochemistry of an *in vitro* culture (rat postnatal day 4) reveals GFAP intermediate filaments underlying cellular morphology. DAPI stained nuclei (Blue).

4. Cell death markers

The standard methods for the detection of cell death in the nervous tissue include anti-Caspase-1 immunohistochemistry and chemical staining for degenerate neurons using Fluorojade B. In the following section, we investigate the co-incidence of these markers with the novel biomarkers of neural damage.

4.1 Caspase 1

A large macromolecular complex, termed the inflammasome, activates cysteine aspartic acid proteases / interleukin-1 beta converting enzyme (caspase-1). The activated caspase-1, a (p20/p10) tetramer, is necessary and sufficient for cleavage/maturation of proinflammatory cytokines interleukin-1beta (IL-1beta) and IL-18 as well as for the induction of apoptosis. Caspases inactivate enzymes, cleave structural proteins and activate other proteases in cell death. These cysteine proteases cleave after aspartic acid and can be activated through an intrinsic pathway involving the mitochondria, through an extrinsic pathway involving membrane receptors, or by an inflammatory response. Inflammation and infection cause potassium channels to open, which causes an efflux of potassium. A low level of intracellular potassium induces binding of caspase 1 to the inflammasome, which leads to controlled inflammation. At normal potassium levels, caspase 1 is inhibited. Traumatic brain injury elicits acute inflammation including caspase-1, that in turn exacerbates primary brain damage (Mariathasan and Monack 2007; Yu and Finlay 2008; de Rivero Vaccari, Lotocki et al. 2009). Therefore, staining for these proteases can be valuable in monitoring and observing cell death and inflammation.

4.2 Fluorojade B stain degenerate neurons

Fluoro-Jade B (FJB) is an anionic fluorochrome capable of selectively staining degenerate neurons in brain tissue. It can be combined with other fluorescent methodologies, such as immunofluorescence. FJB is a tribasic fluorescein derivative with a molecular weight of 445 daltons. It has an emission peak at 550 nm, and excitation peaks at 362 and 390 nm, respectively. The exact mechanism by which FJB stains degenerating neurons is not known. Based on the affinity of the acidic FJB, it can be inferred that a degenerating neuron presumably expresses a strongly basic molecule. FJB is fairly resilient, showing no signs of fading during storage at room temperature for up to 2 years and when fading does occur, it does so in cells and background at the same rate (Schmued, Albertson et al. 1997). For the FJB staining procedure the brain sections air dried at 45-50ºC for 30-60 minutes, rehydrated using successive 5 minute rinses in 100%, 75%, 50%, and 25% ethanol followed by 3 minutes in distilled water, placed in 0.06% potassium permanganate for 15 minutes followed by 2 minutes in distilled water, and then stained in 0.0006% Fluoro-Jade B solution in 0.1% acetic acid for 30 minutes. The slides are air-dried, immersed in xylene and coverslipped with cytoseal.

5. Biomarkers

In medicine, a biomarker is a term used for any molecule that can be an indicator of a particular disease or state of that disease. More specifically, a biomarker indicates a change in expression or state of a protein that correlates with the risk or progression of a disease, or with the susceptibility of the disease state to a given treatment. Biomarkers have characteristic biological properties that can be detected and measured in parts of the body like the blood or tissues. Although the term biomarker is relatively new, biomarkers have been used in pre-clinical research and clinical diagnosis for a long time. The most widely accepted serum biomarkers are creatine kinase isoenzyme MB (CK-MB) and troponin, which have surpassed the electrocardiogram as the standard criterion for the diagnosis of myocardial infarction (Lewandrowski, Chen et al. 2002). Other examples are glucose for the diagnosis and follow up for diabetes, cholesterol values as a biomarker and risk indicator for coronary and vascular disease, C-reactive protein as a marker of inflammation and the prostate specific antigen (PSA) as a marker for prostatic cancer between many other known biomarkers. The use of biomarkers to asses brain injury has been an area of research since late 1970s and early 1980s (Thomas, Palfreyman et al. 1978; Papa, Akinyi et al. 2010). Development of a useful biomarker of brain injury, however, has proven to be more difficult than development of biomarkers for other organ systems for several reasons. Perhaps most important, the brain is a more complex and less homogenous organ and different types of injury can occur to different types of brain cells with variable degrees of severity. In addition, the presence of a blood-brain barrier limits the amount and size of markers that can be detected in blood. Although evaluating the use of serum biomarkers to asses brain injury has been an active area of research for more than 20 years, interest in the use of serum biomarkers to assist in the prediction of outcome after TBI is more recent, with the first studies being published in the early 1990s.

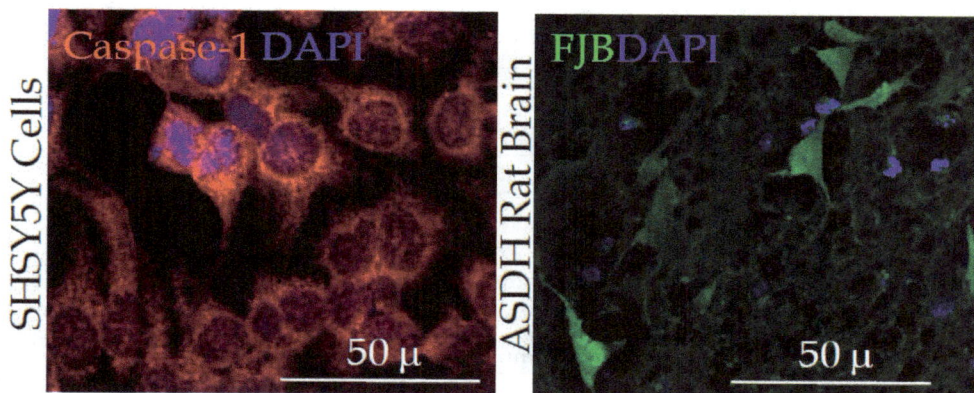

Fig. 3. A confocal image of SHSY5Y neuroblastoma cell line immunostained for Caspase-1 (red) shows cytoplasmic expression (left). Fluorojade B stained degenerate neurons appear fluorescent green in an acute subdural hematoma brain section (right). DAPI stained nuclei (blue). Magnification is shown below.

5.1 Traditional biomarkers

For primary and secondary injury in TBI several classes of biomarkers have been investigated. They include proteins, lipids, metabolites of neurotransmitters, second messengers, ions and glycolytic intermediates. These molecules are cell components that relate specifically with a neurological disease and can be found in body fluids such as cerebral spinal fluid, blood and urine. The presence of the biomarker in these fluids and tissues is detected by antibody-based assays to assess the extent of injury/disease and determine recovery. Several examples of neurological diseases and biomarkers have been reported in the literature such as β-amyloid protein, Tau protein and phosphorylated tau protein in CSF of patients with Alzheimer's disease; or abnormal accumulation of α-synuclein bound to ubiquitin inside the neurons forming Lewy bodies in Parkinson's disease.

Several traditional biomarkers, discovered by top-down approaches (known molecules with functional associations with normal brain or pathology) such as S100β, neuron specific enolase (NSE), tau, myelin basic protein (MBP) are plagued with lack of sensitivity or specificity, thus limiting their clinical utility (Dash, Zhao et al. 2010).

5.2 Novel biomarkers

A more radical bottom-up approach (in which the body fluids of injured patients are screened for molecules and their role/presence in CNS is ascertained later) has recently been developed (see Fig. 4) with advances in molecular biology, specifically in proteomics (Kobeissy, Sadasivan et al. 2008). Despite previous discoveries such as C-reactive protein

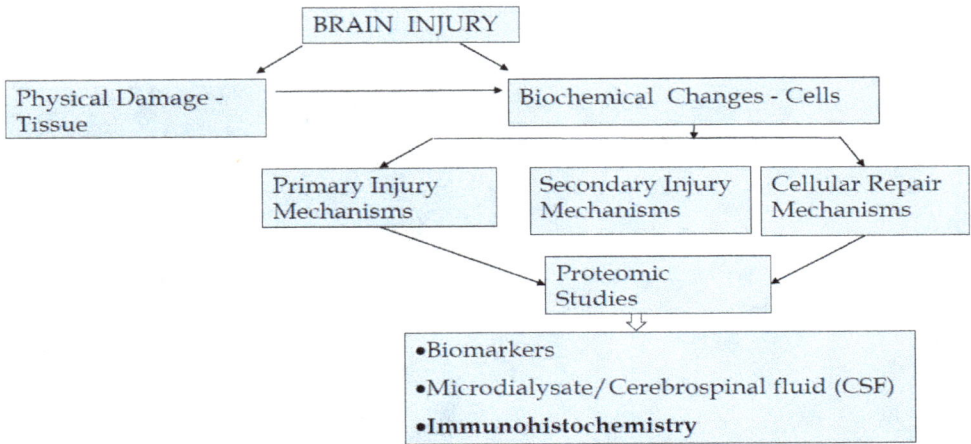

Fig. 4. Benefits of biomarkers: Diagnosis - Prognosis - Therapeutic (Diagram modified from Larner SF., 2008).

and serum amyloid A, which are not very useful as biomarkers due to lack of specificity, some progress has been made by this approach (Dash, Zhao et al. 2010). Pooled naive and injured cortical samples (48 h post injury; rat controlled cortical impact model) were processed and analyzed using a differential neuroproteomics platform. The results included 59 differential protein components of which 21 decreased and 38 increased in abundance after TBI. Proteins which decreased abundance included collapsin response mediator protein 2 (CRMP-2), glyceraldehyde-3-phosphate dehydrogenase, microtubule-associated proteins MAP2A / 2B and hexokinase. Conversely C-reactive protein, transferrin, breakdown products of CRMP-2, synaptotagmin, and αII-spectrin were found to be elevated after TBI (Ottens, Kobeissy et al. 2006). One of the novel biomarkers that increased was Ubiquitin C-terminal hydrolase-L1 (UCH-L1) (Kobeissy, Sadasivan et al. 2008), also known as neuronal-specific protein gene product 9.5 (PGP9.5). UCH-L1 was previously used as a histologic marker for neurons because of its high abundance and specific expression in neurons (Fig. 5). It is present in almost all neurons and averages 1–5% of total soluble brain protein. There are three related enzymes of this class (UCH-L1, UCH-L2 and UCHL3), but only UCH-L1 is highly enriched in the central nervous system. These enzymes are involved in either the addition or removal of ubiquitin from proteins destined for metabolism via the ATP-dependent proteasome pathway (Kobeissy, Ottens et al. 2006; Gong and Leznik 2007). To assess the reliability of UCH-L1 as a potential biomarker for traumatic brain injury (TBI), this study compared cerebrospinal fluid (CSF) levels of UCH-L1 from adult patients with severe TBI to uninjured controls and examined the relationship between these levels and severity of injury, complications and functional outcome. UCH-L1 levels in CSF were assessed using an ELISA in a prospective case control study conducted with 66 patients. Forty one patients with severe TBI, defined by a Glasgow coma scale (GCS) score of <8, who underwent intraventricular intracranial pressure monitoring were compared to 25 controls without TBI requiring CSF drainage for other medical reasons. Ventricular CSF was sampled from each patient at 6, 12, 24, 48, 72, 96, 120, 144, and 168 hrs following TBI. Injury severity was assessed by the GCS score, Marshall Classification on computed tomography and a complicated post injury course. Mortality was assessed at 6 wks and long-term outcome was assessed using the Glasgow outcome score 6 months after injury. TBI patients had significantly elevated CSF levels of UCH-L1 at each time point after injury compared to uninjured controls. Overall mean level of UCH-L1 in TBI patients was 44.2 ng/mL (±7.9) compared with 2.7 ng/mL (±0.7) in controls (p <.001). There were significantly higher levels of UCH-L1 in patients with lower GCS scores at 24 hrs, in those with post injury complications, in those with 6-wk mortality and in those with a poor 6-month dichotomized Glasgow outcome score. These data suggest that this novel biomarker has the potential to determine injury severity in TBI patients. (Hans, Born et al. 1983; Lewandrowski, Chen et al. 2002). However, one question regarding the origin of the biomarker remained unexplored due to the nature of subjects in the study.

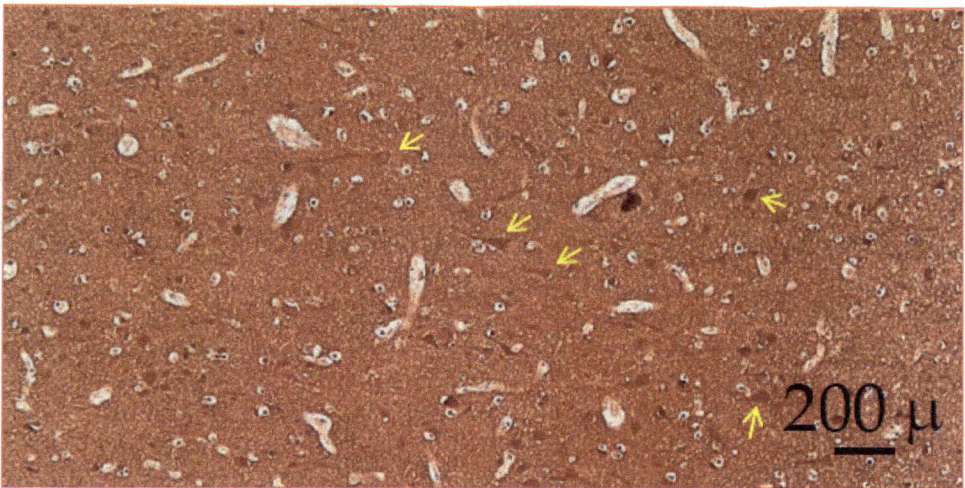

Fig. 5. UCH-L1 staining in normal human brain tissue. Intense brown UCH-L1+ neurons (yellow arrows) can be seen scattered throughout the section.

6. Immunohistochemistry (IHC)

Immunohistochemistry involves application of antibody to antigen-containing tissue and visualization of the antigen-antibody complex using either chromogenic or fluorescent readouts. The data can be collected using appropriate microscopy. Many of the details that are important for a successful immunohistochemical experiment have been extensively discussed elsewhere (Day and Thompson 2010) and in this book. Amplification of the signal is possible in IHC with both chromogenic and fluorescent readouts. The example for chromogenic readout includes the potentially carcinogenic 3-3´diaminobenzidine (DAB) (brown reaction stains in Fig, 6 & 9). With the advent of the tyramide signal amplification (TSA) it has been possible to obtain amplification of fluorescent signals (Raap 1998)(Fig. 6).

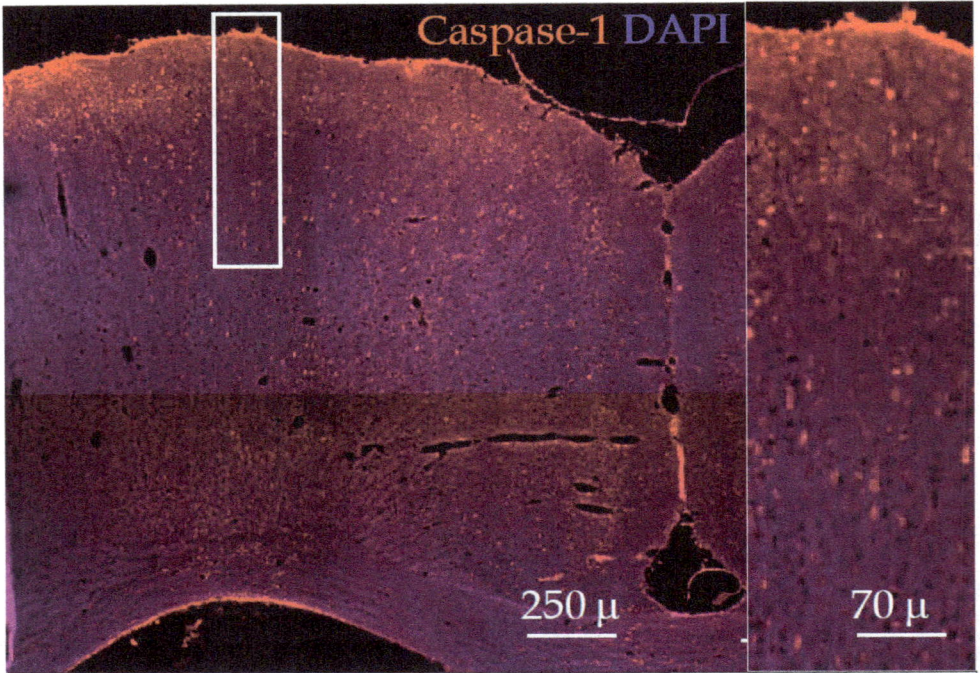

Fig. 6. Caspase-1 immunopositive cells (red) developed using TSA can be seen in cortex 24h following TBI (left). The area in the rectangular box is shown at higher magnification (right) to appreciate the cytoplasmic staining. The micron bar units are shown at the bottom of the image.

7. Factors influencing IHC

In this section we will consider a few factors that influence IHC, specifically in rodent and human brain tissue. More often than not brain tissue for IHC has to be fixed, which involves formation of novel covalent bonds between the amino acid side chains of proteins with aldehydes in order to stabilize the proteins. Fixed tissue is resistant to degradation by bacteria and fungi. Fixation of the brain is achieved by perfusion of the animal with saline, followed by 4% paraformaldehyde (PFA). Following perfusion the brain is rapidly removed and post-fixed in PFA for 6-8h. Excessive time (greater than 6-8h) in PFA can result in increased fluorescent background, which would impede use in IHC. Following post-fixation, brains are cryopreserved in 25% sucrose in phosphate buffered saline. The cryopreserved brains are placed in a Lucite/steel mold which enables accurate sectioning of the brain in blocks. Reproducibility of experiments depends on accurate blocking. The ~23mm brain is sectioned in 8mm blocks A, B, and C. Block B encompasses the hippocampus, the region of interest known to be most vulnerable to TBI. The entire block B can be embedded using media such as embedding matrix EM1 (Thermo) and sectioned on a cryostat, embedded in paraffin wax and sectioned on a microtome, or embedded in gelatin and sectioned on a Vibratome. Each of the embedding strategies is optimal for certain antigens. Tissue sectioning can be achieved in series (which is tedious, but preserves section order), free floating (relatively less tedious but information regarding location is lost). We employed the series strategy to ensure that adjacent sections can be used for different purposes such as chemical and immunostaining. The two series can then be compared and be used to deduce relevant information. The introduction of antigen retrieval has enabled immunohistology to become an integral component of morphologic diagnosis. Antigen retrieval phenomenon involves a Mannich reaction, which occurs with the cross-linking of some proteins. Such cross-linkages can be hydrolyzed by heat or alkalis so that the process of antigen retrieval may be the simple removal of such cross-linked proteins that are sterically interfering with the binding of antibodies to linear protein epitopes in the tissue section. (Leong and Leong 2007). It is important to pay attention to the method of antigen retrieval and to match it with that of the tissue sectioning. For example, formalin fixed paraffin embedded (FFPE) sections can be subjected to antigen retrieval with a citrate buffer of pH6.0, however this buffer is not suitable for paraformaldehyde fixed gelatin embedded tissue, it requires an alkaline pH 9.0 (Shi, Shi et al. 2011). These considerations play a vital role in how the IHC turns out. Double staining by alkaline phophatase and anti-alkaline phosphatase (APAAP) in combination with bromodeoxy uridine (BrdU) and surface antigen with same isotype or different chromogenic reagent combination have aided visualization of antigens that are exclusively distributed (Chaubert, Bertholet et al. 1997). Antibodies should be applied in a manner that minimizes the amount required to achieve the antigen-antibody interaction. We routinely employ the Sequenza cover plate technology (Shandon). Briefly, the brain section containing glass slides are subjected to antigen retrieval and sandwiched by coverplates. The slides are washed and antibody applied in a 120µl volume per slide. The usual methods of acquiring images use a microscope and generate huge image files that are cumbersome to share. Recent developments have allowed digital pathology, which represents an electronic environment for performing pathologic analysis and for managing the information associated with this activity. The utility of digital pathology has already been demonstrated by pathologists in several areas including consensus reviews, quality assurance (Q/A), tissue microarrays (TMAs), education and proficiency testing. The utilization of these tools will be essential for neuropathologists to continue as leaders in diagnostics, translational research and basic science in the 21st century (Guzman and Judkins 2009). The human adrenal glands stained for a low-affinity nerve growth factor were developed with DAB and slide imaged using a Zeiss MIRAX digital slide scanner (Fig 7). Another digital slide system, Aperio, is also widely used to generate and share slides.

8. Immunohistochemistry of biomarkers

In an unpublished study, we investigated the levels of the novel biomarker UCH-L1 in ASDH model. In this study, we present data that suggests the co-incidence of chemical stain for degenerate neurons i.e., FJB and loss of UCH-L1 immunoreactivity in the tissue. By combining the aforementioned methods we found that parts of the brain that contain the degenerate neurons could be the source of novel biomarker in the extracellular fluids. In this study FJB+ cells following ASDH could be observed in the sub-cortex and cornu ammonus 1 (CA1) region of the hippocampus. To determine that the FJB+ cells were indeed degenerate neurons, the brain sections were double stained with NeuN first followed by FJB (Fig. 8).

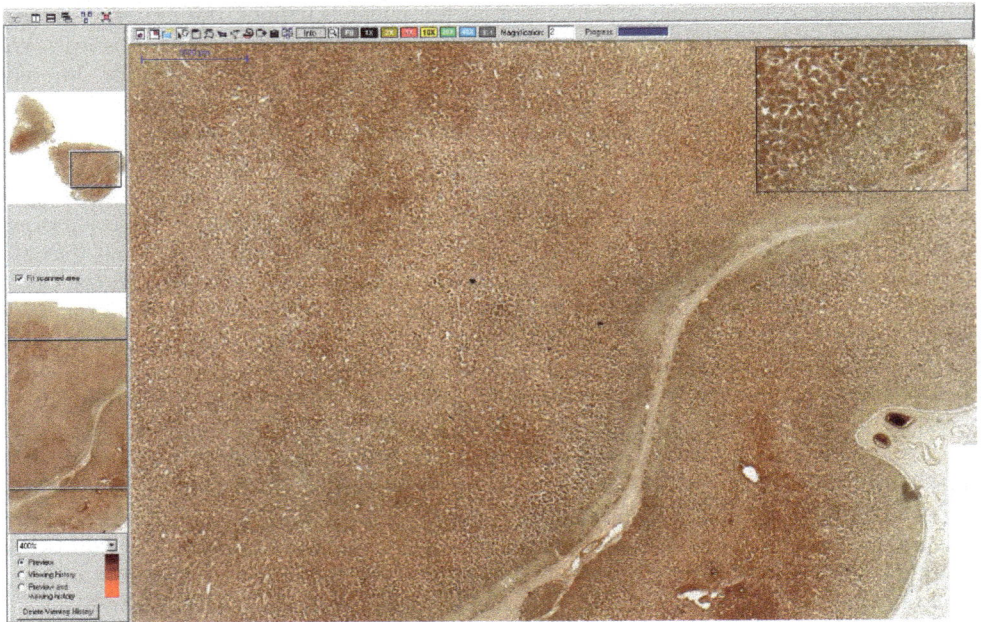

Fig. 7. Digital slide generated by Zeiss Mirax shows the entire slide (top left), the area shown in progressively higher magnification left to right. The scale bar shown on the top is 1000 μm. Brown p75+ adrenal medulla can be seen in the highest magnification (top right). http:/ /www.zeiss.de/mirax.

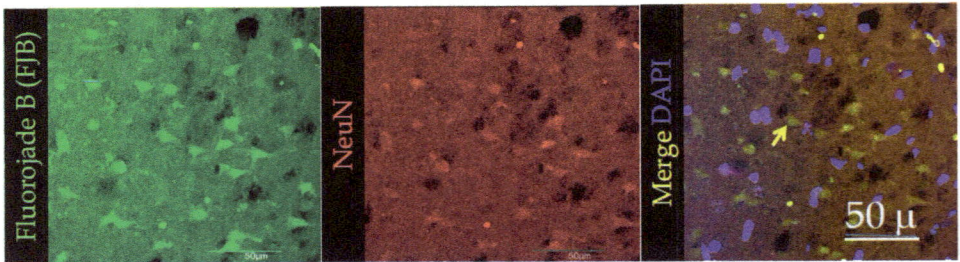

Fig. 8. FJB+ cells are degenerate NeuN+ neurons. FJB+ cells (left) double stained with NeuN (middle). Double stained cells appear yellow in the merged image (left). Absence of nuclear counter stain DAPI (blue) in FJB+ cells indicates neuronal degeneration (yellow arrow). All images are at same magnification, the micron bar units are shown at the bottom.

The degenerate neurons appear yellow upon merging of the confocal images. However this strategy was not possible with UCH-L1, perhaps because FJB protocol is harsh on the UCH-L1 antigen-antibody complex. To circumvent that issue, the FJB staining and UCH-L1 staining were performed on adjacent slides. First, a series of sections were stained for FJB, the distribution of the FJB cells was determined. Next, a few of the adjacent sections were double stained with FJB and NeuN. In this staining the NeuN was developed with TSA (red fluorescence) and followed by FJB (green fluorescence). Despite the harsh conditions in the FJB protocol (with regard to antigen antibody complex stability), the NeuN+FJB+ cells appear yellow due to colocalization of the red and green signals (Fig. 8). We looked for UCH-L1 staining in regions containing FJB+ cells and those devoid of FJB. We observed that the FJB+ area was devoid of UCH-L1 staining; in contrast, robust UCH-L1 staining was present in brain regions that were negative for FJB (Fig. 9). The combination of IHC with serial sectioning made it possible to conclude that the degenerate neurons must be the source of the UCH-L1. Kinetic studies of UCH-L1 release would address the exact sequence of events that are responsible for the appearance of this biomarker in body fluids and its role in the onset of neurodegeneration.

9. Classifications of IHC guidelines

As the applications of IHC crossover into nontraditional areas such as intraoperative neurosurgery, it is important to keep the previously developed guidelines for IHC in perspective in order to avoid mistakes. The Food and Drug Administration (FDA) has guidelines for the industry regarding good guidance practices on conducting unambiguous IHC experiments, reporting IHC data (George 2009), they are summarized below.

Class I IHC's (General Controls). Class I provide the pathologist with adjunctive diagnostic information that may be incorporated into the pathologist's report, but is not ordinarily reported to the clinician as an independent finding. These IHC's are used after the primary diagnosis of tumor is made by conventional histopathology using nonimmunologic histochemical stains such as hematoxylin and eosin.

Class II IHC's (Special Controls/Guidance Document). Class II are intended for the detection and/or measurement of certain target analytes by immunological techniques in order to provide prognostic and predictive data that are not directly confirmed by routine histopathologic internal and external control specimens. The IHC's provide the pathologist with diagnostic that is ordinarily reported as independent diagnostic information to the ordering clinician, and the claims associated with these data are widely accepted and supported by valid scientific evidence.

Class III IHC's (Premarket Approval). These IHC's do not meet the criteria for class I or II, or are IHC's that meet those criteria but raise new issues of safety and effectiveness. Examples are markers used to identify new target analytes in tissues that are claimed to be clinically significant genetic mutations and that cannot be confirmed by conventional histopathologic internal and external control specimens.

Similarly, the Canadian Association of Pathologists recommendations are quality control, risk-assessment, and quality assurance (Torlakovic, Riddell et al. 2010). The aforementioned guidelines have not been set primarily for neurosurgery. With the application of intraoperative immunohistochmical analyses (Uzuka, Aoki et al. 2011) in brain tumor surgeries, virtual biopsy in brain injury diagnosis, there is a need to explore the guidelines for acceptable IHC standards for TBI neurosurgery. Uzuka *et al.*, describe methods and four successfully diagnosed cases. The time for rapid histological diagnosis was 70 minutes. Uzuka *et al.*, suggest that immunohistochemical examination is indicated under the following conditions: (1) preoperative radiologic differential diagnosis includes both high- and low-grade tumors, (2) intraoperative assessment is necessary to determine the extent of excision, and (3) quick and accurate pathological diagnosis is necessary for early initiation of treatment after surgery. In mild TBI cases such as chronic traumatic encephalopathy (CTE), the utility of novel biomarkers such as UCH-L1 is under investigation. Application of robust IHC techniques could revolutionize the brain imaging field. The virtual biopsies could be extended to accurately diagnose mild, moderate and severe TBI. Such improvements could aid assessment of the therapeutic intervention efficacy. Currently, the molecular beacon-based technology is available in oncology to image cancer tissue in live animals (Bhojani, Ranga et al. 2011).

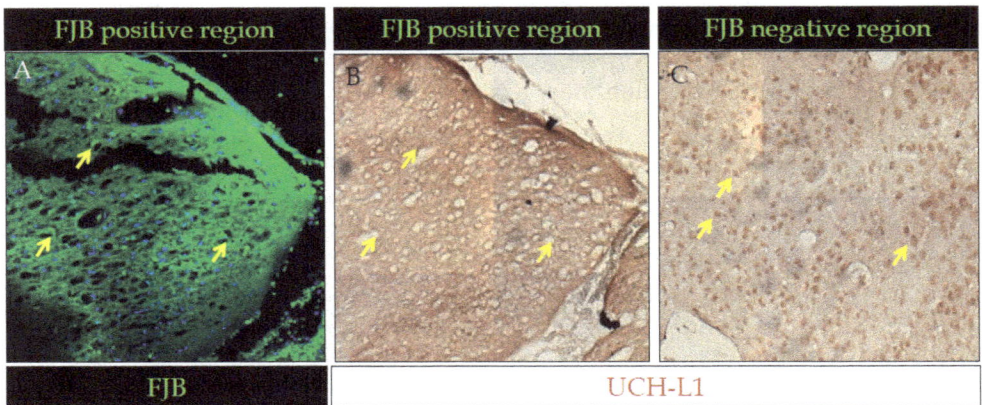

Fig. 9. A confocal image brain section from an ASDH rat shows presence of FJB+ cells (yellow arrows) in microvacuolated tissue (A). A light microscopy image of adjacent slide stained with UCH-L1 (brown) shows absence of immtmoreactivity in the FJB+ region (B). III contrast, robust UCH-L1 immunoreactivity can be seen in FJB negative region within the same section (yellow arrows) in (C). All images are shown at 100x magnification.

Brain-injury homing probes based on their affinity to the novel biomarkers would enhance visualization of the injured area. In silico structure-based drug screening using human UCH-L1 crystal structure data and virtual compound libraries identified one that potentiates the hydrolase activity of UCH-L1, and six that inhibit the activity in enzymatic assays. These compounds may be useful for research on UCH-L1 function, and could lead to candidate therapeutics for UCH-L1-associated diseases (Mitsui, Hirayama et al. 2010).

10. References

Bhojani, M. S., R. Ranga, et al. (2011). "Synthesis and investigation of a radioiodinated F3 peptide analog as a SPECT tumor imaging radioligand." *PLoS One* 6(7): e22418.
Bregy, A. (2010). "Neuroprotection by Sutureless Vascular Tissue Fusion and Artificial Oxygen Carriers."
Cernak, I. (2005). "Animal models of head trauma." *NeuroRx* 2(3): 410-422.
Chaubert, P., M. M. Bertholet, et al. (1997). "Simultaneous double immunoenzymatic labeling: a new procedure for the histopathologic routine." *Mod Pathol* 10(6): 585-591.

Dash, P. K., J. Zhao, et al. (2010). "Biomarkers for the diagnosis, prognosis, and evaluation of treatment efficacy for traumatic brain injury." *Neurotherapeutics* 7(1): 100-114.

Daugherty, W. P., J. E. Levasseur, et al. (2004). "Perfluorocarbon emulsion improves cerebral oxygenation and mitochondrial function after fluid percussion brain injury in rats." *Neurosurgery* 54(5): 1223-1230; discussion 1230.

Day, I. N. and R. J. Thompson (2010). "UCHL1 (PGP 9.5): neuronal biomarker and ubiquitin system protein." *Prog Neurobiol* 90(3): 327-362.

de Rivero Vaccari, J. P., G. Lotocki, et al. (2009). "Therapeutic neutralization of the NLRP1 inflammasome reduces the innate immune response and improves histopathology after traumatic brain injury." *J Cereb Blood Flow Metab* 29(7): 1251-1261.

Dietrich, W. D., O. Alonso, et al. (1994). "Post-traumatic brain hypothermia reduces histopathological damage following concussive brain injury in the rat." *Acta Neuropathol* 87(3): 250-258.

Eng, L. F., R. S. Ghirnikar, et al. (2000). "Glial fibrillary acidic protein: GFAP-thirty-one years (1969-2000)." *Neurochem Res* 25(9-10): 1439-1451.

Gajavelli, S., P. M. Wood, et al. (2004). "BMP signaling initiates a neural crest differentiation program in embryonic rat CNS stem cells." *Exp Neurol* 188(2): 205-223.

George, K. L. (2009). "Immunohistochemical Methods."

Gong, B. and E. Leznik (2007). "The role of ubiquitin C-terminal hydrolase L1 in neurodegenerative disorders." *Drug News Perspect* 20(6): 365-370.

Guzman, M. and A. R. Judkins (2009). "Digital pathology: a tool for 21st century neuropathology." *Brain Pathol* 19(2): 305-316.

Hans, P., J. D. Born, et al. (1983). "Creatine kinase isoenzymes in severe head injury." *J Neurosurg* 58(5): 689-692.

Kobeissy, F. H., A. K. Ottens, et al. (2006). "Novel differential neuroproteomics analysis of traumatic brain injury in rats." *Mol Cell Proteomics* 5(10): 1887-1898.

Kobeissy, F. H., S. Sadasivan, et al. (2008). "Neuroproteomics and systems biology-based discovery of protein biomarkers for traumatic brain injury and clinical validation." *Proteomics Clin Appl* 2(10-11): 1467-1483.

Kwon, T. H., D. Sun, et al. (2005). "Effect of perfluorocarbons on brain oxygenation and ischemic damage in an acute subdural hematoma model in rats." *J Neurosurg* 103(4): 724-730.

Leong, T. Y. and A. S. Leong (2007). "How does antigen retrieval work?" *Adv Anat Pathol* 14(2): 129-131.

Lewandrowski, K., A. Chen, et al. (2002). "Cardiac markers for myocardial infarction. A brief review." *Am J Clin Pathol* 118 Suppl: S93-99.

Mariathasan, S. and D. M. Monack (2007). "Inflammasome adaptors and sensors: intracellular regulators of infection and inflammation." *Nat Rev Immunol* 7(1): 31-40.

Middeldorp, J. and E. M. Hol (2011). "GFAP in health and disease." *Prog Neurobiol* 93(3): 421-443.

Mitsui, T., K. Hirayama, et al. (2010). "Identification of a novel chemical potentiator and inhibitors of UCH-L1 by in silico drug screening." *Neurochem Int* 56(5): 679-686.

Mondello, S., U. Muller, et al. (2011). "Blood-based diagnostics of traumatic brain injuries." *Expert Rev Mol Diagn* 11(1): 65-78.

Mullen, R. J., C. R. Buck, et al. (1992). "NeuN, a neuronal specific nuclear protein in vertebrates." *Development* 116(1): 201-211.

Ottens, A. K., F. H. Kobeissy, et al. (2006). "Neuroproteomics in neurotrauma." *Mass Spectrom Rev* 25(3): 380-408.

Papa, L., L. Akinyi, et al. (2010). "Ubiquitin C-terminal hydrolase is a novel biomarker in humans for severe traumatic brain injury." *Crit Care Med* 38(1): 138-144.

Raap, A. K. (1998). "Advances in fluorescence in situ hybridization." *Mutat Res* 400(1-2): 287-298.

Schmued, L. C., C. Albertson, et al. (1997). "Fluoro-Jade: a novel fluorochrome for the sensitive and reliable histochemical localization of neuronal degeneration." *Brain Res* 751(1): 37-46.

Shandon "The Sequenza Coverplate Technology."

Shi, S. R., Y. Shi, et al. (2011). "Antigen retrieval immunohistochemistry: review and future prospects in research and diagnosis over two decades." *J Histochem Cytochem* 59(1): 13-32.

Thomas, D. G., J. W. Palfreyman, et al. (1978). "Serum-myelin-basic-protein assay in diagnosis and prognosis of patients with head injury." *Lancet* 1(8056): 113-115.

Torlakovic, E. E., R. Riddell, et al. (2010). "Canadian Association of Pathologists-Association canadienne des pathologistes National Standards Committee/Immunohistochemistry: best practice recommendations for standardization of immunohistochemistry tests." *Am J Clin Pathol* 133(3): 354-365.

Uzuka, T., H. Aoki, et al. (2011). "Indication of intraoperative immunohistochemistry for accurate pathological diagnosis of brain tumors." *Brain Tumor Pathol* 28(3): 239-246.

Yu, H. B. and B. B. Finlay (2008). "The caspase-1 inflammasome: a pilot of innate immune responses." *Cell Host Microbe* 4(3): 198-208.

Zhou, Z., D. Sun, et al. (2008). "Perfluorocarbon emulsions improve cognitive recovery after lateral fluid percussion brain injury in rats." *Neurosurgery* 63(4): 799-806; discussion 806-797.

5

Immunocytochemistry of Cytoskeleton Proteins

Arzu Karabay, Şirin Korulu and Ayşegül Yıldız Ünal
Istanbul Technical University
Turkey

1. Introduction

In this chapter, we are going to introduce some of the cytoskeleton proteins and then, discuss about the technical details and specifics of immunocytochemistry of these proteins.

Immunochemistry is a powerful technique to observe tissues (immunohistochemistry) in their surroundings and inner world of cells (immunocytochemistry) in order to understand lives of organisms as close as to their physiological states.

Being the most complex life units, eukaryotic cells have an elaborate infra – structure that is composed of filamentous proteins which all together form the skeleton of cells, "cytoskeleton". The cytoskeletal proteins are not only responsible for the physical structure of the cells, but also are involved in cellular functions requiring rapid reorganization of the cytoskeletal structures. Microfilaments and microtubules are more predominantly involved in dynamic events of the cells, whereas intermediate filaments are mainly important for static structure of the cells. Both microfilaments and microtubules have control on movement within the cell. As the most prominent movement, intracellular protein trafficking is almost a continuous dynamic event, yet this protein trafficking is surpassed by more global dynamic movements of cells such as cell division, differentiation and migration processes. All of these minor and major dynamic events are orchestrated by microfilament and microtubule proteins by taking turns.

2. Cytoskeletal proteins

2.1 Microfilaments

Microfilaments (Fig.1), also known as actin filaments, are flexible tubes composed of two – stranded helical polymers of the actin protein. Each filament has a structural polarity with a plus and minus end. Actin filaments determining the shape of the cell surface are dispersed throughout the cell, and they are necessary for whole cell locomotion, for cells to engulf large particles by phagocytosis and to divide (Lodish & Baltimore, 1995).

2.2 Intermediate filaments

Intermediate filaments are ropelike fibers with many long strands twisted together to provide great tensile strength. They form stable dimers by wrapping around each other in a coiled – coil configuration, and their main function is to provide cells mechanical strength and resistance to shear stress when cells are stretched (Lodish & Baltimore, 1995).

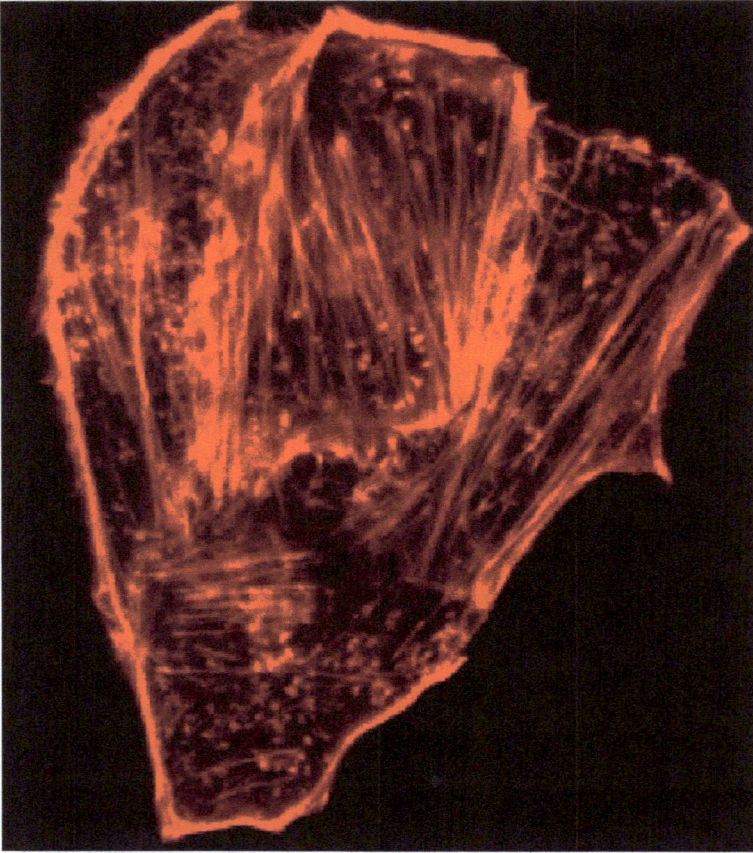

Fig. 1. Immunostaining of actin in fibroblast cell. Filamentous actin was labeled with Alexa Fluor 647 phalloidin. Image was obtained with laser scanning confocal microscope.

2.3 Microtubules

Among the cytoskeletal proteins, microtubules (Fig.2) are thought to have the most important roles, especially in generation of cell shape and polarity, cell division, cell growth and intracellular organelle transport.

Microtubules are polymers of α - and β - tubulin subunits. These subunits are arranged in a cylindrical tube of 24 nm in diameter. There are both lateral and longitudinal interactions between the tubulin heterodimer subunits. These interactions maintain the tubular form of microtubules (Lodish & Baltimore, 1995; Vale & Hartman, 1999).

In addition to α - and β - tubulins, there is a special third type of tubulin, γ - tubulin. It is located in the centrosomal matrix. In animal cells, centrosomes are primary sites for microtubule nucleation, and microtubules are thought to be nucleated from γ - tubulin ring complexes (γ - TuRCs) within the centrosome.

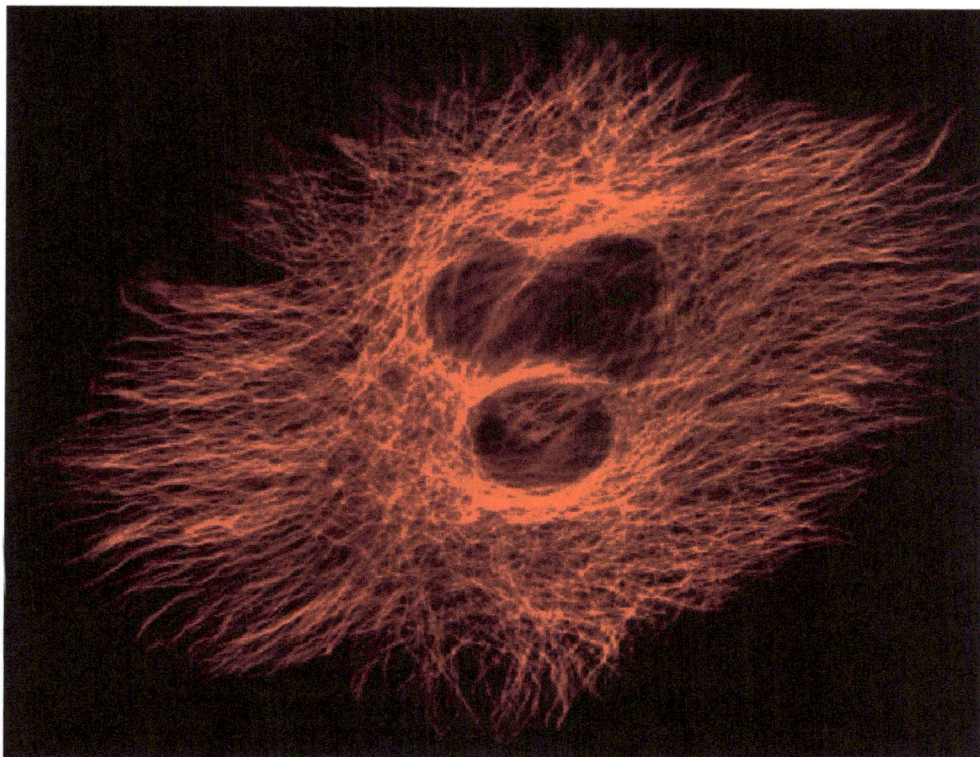

Fig. 2. Immunostaining of microtubules in fibroblast cell. Microtubules were labeled with Cy3 Conjugated mouse anti – beta tubulin monoclonal antibody (Clone TUB 2.1). Image was obtained with laser scanning confocal microscope.

2.4 Microtubule severing proteins

In many eukaryotic cells, minus – ends of microtubules are anchored near the centrosome, whereas the plus – ends are oriented towards the cell periphery. This so called interphase state structure of the cells is changed when the cells are committed to undergo mitosis, and this process requires reorganization of microtubules. In dividing eukaryotic cells, microtubule polymers have to be organized according to the different phases of the cell cycle during mitosis and interphase. This transition from interphase to mitosis involves both orientation reorganization and length changes of microtubule polymers. The formation of specialized mitotic spindle requires severing of relatively long microtubule polymers into shorter pieces by a mechanism called microtubule severing.

In non – dividing, terminally differentiated neurons, microtubule reorganization is much more elaborated in axonal and dendritic differentiation processes (Fig.3). In neurons, microtubules are born at the centrosomes within the cell body, and this birth place is away from their specialized functional areas such as axons and dendrites. These specialized elongated processes require non – centrosomal microtubules to maintain both the structural integrity and the activity of these terminally differentiated cells. Therefore, for axonal and dendritic differentiation, microtubules need to migrate very long distances to reach their final destinations. Recent studies support the idea that microtubule severing is also an important source of non – centrosomal microtubules of neurons (Yu et al., 2008).

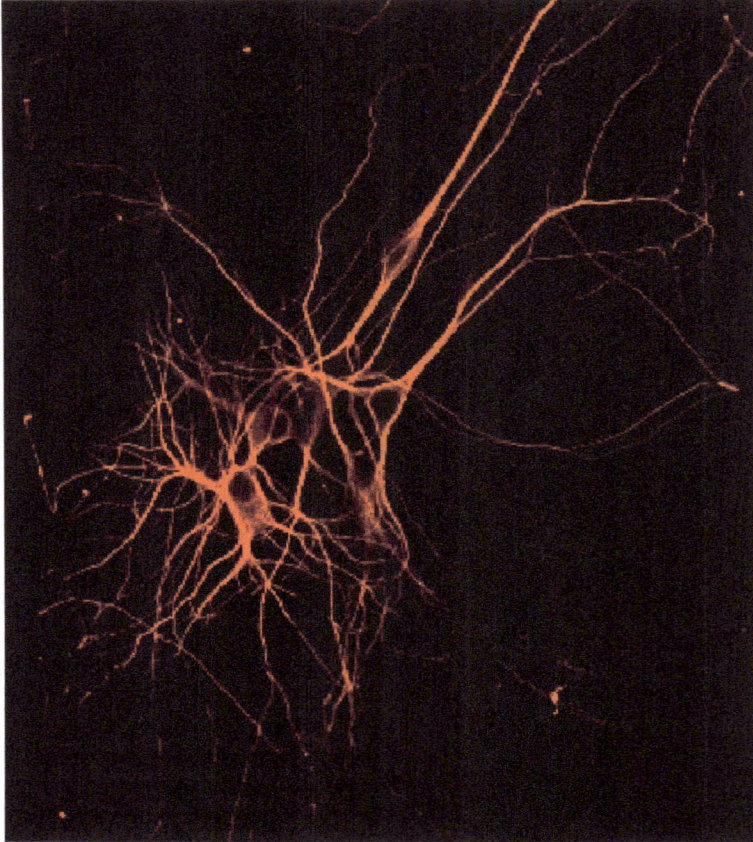

Fig. 3. Immunostaining of microtubules in primary hippocampal neurons. Microtubules were labeled with Cy3 Conjugated mouse anti – beta tubulin monoclonal antibody (Clone TUB 2.1). Image was obtained with laser scanning confocal microscope.

To fulfill the needs of both mitotic and non – mitotic cells, according to the changing conditions, microtubules need not only be regulated by their intrinsic dynamics and by some structural proteins such as MAPs (Microtubule Associated Proteins), but also by molecular motor proteins such as dynein and kinesin and microtubule – severing proteins which regulate the length of the microtubule polymers.

There are mainly two severing proteins in different types of cells, katanin and spastin. Katanin and spastin are members of a large AAA (ATPases Associated with various cellular Activities) protein family. This family proteins play important roles in a number of cellular activities including proteolysis, protein folding, membrane trafficking, cytoskeleton regulation and organelle biogenesis (McNally & Vale, 1993; Vale, 2000). Katanin and spastin are expressed in both mitotic cells and post – mitotic neurons and they function to severe microtubules into shorter pieces to make microtubule transport and reorganization easier in both cell types.

2.4.1 Katanin

Katanin is the most well characterized microtubule – severing protein. It is a microtubule stimulated ATPase, and it forms ring structures when katanin subunits bind to adjacent tubulin subunits on the microtubule wall (McNally, F. et al., 2000).

It is a heterodimer protein consisting of two subunits. 60kD (p60) enzymatic subunit carries out the ATPase and severing reactions. Other subunit, 80kD (p80), localizes katanin to the centrosome and regulates microtubule – severing activity of p60 subunit (Vale & Hartman, 1999; Quarmby et al., 2000).

N – terminal domain of p60 subunit binds microtubules and C – terminal AAA domain affects the binding affinity of the adjacent microtubule – binding domain. This stabilizes p60 rings (Vale & Hartman, 1999).

N – terminal of p80 subunit is composed of WD40 repeat (proline – rich) domain and a C – terminal domain is required for dimerization with catalytic p60 subunit. Studies showed that WD40 repeat domain of p80 subunit is required for spindle pole localization of katanin. WD40 domain probably binds to another spindle pole protein (McNally, K. et al., 2000). Although p60 subunit shows its ATPase and severing activity in the absence of p80 subunit, p80 subunit cannot sever microtubules on its own.

In immunostaining, katanin is co – localized with microtubules. In dividing cells (Fig.4), katanin is localized on the mitotic spindle towards poles that are closer to centrosome regions on mitotic apparatus throughout the cell cycle phases and in neurons (Fig.5), katanin is distributed throughout the cell body and neuronal processes to severe non – centrosomal microtubules to give rise to new axonal and dendritic branches for further elongation.

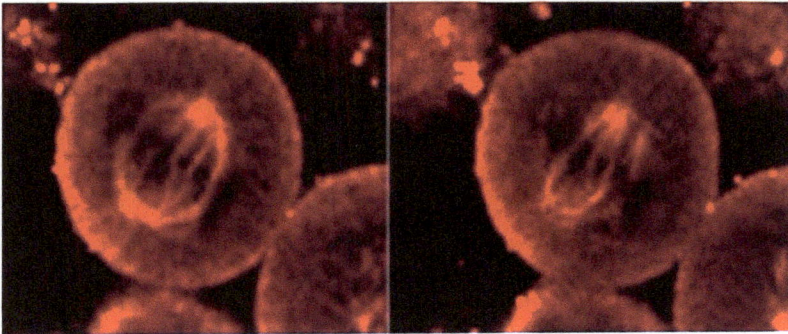

Fig. 4. Immunostaining of p60 katanin in HeLa cells. HeLa cells were labeled with 1G6 mouse p60 katanin primary antibody and visualized with Alexa Fluor 594 goat anti – mouse IgG. Images were obtained with laser scanning confocal microscope.

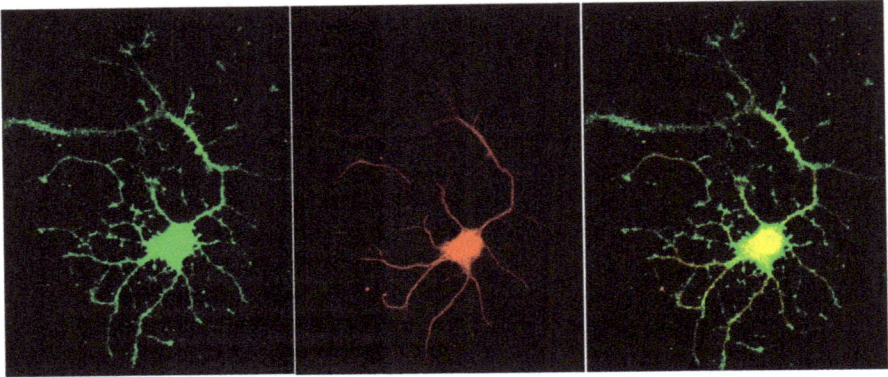

Fig. 5. Immunostaining of microtubules in primary hippocampal neurons. Hippocampal neurons were labeled with two fluorescent dyes. GFP fused p60 katanin was labeled with rabbit anti – GFP primary antibody and visualized with Alexa Fluor 488 goat anti – rabbit IgG (green) and microtubules were labeled with Cy3 Conjugated mouse anti – beta tubulin monoclonal antibody (Clone TUB 2.1) (red). Image on the right panel represents merge of p60 katanin and microtubules (yellow). Image was obtained with laser scanning confocal microscope.

2.4.2 Spastin

Spastin is a member of AAA protein family. It belongs to the meiotic subgroup which also contains proteins involved in vesicle trafficking and microtubule dynamics. Spastin shares great homology with p60 katanin within AAA domain but they do not have homology in their N terminal region.

In dividing cells (Fig.6), spastin is mainly nuclear in interphase cells. It becomes associated with centrosomes, the spindle microtubules, the midzone and finally midbody during cell division (Errico et al., 2004).

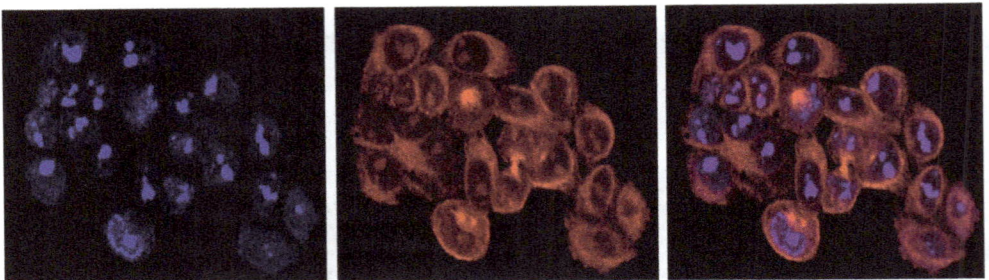

Fig. 6. Immunostaining of spastin in neuroblastoma cells. GFP fused spastin was labeled with rabbit anti – GFP primary antibody and visualized with Alexa Fluor 405 goat anti – rabbit IgG (blue) and microtubules were labeled with Cy3 Conjugated mouse anti – beta tubulin monoclonal antibody (Clone TUB 2.1) (red). Image on the right panel represents merge of spastin and microtubules. Image was obtained with laser scanning confocal microscope.

In post – mitotic neurons (Fig.7), spastin is localized in discrete nuclear domains, but most interestingly detects a specific signal in the neuritis. This signal is characteristically enriched in the distal axon and in the branching regions such as growth cones. Therefore, spastin influences microtubule dynamics in growth cones; thus regulating the stability of axons and axonal transport (Errico et al., 2004).

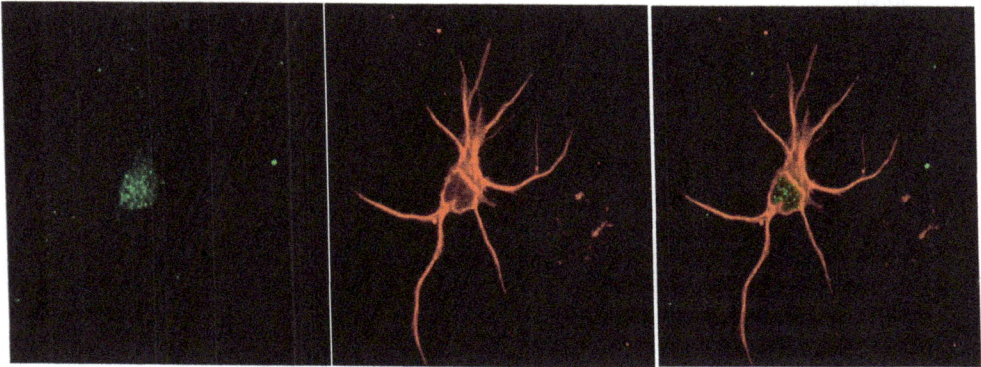

Fig. 7. Immunostaining of spastin in primary hippocampal neurons. Hippocampal neurons were labeled with two fluorescent dyes. GFP fused spastin was labeled with rabbit anti – GFP primary antibody and visualized with Alexa Fluor 488 goat anti – rabbit IgG (green) and microtubules were labeled with Cy3 Conjugated mouse anti – beta tubulin monoclonal antibody (Clone TUB 2.1) (red). Image on the right panel represents merge of spastin and microtubules. Image was obtained with laser scanning confocal microscope.

Among these dynamic polymers of the cytoskeleton, microtubules are the main vital elements of all eukaryotic cytoskeleton in terms of determining cellular architecture and intra – cellular movements. They achieve their functions by interacting with microtubule – related proteins. In order to analyze cellular events occurring via reorganization and dynamic behaviors of microtubules and interacting proteins in accuracy, it is essential to reveal all sub – cellular structures in great detail. Detailed pinpointing of subcellular structures requires having good cytoskeleton related antibodies with high efficiency and specificity.

Although microtubules constitute a long studied area of cell biology, and tubulin antibodies are relatively well studied, there is still a high need and demand for antibodies of microtubule interacting proteins that are more recently being studied. Also, the commercially available cytoskeleton related protein antibodies are usually polyclonal antibodies; and although these antibodies are faster and easier to obtain, they could share epitopes of different proteins, leading to non – specific binding to the other cellular proteins. In addition to this, availability of a polyclonal antibody depends on the life time of the animal. In case of the need to reproduce the same antibody, using another animal with the same antigen could end up as a failed attempt as it would be produced by generation of different clones. Therefore, using these polyclonal antibodies may sometimes be misleading in understanding the cytoskeleton related events, and this gives rise to the need to have monoclonal antibodies with higher specificity and as an endless supply.

However, obtaining a monoclonal antibody specific for a particular protein is very time consuming. Thus, there is limited source of available cytoskeleton related monoclonal antibodies for a specific protein of interest. For instance, there has been no monoclonal p60 katanin antibody available until now, and polyclonal p60 katanin antibodies have been in use (McNally & Thomas, 1998; Karabay et al., 2004). Therefore, the absence of available monoclonal antibody against p60 katanin has been an obstacle in cytoskeleton research, and our lab has recently successfully produced a mouse anti – p60 katanin monoclonal antibody, 1G6, which has been characterized in detail for many immunochemical applications and it has been proven to be a very qualitative antibody in a vast array of organisms (rat, mouse, chicken and human tested) with wide range of applications (Akkor & Karabay, 2010). Successfully obtained 1G6 recognized the endogenous p60 katanin in immunocytochemistry (Fig.8), Western blot analysis (Fig.9) and immunohistochemistry analysis (Fig.10) in addition to ELISA application.

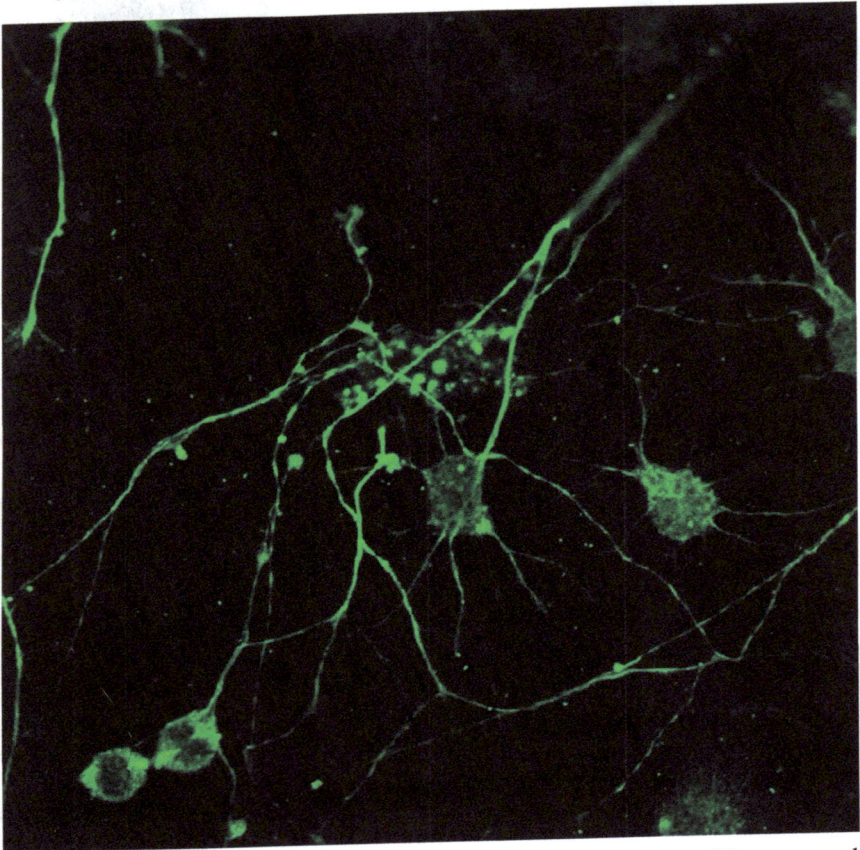

Fig. 8. Immunostaining of p60 katanin in primary hippocampal neurons. Hippocampal neurons were labeled with 1G6 mouse anti – p60 katanin primary antibody and visualized with Alexa Fluor 488 goat anti – mouse IgG. Image was obtained with laser scanning confocal microscope.

Fig. 9. Western blot image indicating p60 katanin in brain extract. Arrowhead points p60 katanin labeled with 1G6 mouse anti – p60 katanin primary antibody and visualized with anti – mouse IgG alkaline phosphatase secondary antibody.

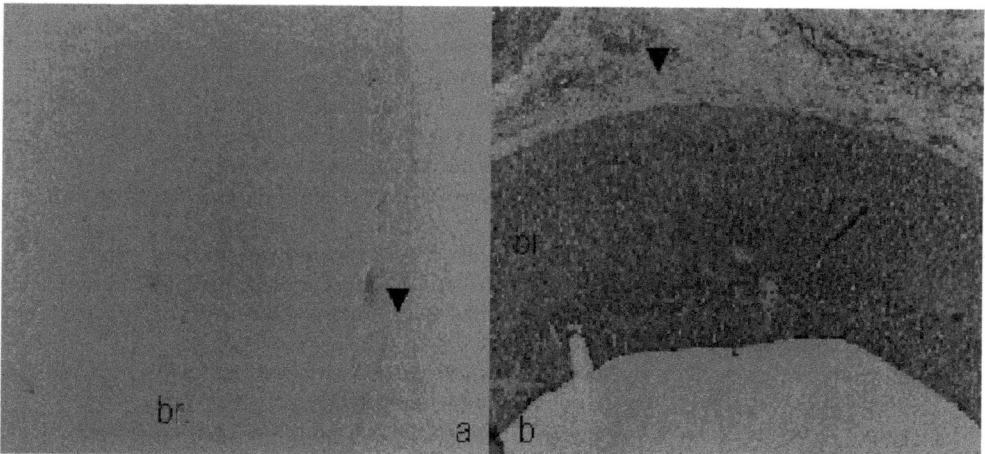

Fig. 10. Immunohistochemistry image showing p60 katanin expression in brain (br) tissue sections of *Gallus gallus* embryo. a) Negative control, b) p60 katanin expressing region. p60 katanin was labeled with 1G6 mouse anti – p60 katanin primary antibody and visualized with anti – mouse IgG alkaline phosphatase secondary antibody. Image was obtained with light microscope.

3. Production of monoclonal antibody against p60 katanin

In our laboratory, we produced the first monoclonal antibody against p60 katanin which is named with its clone name as 1G6. Recombinant p60 katanin protein, which was produced based on a specific region of rat p60 katanin, was expressed in *Escherichia coli* and used as antigen.

p60 katanin monoclonal antibody has been produced by hybridoma technology. In hybridoma technology, hybrid cells are produced from the fusion of B cells and the myeloma cell line (hybridomas). In this technique, the idea is combining these two cells together. In case of p60 katanin hybridomas, B cells have been derived from the lymphatic tissues of recombinant p60 katanin antigen immunized animals and the myeloma cell line have brought the immortality to these B cells when they have been fused together.

The steps of hybridoma technology for 1G6 p60 katanin monoclonal antibody could be summarized briefly as followed: (Akkor & Karabay, 2010).

* BALB/c mice were immunized intra – peritoneally two times with 50 mg of recombinant p60 katanin/mouse at 2 – week intervals.
* After immunization, anti – recombinant p60 katanin antibody response was assayed by indirect ELISA. ELISA plates coated with recombinant p60 katanin antigenic protein were incubated with the mice sera. Anti – recombinant p60 katanin antibody binding reaction was detected by using an alkaline phosphatase conjugated anti – mouse IgG or anti – mouse polyvalent (IgA, IgM, IgG) immunoglobulins as secondary antibody. Although the monoclonal antibodies are more specific compared to polyclonal antibodies, the clones may not always be IgG type, but also IgM or IgE. In our study, we have obtained another anti – p60 katanin antibody, which is an IgM type, but that antibody was not as specific as IgG type 1G6 anti – p60 katanin antibody.
* Mice that developed the IgG response against recombinant p60 katanin were selected for fusion studies. Polyethylenglycol (PEG) was used to fuse the two types of cells.
* Fused cells were cultured in Hypoxanthine Aminopterin Thymidine (HAT) selective medium for selection of hybridomas as the two types of unfused cells would die and remained hybridomas continue to divide in culture plates.
* Hybridoma colonies that produced antibodies against desired recombinant p60 katanin were detected by ELISA.
* In selected wells which had antibody response, there were more than one hybridoma colonies which would result in polyclonal response. By "limited dilution" method, the cells in antibody responding wells were dispersed to new culture plates in order to get hybrid colony produced from single cell.
* Hybridoma colonies that synthesize specific antibody for recombinant p60 katanin antigen was selected by cross – reactivity ELISA method.
* Selected hybridomas were then produced in large quantities and antibodies were purified from their media.

4. Immunocytochemistry of cytoskeleton and cytoskeleton related proteins

Immunocytochemistry is a biochemical technique that applies an antibody to a specific cell protein in the cell. It can be used either to detect whether the protein exists in the sample

(Fig.11, Fig.12), or to highlight the location of the specific protein (Fig.13). Furthermore, this technique even allows discriminating different types of cells in the same culture dish (Fig.14).

If the protein of interest were to be over – expressed in cells by transfecting with the expression vector containing the sequence for the protein of interest, some cells would take up the plasmid whereas others would not. In Fig.11, 3T3 fibroblast cells that were transfected electrophoretically with Kif15 kinesin motor containing expression vector and immunostained for actin and Kif15 could be seen. This double staining would allow discriminating the cells that have taken up the plasmid or not.

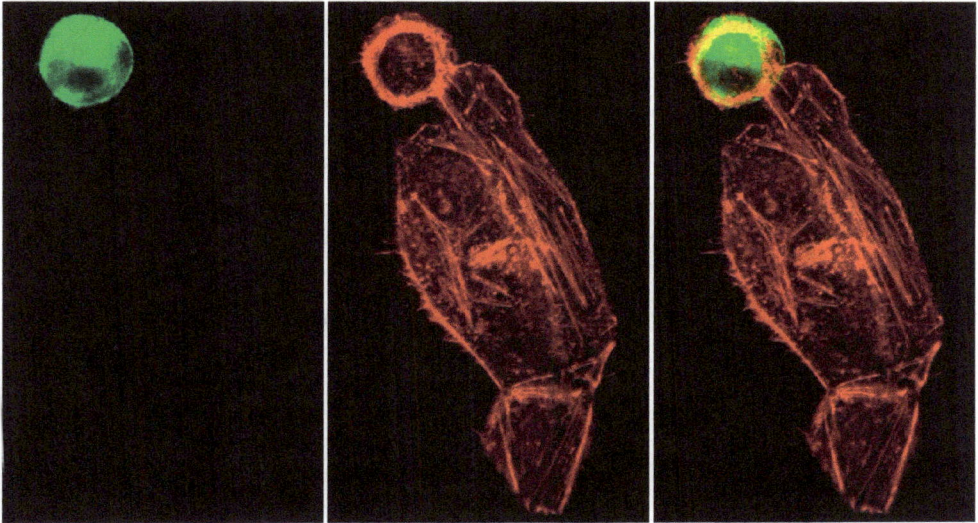

Fig. 11. Immunostaining of Kif15 expressing and non – expressing fibroblast cells. GFP fused Kif15 was labeled with rabbit anti – GFP primary antibody and visualized with Alexa Fluor 488 goat anti – rabbit IgG (green) and filamentous actin was labeled with Alexa Fluor 647 phalloidin (red). Image on the right panel represents merge of Kif15 and actin filaments (yellow). Image was obtained with laser scanning confocal microscope.

In Fig.12, HeLa cells that were electrophoretically transfected with GFP fused p60 katanin expression vector for over – expression of p60 katanin could be seen. The cells were immunostained with rabbit anti – GFP antibody and mouse anti – tubulin antibody. This double staining allowed discriminating the cells that were over – expressing GFP fused p60 katanin.

In Fig.13, primary hippocampal neurons immunostained for cyclin A and neuron specific beta – III tubulin could be seen. The neurons were also stained with DAPI for nuclei. This triple staining would allow observing not only specifically the neurons in the culture, but also the exact location of cyclin A within in the neurons.

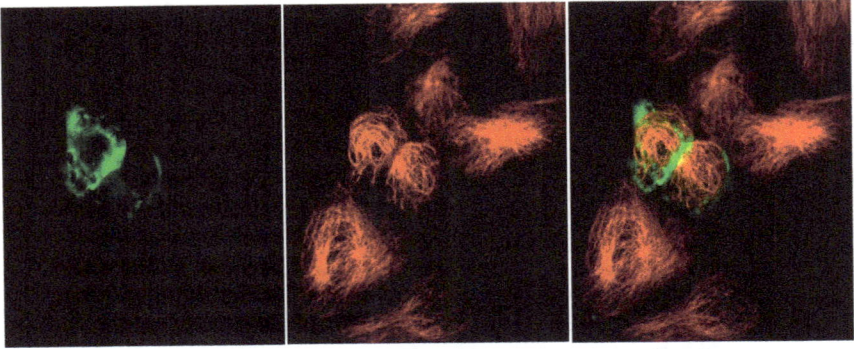

Fig. 12. Immunostaining of p60 katanin expressing and non – expressing HeLa cells. GFP fused p60 katanin was labeled with rabbit anti – GFP primary antibody and visualized with Alexa Fluor 488 goat anti – rabbit IgG (green) and microtubules were labeled with Cy3 conjugated mouse anti – beta tubulin monoclonal antibody (Clone TUB 2.1) (red). Image on the right panel represents merge of p60 katanin and microtubules. Image was obtained with laser scanning confocal microscope.

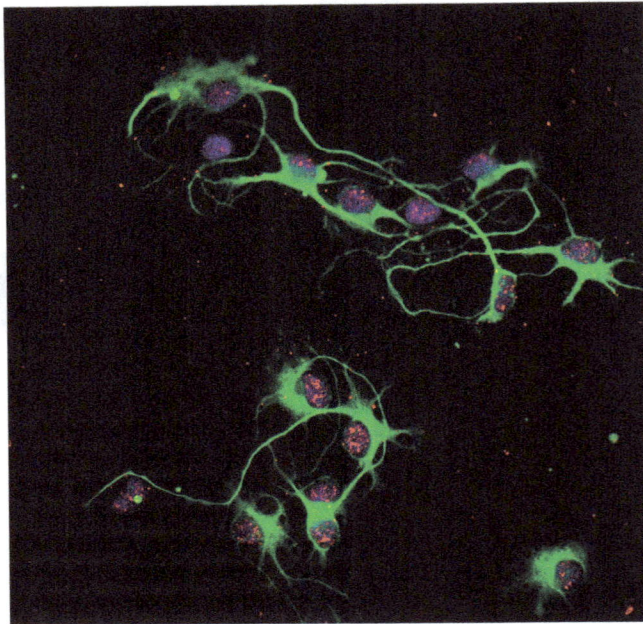

Fig. 13. Immunostaining of cyclin A in primary hippocampal neurons. Hippocampal neurons were labeled with three fluorescent dyes. Cyclin A was labeled with rabbit anti – cyclin A primary antibody and visualized with Alexa Fluor 647 goat anti – rabbit IgG (red); microtubules were labeled with mouse anti – beta III tubulin monoclonal antibody and visualized with Alexa Fluor 488 goat anti – mouse IgG (green) and nucleus was visualized with DAPI (blue). Image was obtained with laser scanning confocal microscope.

Organic solvents such as methanol, ethanol, and acetone denature proteins without any covalent modifications. Their working principle is by removing bound H_2O molecules of the cells. These solvents also remove membrane and some structural lipids, and cells become permeable to antibodies. Therefore, cells do not need to be permeabilized.

For immunostaining of cytoskeletal proteins, the most sensitive and effective fixation method is with aldehydes; glutaraldehyde and paraformaldehyde. Since aldehydes fixation method requires additional permeabilization step, a detergent, Triton – X100 can be used for cell extraction that leads cell permeabilization. Triton X100 is a detergent that is mostly used to extract the membrane from cells so that the cytoskeleton can be accessed. It is usually prepared as 10% solution in ddH_2O and is rotated overnight in the cold room to fully disperse the tick detergent.

Another buffer called "PHEM (Pipes – Hepes – EGTA – MgCl₂) is also necessary for microtubule stabilization. Stock solution (2X PHEM) can be prepared as indicated in the Table 4 and buffer pH is adjusted to 6,9 with NaOH.

| 18,14 g Pipes |
| 5,96 g Hepes |
| 3,8 g EGTA |
| 0,41 g MgCl2 |
| up to 500 ml ddH₂O |

Table 4. 2X PHEM Buffer Ingredients.

Cell permeabilization and fixation can either be performed separately or together in one step as termed fixation/co – extraction depending on the design of the study. For instance, in our p60 katanin – microtubule co – localization studies we preferred simultaneous fixation/co – extraction method. Since p60 katanin severs microtubules into short pieces, fixing the cells first would clutter inner side of the cells with short pieces/subunits of tubulin and therefore would cause to have higher background with the antibody staining. Whereas, if the fixation/co – extraction steps were performed simultaneously, it would allow some of the tubulin subunits to be extracted from the cell and therefore would give a cleaner staining while keeping some of the short microtubule pieces inside the cells. On the other hand, if the extraction step were to be done at first, as it would cause most of the short pieces to leave the cell, it may not be possible to see short microtubule pieces that are created by katanin. Therefore as in the case of the examples, one should decide about the methodology of the cell permeabilization and fixation depending on the need of the question asked.

Below you can see some examples for different cell fixation protocols commonly used in the fixation of microtubules and microtubule related proteins in dividing cells.

4.2.1 Paraformaldehyde fixation

- To prepare paraformaldehyde fixation solution, ingredients are mixed together and pH is adjusted to 7,3 by NaOH. Then, the fixation solution needs to be heated to 60°C with stirring it for overnight.
- To perform paraformaldehyde fixation procedure, fixation solution is warmed in 37°C water bath.

- Petri dishes containing the cultured cells are filled with fixation solution and incubated at room temperature for 15 minutes.
- Fixed cells are rinsed with 1X Phosphate Buffered Saline (PBS) 3 times for 5 minutes.
- Following fixation; permeabilization step can be performed by keeping cells in 0,1% Triton X100 for 10 minutes at room temperature.
- Cells are rinsed with PBS 3 times for 5 minutes.
- Free aldehyde groups which are formed during the fixation procedure may decrease antigenic site accessibility. To prevent this, Sodium Borohydrate treatment has to be performed. 2 mg/ ml sodium borohydrate solution is prepared in PBS. Cells are quenched 2 times for 15 minutes with 2 mg/ ml sodium borohydrate solution.
- Following final rinse with PBS 3 times for 5 minutes, cells can be kept at 4°C until the day of immunostaining.

Chemical	Final Concentration
Paraformaldehyde	4%
PHEM	1X
ddH$_2$O	up to final volume

Table 5. Fixative ingredients for paraformaldehyde fixation.

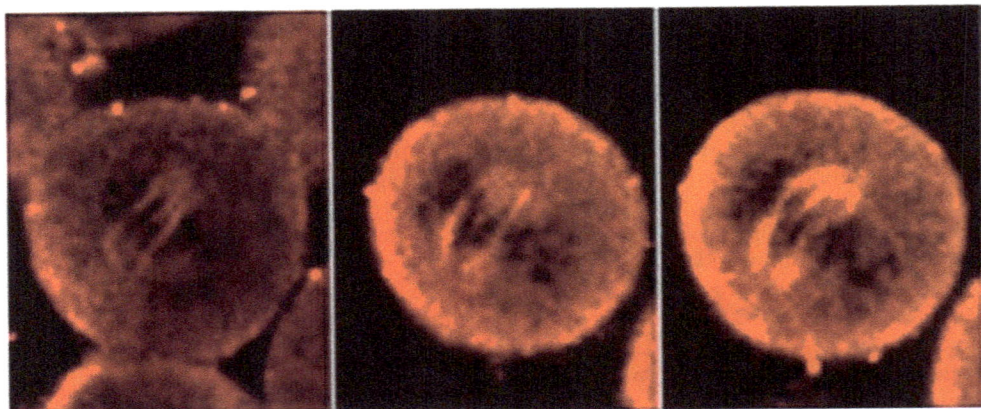

Fig. 15. Immunostaining of p60 katanin in paraformaldehyde fixed HeLa cells. p60 katanin was labeled with 1G6 mouse anti – p60 katanin primary antibody and visualized with Alexa Fluor 546 goat anti – mouse IgG (red). Images were obtained with laser scanning confocal microscope.

4.2.2 Glutaraldehyde fixation

- The glutaraldehyde fixation solution is warmed in 37°C water bath.
- Petri dishes containing the cultured cells are filled with fixation solution and incubated at room temperature for 15 minutes.
- Fixed cells are rinsed with 1XPBS 3 times for 5 minutes.
- Following fixation, permeabilization step can be performed by keeping cells in 0,1% Triton X100 for 10 minutes at room temperature.

- Cells are rinsed with PBS 3 times for 5 minutes.
- Cells are quenched 2 times for 15 minutes with 2 mg/ml sodium borohydrate solution.
- Following final rinse with PBS 3 times for 5 minutes, cells can be kept at 4°C until the day of immunostaining.

0,2%Glutaraldehyde
10µM Taxol
1X PHEM
up to final volume – ddH$_2$O

Table 6. Fixative ingredients for glutaraldehyde fixation.

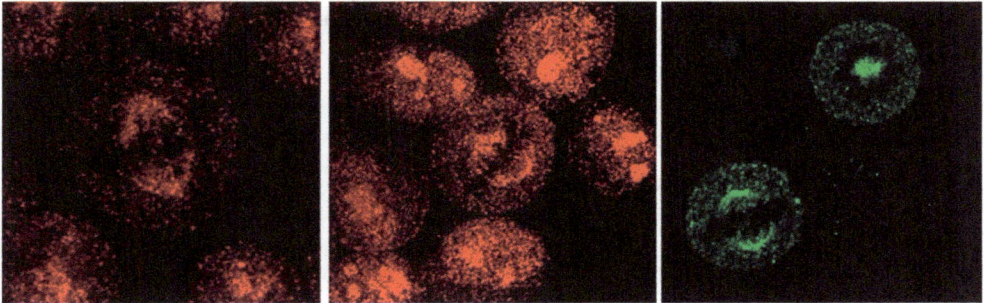

Fig. 16. Immunostaining of p60 katanin in glutaraldehyde fixed HeLa cells. p60 katanin was labeled with 1G6 mouse anti – p60 katanin primary antibody and visualized with Alexa Fluor 546 goat anti – mouse IgG (red) and Alexa Fluor 488 goat anti – mouse IgG (green). Images were obtained with laser scanning confocal microscope.

4.2.3 Methanol fixation

Since organic solvents also permeabilize cell membrane, additional protein extraction step is not required in methanol fixation method. Addition of nearly 200 µl of ice cold methanol per slide and placing at – 20°C for 10 minutes is enough to obtain fixation of the cells. Cells can be rinsed with PBS 3 times for 5 minutes and are ready for blocking step.

Exact and sharp location identification of p60 katanin protein is crucial for some critical places such as on mitotic spindle and on branch points of axons and dendrites. Since good protein visualization is obtained upon optimum fixation method and optimum fixation method changes depending on the epitope and antibody specificity, we have tested our home-made 1G6 antibody for p60 katanin under different fixation methods including paraformaldehyde fixation (Fig.15), glutaraldehyde fixation (Fig.16), and methanol fixation (Fig.17). 1G6 gave the best results with paraformaldehyde, whereas other fixation methods caused fuzzy appearance of the p60 katanin protein.

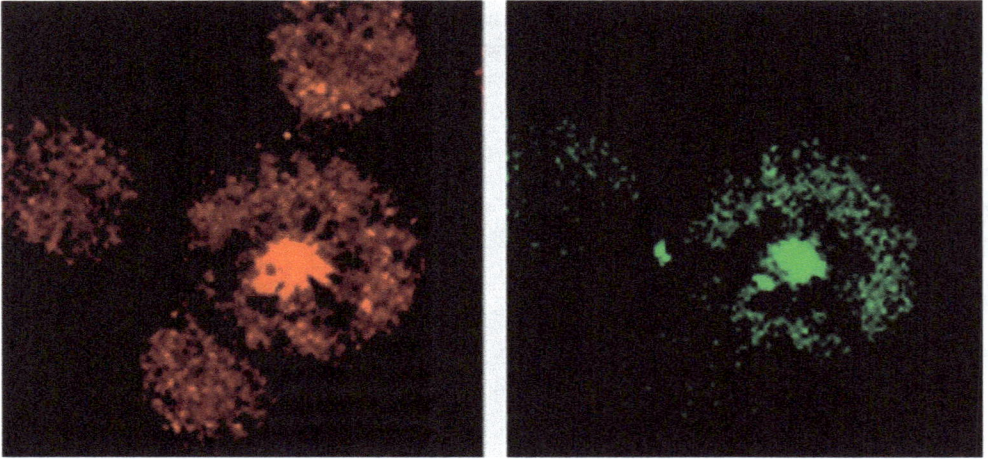

Fig. 17. Immunostaining of p60 katanin in methanol fixed HeLa cells. p60 katanin was labeled with 1G6 mouse anti – p60 katanin primary antibody and visualized with Alexa Fluor 546 goat anti – mouse IgG (red) and Alexa Fluor 488 goat anti – mouse IgG (green). Images were obtained with laser scanning confocal microscope.

4.3 Blocking for immunostaining

Before the antibody application part of the immunostaining procedure, cells should be blocked with an appropriate blocking solution to reduce nonspecific bindings of antibodies. To prepare blocking solution, 10% appropriate serum (goat or donkey depending on the type of secondary antibody) and 10 mg/ml Bovine Serum Albumin (BSA) is dissolved in PBS. Cells are then incubated in this blocking solution at room temperature for 1 hour or any optimized time period.

4.4 Immunostaining

Antibodies are the main players of immunostaining procedure. Antibodies either can be applied in a single step or in two steps. In single – step method, antibodies are directly conjugated to a colored agent, fluorochrome; whereas in two – step method the first antibody does not have the chemical structure to support a colored agent. In this case, following the application of the primary (colorless) antibody, a secondary antibody linked to a colored agent is applied, where the secondary antibody binds to the primary. The colored agent can then be visualized under the microscope. Comparing to the single – step method two – step method has some advantages. Variety of coloring agents can be conjugated to any given type of secondary antibody which gives chance to visualize different types of proteins at the same time.

Antibodies are diluted to obtain working stock depending on the manufacturer's instructions, but they usually require optimization studies to find the exact dilution ratio for the antibody of interest. To obtain clear visualization, any precipitates which may be formed during storage need to be cleaned by centrifuging the diluted antibodies at 10.000 rpm for 10 minutes at 4°C before use. Below you can find a sample for two – step immunostaining procedure.

- Diluted primary antibody solution is put on the slide containing fixed cells.
- Primary antibody containing cells are incubated at 4°C for overnight. During this staining period, the cover glasses are placed in a "humidified chamber," in a large Petri dish containing moistened filter paper, which minimizes evaporation of the antibody.
- Primary antibody is removed by vacuum carefully and dishes are washed by using PBS 3 times for 5 minutes each.
- Cells are then blocked again for 1 hour at room temperature with appropriate blocking solution.
- Secondary antibody is diluted and centrifuged in the same way as primary antibody.
- Diluted secondary antibody solution is put on the slide and cells are incubated for 1 hour in 37°C incubator. Secondary antibody application requires working in dark.
- Cells are washed with PBS 3 times for 5 minutes each.
- Finally 6 – 7 drops of anti – fade mounting medium (Table 7) is added on the slides in order to prevent rapid fading of fluorescent signals.
- Cover slips are placed on the cells and fixed from the edges by using nail polisher.
- Cells are visualized and analyzed by using fluorescent or confocal microscopy.

| 0,106 gr N – Propyl gallate |
| 5 ml PBS |
| 45 ml glycerol |

Table 7. Ingredients for mounting medium.

5. Conclusion

Immunocytochemistry is a powerful biochemical technique to identify the inner worlds of cells. Capturing the intracellular architecture and dynamics of cells as close as to their physiological state would require visualization and image analysis with state of the art techniques upon immunostaining with highly specific monoclonal antibodies and optimized fixation.

Therefore, there should be much effort to obtain highly specific monoclonal antibodies for a wide range of applications, and antibodies on the market would require much more vigorous characterization steps before they are made commercially available.

6. Acknowledgement

We would like to thank all the past and present members of Karabay Molecular Neurobiology Lab for their contribution to all the work presented here.

7. References

Akkor, M., Karabay, A. (2010). Production and characterization of a monoclonal antibody against P60 – katanin, *Hybridoma*, 6:531 – 7, 1554 – 0014

Errico, A., Claudiani, P., D'Addio, M., Rugarli, E.I. (2004). Spastin interacts with the centrosomal protein NA14, and is enriched in the spindle pole, the midbody and the distal axon. *Human Molecular Genetics*, 13, 182121–182132, 0964 – 6906

Karabay, A., Yu, W., Solowska, J. M., Baird, D. H., Baas, P. W. (2004). Axonal growth is sensitive to the levels of katanin, a protein that severs microtubules, *Journal of Neuroscience*, 24, 5778–5788, 0270 – 6474

Lodish, H. & Baltimore, D., (1995). *Molecular Cell Biology* (3rd edition). Scientific American Books, 0716776014, New York

McNally, F. J. & Vale, R. D. (1993). Identification of Katanin, an ATPase That Severs and Disassembles Stable Microtubules, *Cell*, 75, 419 – 429, 0092 – 8674

McNally, F. J. & Thomas, S. (1998). Katanin is responsible for the M – phase microtubule – severing activity in Xenopus eggs, *Molecular Biology of the Cell*, 9, 1847–1861, 1059 – 1524

McNally, F. J. (2000). Capturing a Ring of Samurai, *Nature Cell Biology*, 2: E4 – E7, 1465 – 7392

McNally, K. P., Bazirgan, O. A., McNally, F. J. (2000). Two domains of p80 katanin regulate severing and spindle pole targeting by p60 katanin, *Journal of Cell Science*, 113, 1623 – 1633, 0021 – 9533

Quarmby, L. (2000). Cellular Samurai: katanin and the severing of microtubules, *Journal of the Cell Science*, 113, 2821 – 2827, 0021 – 9533

Yu, W., Qiang, L., Solowska, J.M., Karabay, A., Korulu, S., Baas, P.W. (2008). The microtubule – severing proteins spastin and katanin participate differently in the formation of axonal branches, *Molecular Biology of the Cell*, 4:1485 – 98, 1059 – 1524

Vale, R. D. & Hartman, J. J. (1999). Microtubule Disassembly by ATP – Dependent Oligomerization of the AAA Enzyme Katanin, *Science*, 286, 782 – 784, 0036 – 8075

Vale, R. D. (2000). AAA Proteins: Lords of the Ring, *Journal of Cell Biology*, 150, F13 – F19, 0021 – 9525

6

Immunocytochemical Tools Reveal a New Research Field Between the Boundaries of Immunology and Reproductive Biology in Teleosts

Alfonsa García-Ayala[1] and Elena Chaves-Pozo[2]
[1]*Department of Cell Biology and Histology, Faculty of Biology,*
University of Murcia, Murcia
[2]*Centro Oceanográfico de Murcia, Instituto Español de Oceanografía (IEO),*
Carretera de la Azohía s/n. Puerto de Mazarrón, Murcia,
Spain

1. Introduction

Since the specific binding of the antigen and immunoglobulin was discovered, multiple applications have been developed to asses different issues such as: (i) localization of cells and molecules in different cell compartments, (ii) characterization of functional subsets of cells, (iii) identification of different cell types, (iv) and specific blocking or enhancement of receptors or other protein functions (Ruigrok et al., 2011). However, the array of available antibodies specific to fish species has limiting those experiments mostly in lower vertebrates and invertebrates. Mulero and colleagues, in the heart of our research group, has obtained some specific antibodies against several leukocytes and cytokines using different approaches. Moreover, we have developed some immunocytochemical techniques that allowed us to identify the immune cells in the gonads and to study their main activities and their regulation by different molecules or physiological processes in a teleost species, the gilthead seabream (*Sparus aurata* L.). The gilthead seabream is a sequential hermaphrodite species that, in the western Mediterranean area, develops as male during the first two reproductive cycles, while from the third reproductive cycle onwards the population divides into males and females (Chaves-Pozo et al., 2005a; Liarte et al., 2007) and due to the morphological arrangement of its bisexual gonad, it´s an interesting model for studying the mechanisms involved in the regulation of the gonads. This chapter deals with the antibodies and the techniques used to set up a novel research line in fish, such as the study of the immune-reproductive interactions which are essential for normal reproductive physiology. The antibodies have been obtained using three different approaches: (i) immunizing mice with cells, (ii) immunizing rabbits with recombinant proteins, or (iii) with synthetic peptides. Moreover, we have also used commercial antibodies in order to achieve information about the renewal of the cells in the gonad, especially analysing processes such as cell proliferation and apoptosis. The results obtained allowed us the identification and functional characterization of the leukocytes and immune molecules located in the gilthead seabream gonads.

2. Characterization of immune cells and molecules in the gonads of gilthead seabream by using immunocytochemical approaches

We have used antibodies and different immunocytochemical techniques to characterize the immune cells and molecules present in the gonads of the gilthead seabream throughout their lifespan as well as the regulatory mechanisms involved in the immune regulation of the reproductive functions. Notably, throughout this research line we have clearly demonstrated that, similar as occurs in mammals, the presence of immune cells and molecules are strictly regulated in the fish gonad and they are essential for the reproductive physiology. To reach this aim we have used several antibodies: one monoclonal antibody (G7) that is specific to gilthead seabream acidophilic granulocytes (Sepulcre et al., 2002) and three polyclonal antibodies: two against interleukin (Il)1β, obtained from the serum of an immunized rabbit with gilthead seabream recombinant Il1β (Pelegrín et al., 2004) or with the C-terminus of the gilthead seabream Il1β (RRHRIFKFLPPKPEVEGGEC), and one against macrophage colony stimulating factor receptor (Mcsfr), obtained from the serum of an immunized rabbit with a synthetic peptide corresponding to an internal epitope of the extracellular domain of the gilthead seabream Mcsfr (SLRVVRKEGEDYLLPC). The two last antibodies were produced by Pacific Immunology and characterized by Chaves-Pozo et al. (2008a) and Mulero et al. (2008), respectively. We have also used three commercial antibodies: two for the assessment of cell proliferation, anti-5′-Bromodesoxyuridine (BrdU) (Caltag) and anti-proliferating cell nuclear antigen (PCNA) (Dako) and an anti-B lymphocytes (Aquatic Diagnosis) which immunostained fish cells (Cabas et al., 2011; Chaves-Pozo et al., 2005a; Sepulcre et al., 2011).

First of all, we described the reproductive cycles of gilthead seabream males by describing the cell types present and their proliferative and apoptotic status in each stage of the reproductive cycle. For that, we used haematoxylin-eosin and Mallory trichromic staining, the anti-BrdU antibody to assess cell proliferation and the *in situ* detection of fragmented DNA (TUNEL) technique to assess apoptosis. Thus, the first two reproductive cycles of the gilthead seabream were divided in four gonadal stages: gametogenic activity, spawning, post-spawning and resting or testicular involution prior to sex change. In these reproductive cycles, spermatogenesis, spawning and post-spawning stages showed similar features. Thus, the early gametogenetic cells (spermatogonia stem cells and primary spermatogonia) and Sertoli cells were always present in the testis and proliferated throughout the year at variable rates, which depended on the reproductive stage. The late gametogenetic cells (spermatocytes, spermatids and spermatozoa) appeared in the last stages of the spermatogenesis and spawning (Figure 1).

Regarding proliferation, the highly proliferative activity observed during spermatogenesis contrasted with the scarce proliferative activity of early gametogenetic cells observed during spawning (Figure 2). During post-spawning the proliferative activity was resumed, although necrotic areas and apoptotic germ cells were also observed (Chaves-Pozo et al., 2005a; Liarte et al., 2007). Interestingly, the last stages of each reproductive cycle (resting and testicular involution, respectively) differed completely. Thus, compared with what happens in post-spawning, the resting stage was characterized by an increase in the number of proliferative cells and no apoptotic cells (Chaves-Pozo et al., 2005a), while, during the testicular involution stage, the number of proliferative cells were similar and the number of apoptotic cells increased as did the size of the necrotic areas (Liarte et al., 2007). The ovarian

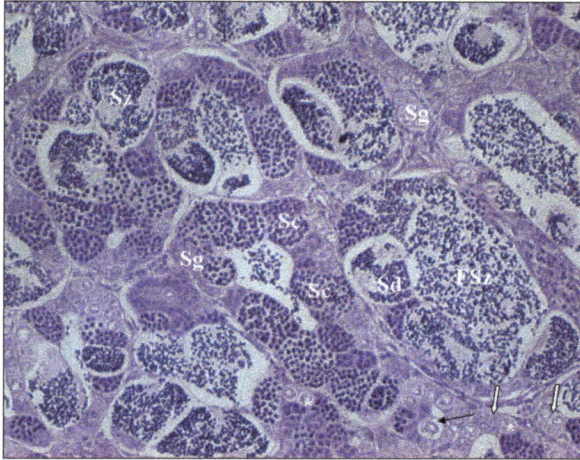

Fig. 1. Section of gilthead seabream testis in the spermatogenesis stage stained with
haematoxylin-eosin. During middle-late spermatogenesis all gametogenic cells are present:
Spermatogonia stem cells (arrow), cysts of primary spermatonia (asterisks), A and B
spermagonia (white arrows) spermatogonia (Sg), spermatocytes (Sc), spermatids (Sd), and
spermatozoa (Sz). Free spermatozoa (FSz) are also seen in the lumen of the tubules.
Magnification x 200.

area showed proliferative activity during both resting and testicular involution stages, at a
rate that did not differed from the proliferative activity that have been described during
each resting stage of the male phase in several sparid species (Micale et al., 1987). However,
only during testicular involution the immature oocytes (pre-perinucleolar and
perinucleolar) developed and the first vitellogenic oocytes appeared (Liarte et al., 2007).

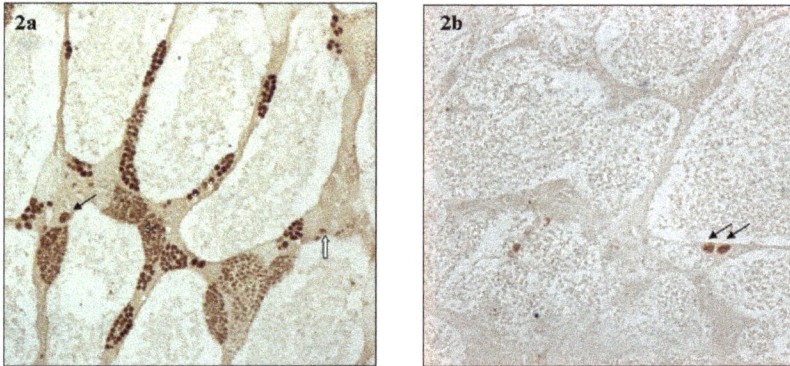

Fig. 2. Sections of gilthead seabream testis in spermatogenesis (a) and spawning (b) stages
immunostained with anti-BrdU. During spermatogenesis the proliferative activity was high
and spermatogonia stem cells (arrow), Sertoli cells (white arrow) and cysts of spermagonia
(white asterisks) and spermatocytes (Sc) were immunostained. During spawning scarce
spermatogonia stem cells (arrows) proliferated. Magnification x 200.

In some teleost species, the presence of macrophages, after the shedding of spermatozoa, has been observed in the testis by conventional microscopy (Besseau et al., 1994). Moreover, Sertoli cells, alone or together with macrophages, have been observed to be involved in germ cell elimination (Scott and Sumpter, 1989). In the gilthead seabream, we have demonstrated the presence of three types of leukocytes (acidophilic granulocytes, macrophages and lymphocytes) in the gonad using immunocytochemical approaches. By using the G7 antibody in paraffin embedded sections (Figure 3), flow cytometry and cell sorting techniques we have characterized the role of acidophilic granulocytes inside the gonad. Firstly, this cell type actively migrate from the head-kidney, where they are produced, into the gonad in healthy conditions since they were not able to proliferate inside the gonad (Chaves-Pozo et al., 2003) in contrast to what happened in condrictian fish species in which the gonad is an haematopoietic organ (Zapata et al., 1996). Secondly, the infiltration of acidophilic granulocytes depended on the reproductive cycle stage of the specimens and it was strongly regulated by soluble factors produced by the gonad (Chaves-Pozo et al., 2003, 2005a, 2005b). Moreover, this migratory influx increased the amount of acidophilic granulocytes in the testis at specific stages (post-spawning and testicular involution) coinciding with a very low proliferative activity of germ cells and with the appearance of necrotic areas and apoptotic germ cells (Chaves-Pozo et al., 2003, 2005a,b; Liarte et al., 2007).

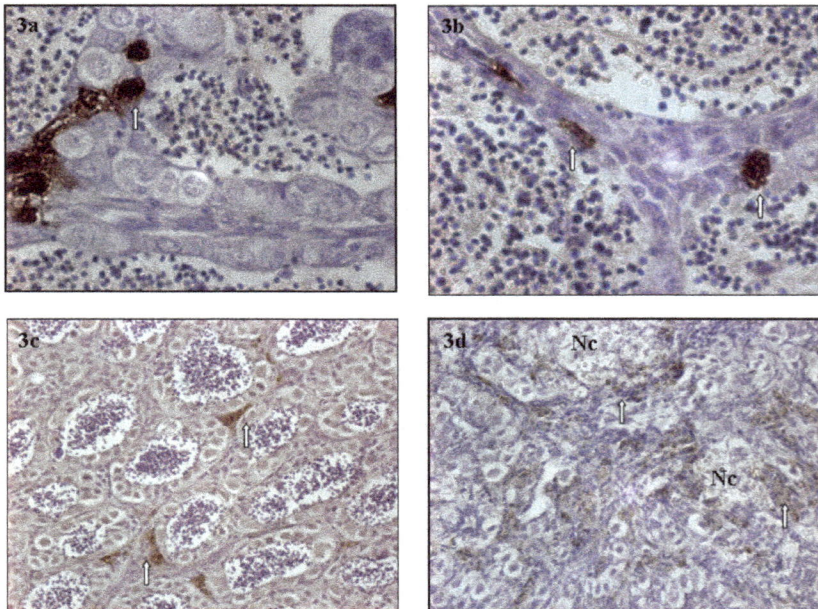

Fig. 3. Sections of gilthead seabream testis in the spawning (a,b) and post-spawning (c,d) stages immunostained with a monoclonal antibody specific to gilthead seabream acidophilic granulocytes. Acidophilic granulocytes (white arrows) are seen in the seminal epithelium in contact with germ cells during spawning (a,b) and post-spawning (c). Moreover, in some areas cluster of acidophilic granulocytes were observed surrounded the necrotic areas (Nc) during post-spawning (d). Magnification x 400 (a,b) and x 200 (c,d).

Immunocytochemical Tools Reveal a New Research Field Between the Boundaries of Immunology and
Reproductive Biology in Teleosts

121

Interestingly, the ultrastructure of these testicular acidophilic granulocytes at the testicular involution stage strongly suggests their involvement in the renewal process that takes place in the gonad at post-spawning and testicular involution stages (Liarte et al., 2007). The use of the G7 antibody has also allowed enriching head-kidney (hkG7+ cells) and testis (tG7+ cells) cell suspensions by magnetic activated cell sorting (MACS). Those enriched cell suspensions together with total testicular and head-kidney cell suspensions have been used in flow cytometry assays to asses phagocytosis (Esteban et al., 1998), reactive oxygen intermediates (ROI) production (Banati et al., 1994) and metalloproteinases activity (Leber and Balkwill, 1997). Combining those techniques, it has been demonstrated that head-kidney acidophilic granulocytes infiltrate the gonad in response to gonadal and hormonal stimulus and showed completely different functional capabilities than their head-kidney counterparts (Chaves-Pozo et al., 2005b, 2008a,b): impaired phagocytic and ROI production activities although they were the majority of cells which were able to produce ROIs in the testis upon phorbol myristate acetate (PMA) stimulation (Chaves-Pozo et al., 2005b). Moreover, testicular acidophilic granulocytes also showed a completely different array of metalloproteinases activity than their head-kidney counterparts (Chaves-Pozo et al., 2008b). These data demonstrated that the activities of acidophilic granulocytes were modified by the microenvironment of the testis once they infiltrate the organ (Figure 4).

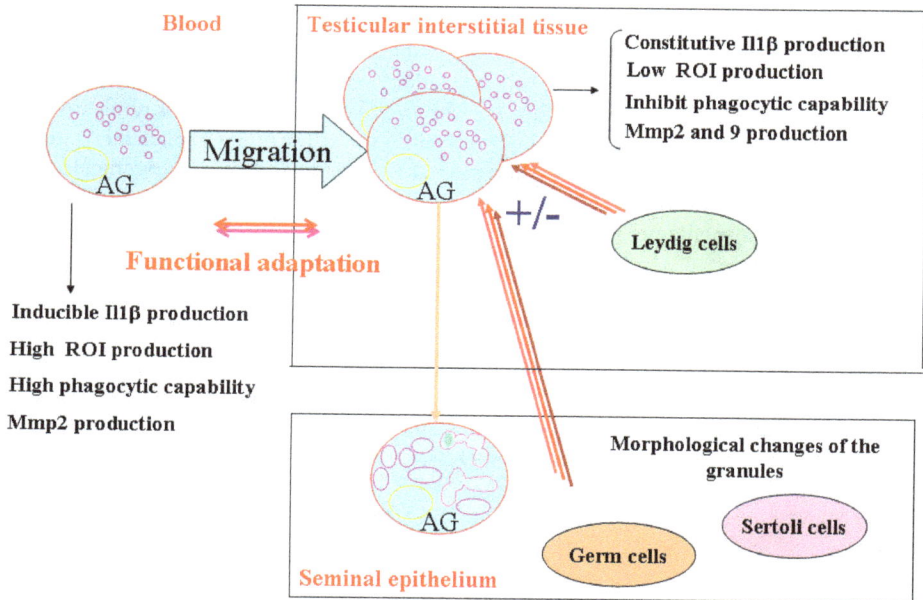

Fig. 4. A diagram of the functional adaptation of acidophilic granulocytes (AG) once they infiltrate the testis. Il1β, interleukin 1β; Mmp, metalloproteinases.

In the other hand flow cytometry techniques and electron microscopy combined with methanol fixed blood smears and paraffin embedded tissue subjected to immunofluorescence and peroxidase-antiperoxidase methods using an anti-Il1β and G7 immunostaining analysis revealed that the acidophilic granulocytes in the gilthead seabream are the first cell type to response to an infection mobilizing to the site of infection, phagocyting the pathogen and re-circulating to the head-kidney and spleen to act as antigen presenting cells (Chaves-Pozo et al., 2004a, 2005c). Interestingly, blood circulating acidophilic granulocytes only contained Il1β upon infection, however testicular acidophilic granulocytes constitutively produced Il1β (Chaves-Pozo et al., 2003, 2004a). Moreover, in the testis some spermatogonia, spermatocytes and scattered interstitial cells also produced Il1β (Chaves-Pozo et al., 2008a). In mammals Il1β has been described to be involved in germ cells proliferation in the testis (Syed et al., 1995), role that might also developed in fish as only the spermatogonia present in spermatogenesis and spawning stages, where the proliferative rate of spermatogonia are high, were immunostained with anti-Il1β antibody while at post-spawning and involution stages, where the proliferative rate of spermatogonia were low, scarce spermatogonia were immunostained (Chaves-Pozo et al., 2005a, 2008b).

In the mid-time we analyzed the sex-steroid hormone levels (17β-estradiol, testosterone and 11-ketotestosterone) in gilthead seabream males as they are important for understanding the dynamic of cell populations in the gonads. The changes in serum levels suggested that each steroid play a different and specific role in the testicular physiology being the 17β-estradiol which mainly orchestrates the testicular regression process that occurred in both post-spawning and testicular involution stages (Chaves-Pozo et al., 2008c). Interestingly, when we increased the 17β-estradiol serum levels experimentally in spermatogenically active males and then, performed head-kidney and testicular sections immunostained with G7, we observed a clear migration of the acidophilic granulocytes from the head-kidney to the testis (Chaves-Pozo et al., 2007). In the same way, the dietary intake of 17α-ethynilestradiol, a pharmaceutical compound used for oral contraceptives and hormone replacement therapy, also promoted the infiltration of acidophilic granulocytes into the spermatogenically active testis, while cell proliferation (analyzed using the anti-PCNA serum) was impaired at the beginning of the treatment and stimulated in Sertoli cells and spermatogonia later on (Cabas et al., 2011).

Regarding macrophages, the anti-Mcsfr revealed that this cell type was scattered in the testis and did not showed any cycling modifications in numbers, not during the first two reproductive cycles neither after 17α-ethynilestradiol dietary intake (Cabas et al., 2011; Chaves-Pozo et al., 2008a). Moreover, Sertoli cells also presented this molecule in their membrane, as other tissue specific macrophages such as those located in the spleen, liver or gills (Mulero et al., 2008). These data together with the fact that the testicular cells which were not recognized by the G7 antibody (G7- cells) have a huge phagocytic capability as revealed by an *in vitro* challenge with bacteria and subsequent electron microscopy analysis (Chaves-Pozo et al., 2004b, 2005b), support the hypothesis that Sertoli cells are the main phagocytic cells in the testis and in contrast to what occurs in mammals, teleosts Sertoli cells have proliferative capability (Chaves-Pozo et al., 2005a; Schulz et al., 2010). The origin of the new Sertoli cells that restart spermatogenesis during the following reproductive cycle has been widely discussed and is still a matter of controversy. Our data demonstrated that after releasing the spermatozoa to the lumen of the tubules, the Sertoli cells remain forming the epithelium of the tubules where they phagocytose the remaining spermatozoa after spawning (Chaves-Pozo et al., 2005a, 2008a). Similarly when a post-spawning stage was induced by dietary intake of 17α-ethynilestradiol, the Sertoli cells released from the open

cyst remain forming the epithelium of the cysts as the anti-mcsfr staining of this cells
suggest (Cabas et al., 2011).

In contrast to what happened with macrophages, B lymphocytes showed the same
behaviour as acidophilic granulocytes upon estrogen treatments. B lymphocytes,
determined as IgM producing cells, were normally presented in the interstitial tissue of the
testis and infiltrated the testis and the ovary upon 17α-ethynilestradiol dietary intake,
appearing in the interstitial tissue and in the lumen of the tubules between the spermatozoa.
Moreover, secreting IgM was increased in the gonad blood vessels of treated fish (Cabas et
al., 2011).

All this data suggested that in adult fish gonads, leukocytes are essential for sperm
production and cell renewal after spawning. Some immunocytochemical methods were also
used to asses the role of leukocytes during the ontogeny of the gonad. For that, gilthead
seabream larvae were sampled every 2 days from 1 to 83 days post-hatching (dph) and
juveniles were sampled at 92, 103, 111, 131 and 180 dph (Chaves-Pozo et al., 2009). Our data
suggested that leukocytes also have a prominent role during the ontogeny of the gonad.
Thus, acidophilic granulocytes were present in large amounts in the undifferentiated gonad
from 111 days dph until the newly formed gonad at 180 dph, as also occurs in *Cichlasoma
dimerus* testis (Meijide et al., 2005); afterwards the amount of acidophilic granulocytes
decreased and they began to be scattered as occurs in the spermatogenesis stage of mature
males (Chaves-Pozo et al., 2009). Moreover, in 180 dph juveniles, acidophilic granulocytes
and mesenchymal cells and a few germ cells produced Il1β, while in 270 dph males, in
which the spermatogenesis process is active, Il1β was only located in acidophilic
granulocytes and spermatocytes as occurs in one year old fish (Chaves-Pozo et al., 2008a).
Our observations of the gonadal primordial support the hypothesis that Il1β is a testicular
germ cell growth factor, as is also the case in mammals, since Il1β is not produced in the
developing ovarian area (Khan and Rai, 2007; Parvinen et al., 1991; Pollanen et al., 1989).
The appearance of Mcsfr was only observed in the well-developed testicular areas of 240-300
dph males, produced by the same cell types as in the adult testis (Chaves-Pozo et al., 2008a).

3. Production of fish-specific antibodies

3.1 Fish-specific antibodies

There are not many available antibodies which specifically recognize immune cells and
immune molecules of fish species with commercial interest. Moreover, the high diversity of
the teleost group makes it difficult to obtain antibodies against all different species.
Therefore, in the previous immunocytochemical studies carried out, antibodies obtained in
other fish and even in mammals, have been used when they showed some cross-reaction;
however, not all antibodies against fish molecules cross-react between fish species. All this,
together with the economic interest of commercial fish species, have led to the development
of specific antibodies. Especially in the gilthead seabream, numerous antibodies have been
obtained in the recent years, particularly those specific for studying the endocrine regulation
of multiple biological processes such as reproduction, growth, metabolism, etc. (Abad et al.,
1992; García-Ayala et al., 2003; Modig et al., 2008; Morgado et al., 2007; Picchietti et al., 2007;
Pinto et al., 2009; Pirone et al., 2008; Radaelli et al., 2003, 2005; Santos et al., 2001; Villaplana
et al., 1997). Regarding the gilthead seabream immune response, an array of monoclonal and

polyclonal antibodies have been developed for leukocyte markers (Sepulcre et al., 2002), immune molecules (Corrales et al., 2010; García-Castillo et al., 2004; Pelegrín et al., 2004) or immunoglobulins (Picchietti et al., 2006).

3.2 Production of antibodies

Antibodies are serum immunoglobulins produced by B lymphocytes in response to foreign proteins, called antigens. The short amino acid sequence of the antigen that the antibody recognizes and binds to is called epitope. Antibodies have binding specificity for particular antigens and function as markers, being of enormous utility in experimental biology, medicine, biomedical research, diagnostic testing, and therapy (Leenaars and Hendriksen, 2005).

3.2.1 Monoclonal antibodies

The production of monoclonal antibodies was pioneered by Kohler and Milstein (1975). This production is based in the fact that each B cell in an organism synthesizes only one type of antibody. The culture of a clone of B cells, derived from a single ancestral B cell, produce an unique and specific antibody that would allow to molecular biologists to harvest this specific antibody and to have substantial amounts of it. This population of B cells would be correctly described as monoclonal, and the antibodies produced by this population of B cells are called monoclonal antibodies. Monoclonal antibodies are homogeneous, of defined specificity and can be produced in unlimited quantities, making them a powerful immunological tool to biomedical and experimental studies. Thus, monoclonal antibodies have been and will continue to be important for the identification of proteins, carbohydrates, and nucleic acids. Its use has led to the elucidation of many molecules that control cell physiology, advancing our knowledge of the relationship between molecular structure and function (Lipman et al., 2005). In order to isolate a B cell population producing a certain antibody, we first have to induce the production of such B cells in a mouse. The antigen could be a purified protein, a recombinant protein, a small synthetic peptide or even cells. After several injections of the antigen every 2-3 weeks, a sample of B cells is extracted from the spleen of the immunized mouse and added to a culture of myeloma cells (cancer cells). The fusion of cells to each other is induced using polyethylene glycol, a virus or by electroporation, and hybridomas are formed (Yokoyama, 2001). The next step consisted in the selection of the hybridomas using the presence or absence of hypoxanthine-guanine phosphoribosyl transferase (HGPRT), an enzyme involved in the synthesis of nucleotides from the amino acid, hypoxanthine. In the next stage, the culture of the hybridomas is grown in hypoxanthine-aminopterin-thymine (HAT) medium, which can sustain only the growth of HGPRT positive cells. Thus, the myeloma cells that fuse with another myeloma cells or do not fuse at all die in the HAT medium since they are HGPRT negative. The B cells that fuse with another B cells or do not fuse at all die because they do not have the capacity to divide indefinitely. Only hybridomas between B cells and myeloma cells survive, being both HGPRT positive and cancerous. As the B cells obtained from the spleen of the immunized mouse are heterogenous (i.e. they do not all produce the same antibody) the hybridomas obtained does not produce a single antibody. Moreover, some hybridomas lost their capability to produce antibodies due to the fact that they are initially tetraploid, having been formed by the fusion of two diploid cells. However, the extra chromosomes are somehow lost in subsequent divisions in a random manner. In this context, we need to clone the different antibodies produced hybridomas and the screening of the clones obtained is required to decide which hybridoma cells are producing the desired antibody. The screening method has to be choosed depending on the posterior use of the antibody to really obtain a worthless tool for our studies. There are several methods of screening (Yokoyama, 2001) and one of the most used, for developing monoclonal antibodies valid for cell markers, is the staining of the interested cells with the different hybridoma cultured media and the subsequent flow cytometry analysis. In other cases we can also used other techniques such as sodium dodecyl sulfate-polyacrylamide gel electrophoresis (SDS-PAGE) and Western blots using the epitope of the antibody labelled with radioactivity or immunofluorescence as probe. Once we are sure that a certain hybridoma is producing the right antibody, we can culture this hybridoma indefinitely and harvest monoclonal antibodies from the medium. Under the appropriate conditions, monoclonal antibodies-producing hybridomas survive indefinitely, so the continuous production of monoclonal antibodies is associated with the use of fewer animals, especially when production involves the use of *in vitro* methods.

The G7 antibody was produced following the method developed by Kohler and Milstein (1975) using 10^7 seabream peritoneal exudate cells as antigen to immunize Balb/c mice (Sepulcre et al., 2002). The G7 was a very multi-functional antibody, able to recognise its specific epitope under denature protein conditions after fixation as in immunocytochemical techniques or under native protein conditions as in cell live staining protocols useful to flow cytometry or MACS assays (Sepulcre et al., 2002).

3.2.2 Polyclonal antibodies

In contrast, antibodies obtained from the blood serum of an immunized animal are called polyclonal antibodies. They consisted in a variety of antibodies that binds to different epitopes of a target antigen, and are therefore less specific that the monoclonal antibodies. The production of polyclonal antibodies can be divided into two or three stages; (i) immunization of the animal, (ii) blood collection and, when needed, (iii) purification of the antibodies presented in serum. In brief, the researcher needs to choose the animal species to be immunized which depends, at least in part, on the amount of antiserum needed and the ease for obtaining blood samples. Several parameters such as the immune status of the immunized animals, the nature of the antigen, the intended purpose of the antiserum produced, the toxicity of the antigen preparation or the adjuvant (a substance which enhance the immune response) and the routes of application might influence the immunization protocol (review in Leenaars et al., 1999). The antisera obtained from the serum of immunized animals contain different classes and subclasses of immunoglobulins with different affinities for the injected antigens. Moreover, the immunized animals may have natural antibodies in their serum that may cross-react or non-specifically bind with the proteins of the tissue. However, many polyclonal antibodies have recently been generated to synthetic peptides and epitopes, reaching specific antibodies that are easily purified by affinity chromatography with resins bound to the peptides (Yamashita, 2007). We have used three polyclonal antibodies, two against gilthead seabream Il1β and one against gilthead seabream Mcsfr. Thus, Pelegrín et al. (2004) produced a serum against Il1β injecting on rabbits the recombinant His6-tagged sbIl1β excised of SDS-PAGE gels reversibly stained with Zn–imidazole negative staining and following standard protocols (Ausubel et al., 1995). Although the antibody allowed the performance of several techniques such as

Western Blot and immunocytochemistry, a more effective antibody was later on developed by injection of a synthetic peptide corresponding to the C-terminus of the gilthead seabream Il1β protein into rabbits (Pacific Immunology). Using a synthetic peptide corresponding to an internal epitope of the gilthead seabream Mcsfr, a specific polyclonal antibody against gilthead seabream macrophages was also developed (Pacific Immunology) and characterized (Mulero et al., 2008). In this case, the serum obtained recognise the antigen under denature conditions; however, it was not useful for live cell staining techniques such as flow cytometry or MACS.

4. Overview of immuno-detection techniques

Although the different immunocytochemical techniques in which these antibodies are used seem to be quite different, all of them are based on the same principle. A mixture of antigens bound to solid phase (a tissue section, a cell in suspension, a membrane, etc.) is exposed to the primary antibody which is directed against the antigen of interest (BrdU, PCNA, acidophilic granulocytes, Il1β, Mcsfr, IgM). After the incubation of the primary antibody, the unbound antibody is removed and different approaches would be followed to detect the antigen-antibody reaction. In the indirect immunocytochemical method (Sternberger, 1986) the bound primary antibody is bounded with a secondary antibody, directed against the primary antibody host species, and tagged with an enzyme (i.e. peroxidase) that, later on, allowed the detection of the immune reaction, after removing the unbound secondary antibody. In the presence of hydrogen peroxide (H_2O_2), the enzyme, peroxidase, is able to convert 3,3′-diaminobenzidine tetrahydrochloride, (DAB) to an insoluble brown reaction product producing water at the same time. Moreover other enzymes (i.e. alkaline phosphatase) or fluorochromes can be conjugated to the secondary antibody, in the last cases those techniques are usually called immmunofluorescence techniques. There are also some methods that allowed a greater amplification of the signal by using a third antibody that specifically bound to the second antibody and to a complex of enzyme molecules such as peroxidase-antiperoxidase (peroxidase anti-peroxidase, PAP, techniques) (Sternberger, 1986) or by using a secondary antibody tagged with a molecule such as biotin that is able to bind a complex with multiple enzymes such as avidine-peroxidase or streptavidine-peroxidase complex (avidine/streptavidine-biotin techniques) (see Figure 5).

The great variability of the immunocytochemical techniques carried out depended on the sample (a tissue section, a cell suspension, a membrane or a plate with proteins or nucleic acid, etc.), the molecule conjugated to the secondary or tertiary antibody (an enzyme, a fluorochrome, radioactive molecule, colloidal gold or even a molecule that can be complex with several molecules of enzymes in a third stage) and the method by which we observed or measured the results (light or electron microscopy, flow cytometry, cell sorting, light or radioactive detector, fluorometer, etc.). Moreover, it is also possible to localize two different antigens on the same sample. In those cases the species in which the primary antibodies were performed became highly important; since all the primary antibodies used at the same time must be performed in different species to avoid cross-reactions of the secondary antibodies (see Figure 6). Although it is possible to achieve different methods to locate several antigens on the same sample using enzymes as the conjugated molecule, the most used molecules are different fluorochromes which emissions are performed in different colours that can be independently detected (see Figure 6).

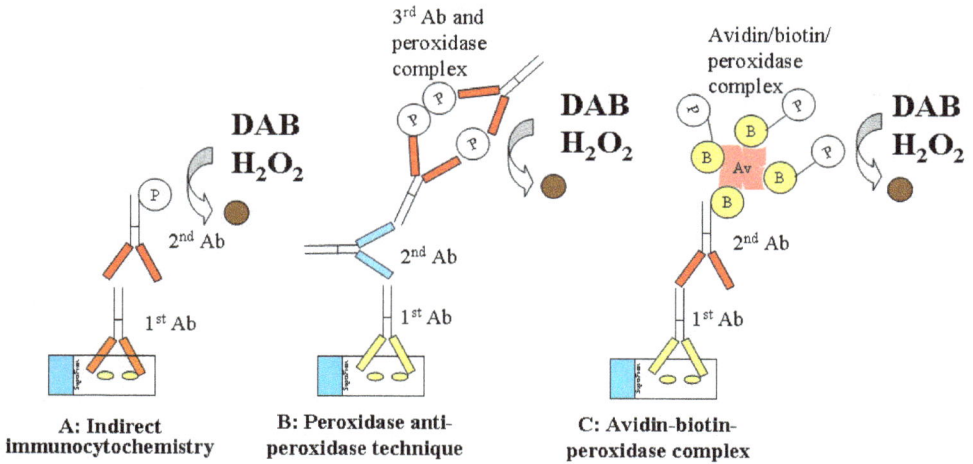

Fig. 5. A diagram of the indirect (A), peroxidase anti-peroxidase (B) and avidin-biotin-peroxidase (C) techniques performed on tissue sections. Ab, antibody, P, peroxidase; DAB, 3,3'-diaminobenzidine tetrahydrochloride; H_2O_2; hydrogen peroxide, B, biotine; Av, avidine; ●, brown staining.

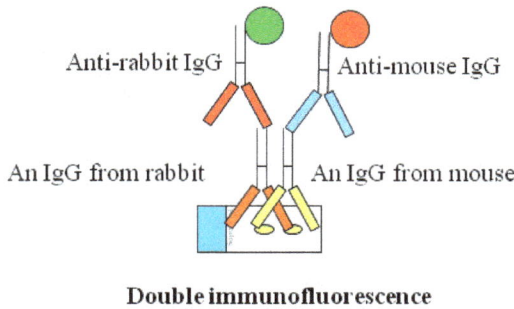

Double immunofluorescence

Fig. 6. A diagram of the double immunofluorescence immunoreaction applied on a tissue section. IgG, immunoglobulin G.

In order to determine which technique is applicable using a specific antibody, it is also worth bearing in mind that not all antibodies detect the antigen in multiple conformations. Thus, there are antibodies which detect their antigens in its native or denaturated conformations, others that only detected their antigens whether they are denatured in a very specific way (i.e. using a specific fixative but not others) and there are a small number of antibodies that are able to recognise its specific antigen in both, native or denature conditions.

Taking all this into account the first step to start working with antibodies is to determine which technique is allowed by our specific antibody characteristics and between them, which one will lead to results that fulfil our research questions.

4.1 Choice of fixatives, antigen retrieval methods, and immunohistochemical procedures

One of the most common applications of the antibodies is the localization, on tissue sections, the cell types that produce a determined antigen. The method relies on proper fixation of tissue to retain the antigen and to preserve cellular morphology. A lot has been written on the chemical characteristics of the different fixatives and how they work (for review see Hopwood, 1977). In general, a fixative can perfectly preserve some molecules on a tissue while others molecules might suffer degradation. So, it is worth knowing the chemical nature of the epitope that we are interested in detecting. In the fixation of tissues, the most important reactions are probably those which stabilize the proteins. Thus, we can talk about crosslinking (aldehydes, cardodiimides), oxidising (osmium tetroxide, potassium permanganate) or protein-denaturing fixatives (acetic acid, methyl alcohol, ethyl alcohol). Moreover, the combination of some of these fixatives allowed obtaining fixative solutions that preserved specific cell components. Here, we included brief information about the fixatives that we have used in our studies and their chemical characteristics (Table 1).

	Methanol	Paraformaldehyde	Acetic acid	Picric acid
Volume changes	decreased	not affected	increased	decreased
Effect on proteins	denatured	cross-linked	denatured	unknown
Effect on lipids	extracted	cross-linked	not affected	not affected
Effect on nucleic acids	not affected	well conserved	coagulated	hydrolysis
Effect on carbohydrates	well conserved	not affected	not affected	not affected

Table 1. Some chemical properties of the fixatives used in the immunocytochemical characterization of the gonad of gilthead seabream (for review see Hopwood, 1977).

The permeability of the tissue to the antibody must be also taking into account. In most of the cases, the use of phosphate-buffer saline (PBS) with detergents during the washes performed before and between primary and secondary antibodies incubation is enough to obtain an optimal immunoreaction. However, in some cases, a heavier antigen retrieval treatment which partially denatures fixed proteins exposing the epitope to the antibodies is used before the primary antibody incubation. The efficacy of those treatments depends on tissue fixation strength, the characteristics of the epitope and the welfare of morphology needed. Those treatments varied according with their strength from the incubation with a heated antigen retrieval solution to a treatment with some proteases, including microwaves in some cases. However, those treatments were not free of charge, since a lost of morphology will always be produced whenever an increase of permeability is achieved. In table 2, we have included some of the lower strength treatments, however for review see Yamashita (2007).

Immunocytochemical Tools Reveal a New Research Field Between the Boundaries of Immunology and
Reproductive Biology in Teleosts

129

Treatment	Reagent	Time (minutes)	Heat	Morphology effect	Temperature
1	3mM sodium citrate with 0.1% Triton X-100	8		Poor effects	Room temperature
2	0.01M sodium citrate pH 6.0	5	650-700 W microwaves	High effects	
3	0.01M sodium citrate pH 6.0	5	650-700 W microwaves	High effects	
4	1mM EDTA with 0.05% Tween 20 pH 8.0	20	Heated bath	Medium effects	95°C
5	1mM EDTA with 0.05% Tween 20	10		Poor effects	Room temperature
6	0.01M sodium citrate pH 6.0	5	Heated bath	Medium effects	98°C
6	0.01M sodium citrate pH 6.0	10		Poor effects	Room temperature
7	10-20µg/ml proteinase K in 10mM Tris-HCl pH 7.4	15-30	Heated bath	High effects	21-37°C
8	0.1M sodium citrate pH 6.0	1	750 W microwaves	High effects	
10	0.05% tripsine in 20mM Tris-HCl pH 7.4 with 130mM NaCl	5		High effects	Room temperature
11	1mM EDTA with 0.05% Tween 20	20	Heated bath	Medium effects	98°C
12	0.1M sodium citrate pH 6.0	1	800 W microwaves	High effects	

Table 2. Some antigen retrieval treatments which have different effects on tissue morphology (for review see Yamashita, 2007).

In our research, we firstly determined the fixative and the embedded method appropriate to specifically detect the gilthead seabream acidophilic granulocytes using the G7 antibody in order to clearly determine where the acidophilic granulocytes were located in the gonad. We have used Bouin's fluid at 4°C for 24 hours (Chaves-Pozo et al., 2003). However, when we combined two antibodies (G7 and anti-Il1β or G7 and anti-BrdU) to determine whether testicular acidophilic granulocytes were able to constitutively produce Il1β or whether they were able to proliferate, the fixative used was 4% paraformaldehyde that allowed the reaction of all antibodies although the preservation of the gonad morphology was worse. In the case of the BrdU detection, a good fixative of DNA such as aldehyde (see Table 1) will be mandatory to optimally detect all the proliferative cells taking into account that BrdU is a molecule that must be injected, previously, to the specimens or incubated with the cell suspensions, to allow its incorporation in the DNA of all proliferating cells of the gonad or the cell suspension.

Regarding the antigen retrieval, we applied different treatments to each antibody in order to get an optimal balance between the immunostaining and the preservation of the morphology. A special attention needed the detection of BrdU inside the DNA of the cells due to the fact that a special treatment to denature the proteins that surround the DNA is mandatory. However, as the anti-BrdU is a commercial antibody, each manufacture provided the best protocol. Although, all of them included acid and heat treatments. Those treatments together with the needed to use paraphormaldehyde as fixative, avoid the obtaining of a good morphology in the anti-BrdU immunostained sections. The rabbit polyclonal anti-PCNA serum cross-react with PCNA from all vertebrate species investigated so far, including fish (Kilemade et al., 2002) and allowed Bouin's fluid as fixative which preserved a better morphology of the tissue. Moreover, heavy treatments of the sections are not needed as the PCNA is a protein presented in the nuclei and cytoplasm at specific stages of the cell cycle which is available for the antibody using standard protocols. However, it is important to keep in mind that the amount of proliferating cells detected by both antibodies, anti-BrdU and anti-PCNA, might not be the same in a determined tissue since anti-BrdU label the cells which are synthesizing DNA, while PCNA is present in the cell at the four phases of the cell cycle (i) G1, where cells grow in size, assess their metabolic status and get ready to divide; (ii) Synthesis, where the actual genome duplication takes place; (iii) G2 where cells check for completion of DNA replication and prepare to divide, and (iv) Mitosis where mitosis and cytodieresis take place. In the other hand, an advantage of the anti-PCNA immunostaining is that this antibody does not depend on previous labelling step which might present difficulties depending in the animal species study (Kubben et al., 1994; Maga and Hubscher, 2003). In order to make easy the understood of all the different protocols applied and their theoretical motivation, we have included all their main stages in Table 3.

As previously discussed (see point 2), there are several methods to determine whether two different antigens reacted with the same cell. In the case of G7 and anti-Il1β, both antibodies were developed in different species so, in theory, we could use both over the same section and apply two secondary antibodies tagged with different fluorochromes (see Figure 2). However, the testis has a well developed interstitial tissue mainly formed by collagen which notably increases the autofluorescence of the tissue and disguises the specific fluorescence. In our case we could only used fluorescence microscopy when blood was processed (Chaves-Pozo et al., 2004a). In the case of the testis we opted to perform serial sections of the gonad and used adjacent sections to perform each immunoreaction separately, as also occurred with the G7 and anti-BrdU antibodies which were both developed in mouse. Thus, we demonstrated that acidophilic granulocytes present in testis did not proliferate and constitutively produced the cytokine Il1β, in contrast to what happened in other tissues such as blood or head-kidney where this cell type only produced Il1β upon infection (Chaves-Pozo et al., 2003, 2004a).

Sample	Tissue sections					Blood smears	Cell suspension
Fixation	Bouin's fluid	Paraformaldehyde or Bouin's fluid	Paraformaldehyde	Bouin's fluid	Bouin's fluid	Methanol	
Dewaxing	Xilol	Xilol	Xilol	Xilol	Xilol		
Rehydration	Decreased dilutions of alcohol in water	Decreased dilutions of alcohol in water	Decreased dilutions of alcohol in water	Decreased dilutions of alcohol in water	Decreased dilutions of alcohol in water	Decreased dilutions of alcohol in water	
Permeabilizing			1% periodic acid in water at 60 °C for 30 minutes	1 mM EDTA and 0.05% Tween 20 buffer (pH: 8) at 95°C during 20 minutes			
Washes	water	water	water	water	water	water	
Endogenous enzyme quenching solution	peroxidase quenching solution (H_2O_2 in methanol, 1:9)	peroxidase quenching solution (H_2O_2 in methanol, 1:9)	peroxidase quenching solution (H_2O_2 in methanol, 1:9)	peroxidase quenching solution (H_2O_2 in methanol, 1:9)	peroxidase quenching solution (H_2O_2 in methanol, 1:9)	peroxidase quenching solution (H_2O_2 in methanol, 1:9)	
Washes	Coons buffer and CBT	Coons buffer and CBT	PBS	PBS and PBT	PBS and PBT	Coons buffer and CBT	
Blocking	5% BSA in PBS	5% BSA in PBS	5% BSA in PBS	5% BSA in PBS	5% BSA in PBS	5% BSA in PBS	5% fetal calf serum and 2 mM EDTA in PBS at 4°C
Washes	Coons buffer and CBT	Coons buffer and CBT	PBS	PBS and PBT	PBS and PBT	Coons buffer and CBT	2% fetal calf serum and 0.05% sodium azide in PBS at 4°C
Primary antibody	G7	Anti-IIIβ	Anti-BrdU (Becton Dickinson)	Anti-Mcsfr (Pacific Immunology)	Anti-lymphocytes B (Aquatic Diagnostic)	G7 and anti-IIIβ	G7 at 4°C
Washes	Coons buffer and CBT	Coons buffer and CBT	PBS	PBS and PBT	PBS and PBT	Coons buffer and CBT	2% fetal calf serum and 0.05% sodium azide in PBS at 4°C

Table 3. Part I

Table 3. Part II

	Anti-mouse IgG conjugated with peroxidase (Sigma-Aldrich)	Anti-rabbit IgG (Dako)	Anti-mouse IgG conjugated with peroxidase (Dako)	Anti-rabbit IgG conjugated with biotin (Dako)	Anti-mouse IgG conjugated with biotin (Sigma-Aldrich)	Anti-mouse IgG conjugated with fluorescein isothiocyanate (Sigma) and anti-rabbitt IgG conjugated with tetramethylrhodamine isothiocyanate (Dako)	Anti-mouse IgG conjugated with fluorescein isothiocyanate (Sigma) at 4°C	Anti-mouse IgG conjugated with micromagnetic Beads (Miltenyi Biotec) at 4°C
Secondary antibody								
Washes	Coons buffer and CBT	Coons buffer and CBT	PBS	PBS and PBT	PBS and PBT	Coons buffer and CBT	2% fetal calf serum and 0.05% sodium azide in PBS at 4°C	5% fetal calf serum and 2 mM EDTA in PBS at 4°C
Tertiary antibody or complex		Peroxidase-anti-peroxidase complex (Dako)		Avidin-biotin-peroxidase complex (Vectastain ABC kit)	Avidin-biotin-peroxidase complex (Vectastain ABC kit)			
Washes		Coons buffer and CBT		PBS and PBT	PBS and PBT			
Incubation with the substrate	2 mM DAB with 0.05% H_2O_2 in PBS	2 mM DAB with 0.05% H_2O_2 in PBS	2 mM DAB with 0.05% H_2O_2 in PBS	2 mM DAB with 0.05% H_2O_2 in PBS	2 mM DAB with 0.05% H_2O_2 in PBS			
Dehydration	Increased dilutions of alcohol in water	Increased dilutions of alcohol in water	Increased dilutions of alcohol in water	Increased dilutions of alcohol in water	Increased dilutions of alcohol in water			
Clearance	Xilol	Xilol	Xilol	Xilol	Xilol			
Mounting	DPX	DPX	DPX	DPX	DPX			
Analyze	Light microscopy	Light microscopy	Light microscopy	Light microscopy	Light microscopy	Light (fluorescence) microscopy	Flow cytometry	Purified by MACS

Table 3. A comparative overview of the different applications of the antibodies that allowed the identification and functional characterization of the different leukocytes presented in the gilthead seabream gonads and some of their functional activities. PBS, phosphate buffer saline; PBT, PBS with BSA and Triton-X100; CBT, Coons buffer with BSA and Triton-X100; BSA, bovine serum albumine; DAB, 3,3'-diaminobenzidine tetrahydrochloride; DPX, Mounting Media; MACS, magnetic activating cell sorting. For buffer recipes see appendix I.

4.2 Flow cytometry and magnetic activating cell sorting (MACS)

The detection of an antigen in a tissue section allowed its localization on the different cell types of the tissue. However, the quantification of the cells marked with our antibodies is difficult to perform on tissue sections as the cells (acidophilic granulocytes) are located in large clusters in the interstitial tissue of the testis (Chaves-Pozo et al., 2003). Thus, in order to quantify the amount of testicular acidophilic granulocytes in the different stages of the reproductive cycles we used flow cytometry. To fulfil the requirements of flow cytometry detection, a homogenous cell suspension of the tissue is mandatory together with an antibody that recognises a protein which is stacks in the membrane and in its native conformation (review in Shapiro, 2005) in order to analyse the morphological characteristics (granularity and size) of the cells that specifically immunoreacted with our antibody. The protocol to dissociate gilthead seabream testis was described in detail in Chaves-Pozo et al. (2004b). Once a homogenous testicular cell suspension was obtained, a specific cell type can be labelled using a specific antibody. We have used the G7 antibody which specifically immunoreacted with acidophilic granulocytes and detected a membrane protein in its native or denature conformations. Since the antibody was able to bind to the epitope in the native conformation, a fixative was not needed, although the staining must be performed at 4°C to avoid internalization of the protein-antibody complex (see Table 3). Analyzing the fluorescence of the immunostained cells by flow cytometry, we have determined that there was an increase in the amount of acidophilic granulocytes present in the testis at specific stages of the reproductive cycle (post-spawning and testicular involution stages) reaching the 6 and 10% of the total cells, respectively (Chaves-Pozo et al., 2005b; Liarte et al., 2007). This data, together with the fact that the acidophilic granulocytes did not proliferate in the testis, demonstrated the infiltration of this cell type from the blood stream to the testis at these specific stages of the reproductive cycle, suggesting a role on fish reproduction of these immune cells (Chaves-Pozo et al., 2003, 2005a,b).

The fact that the G7 antibody recognised its epitope on native proteins allowed us to combine the immunostaining of acidophilic granulocytes with other functional assays *in vitro* such as the phagocytic assay on cell suspensions. In that sense we have first performed the phagocytic assay (review in Esteban et al., 1998), and then the immunostaining keeping the cells at 4°C to avoid digestion of the phagocyted-fluorescence-bacteria and internalization of the antigen-antibody complex at the membrane. Thus, we have observed that the 95% of the head-kidney acidophilic granulocytes had phagocytic capability in contrast with the 1% of the testicular acidophilic granulocytes (Chaves-Pozo et al., 2005b; Sepulcre et al., 2002). However, most of the functional activities of the immune response can not be combine with a posterior antibody immunostaining, so the purification of the testicular acidophilic granulocytes were mandatory to really understand the role of the acidophilic granulocytes once they migrated into the gonad. The G7 also allowed applying the MACS techniques using micromagnetic-beads to specific purified acidophilic granulocytes guaranteeing the viability and functional characteristics of these cells as occurred in mammals (Miltenyi et al., 1990) (Figure 7).

Thus, we demonstrated that the main functional activities of acidophilic granulocytes are inhibited once they migrated into the testis (Chaves-Pozo et al., 2005b). As previously discussed (see point 4) each antibody allowed a fixed number of techniques that depended on its ability to react with its specific epitope under native or denatured conditions. The parameters that determined this ability of the antibodies are not completely known.

However, the use of native or denatured antigens to produce the antibodies and whether they are polyclonal or monoclonal will lead to more multifunctional antibodies.

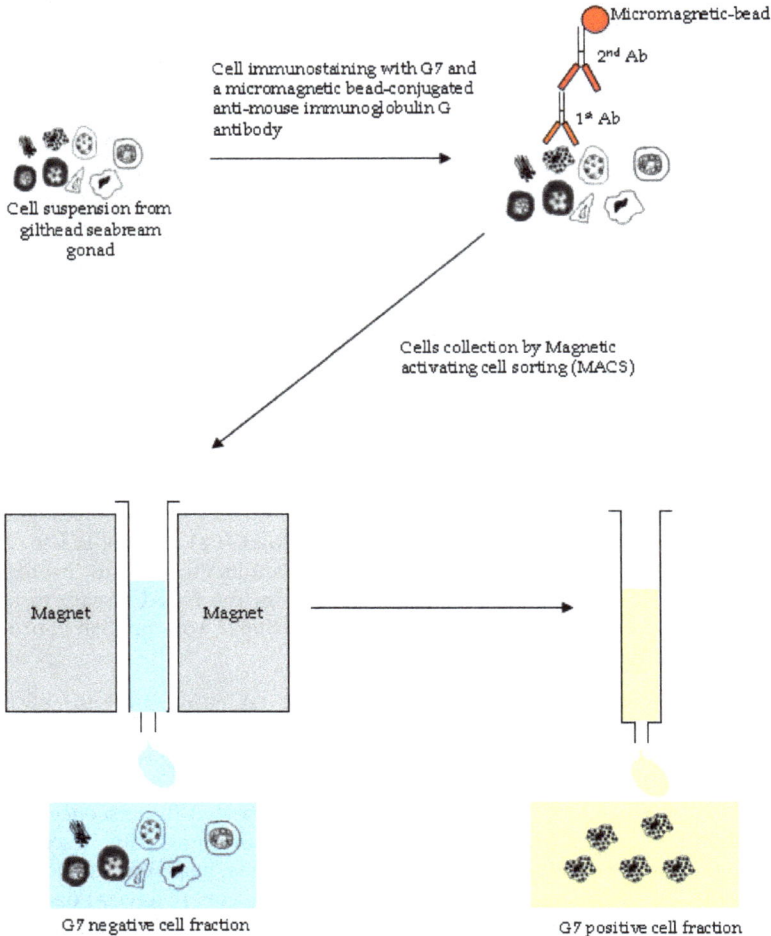

Fig. 7. A diagram for obtaining acidophilic granuloyctes (G7 positive cells) enriched cell fraction from a testicular cell suspension by the immunostaining with G7 and the subsequent application of the MACS technique. The G7 negative cell fraction included the rest of the gonadal cells. Ab, antibody.

5. Conclusions

The availability of specific antibodies has allowed us to characterize the immune cells present in the gonad of the gilthead seabream as well as to determine some of the molecules involved in the immune regulation of the reproductive physiology. Nowadays, the knowledge of the sequence of the proteins allowed the design of specific epitopes sequences that in turn can be used to develop specific antibodies against any molecule of fish with scientific and commercial purposes. Although the characteristics of the antibodies will determine the usefulness of these antibodies in immunocytochemical studies, it is important to bear in mind that some modifications of the protocols, mainly detecting molecules on tissue sections, such as fixation, antigen retrieval, the method of detection of the reaction and so on, might allow the obtaining of good specific immunoreactions.

6. Acknowledgements

This work was supported by the Fundación Séneca, Coordination Center for Research, CARM (proyect 04538/GERM/06 to A.G.A), the Spanish Ministry of Science and Innovation (contract RYC-2009-05451 to E.C.P., project AGL2008-04575-C02-01 to A.G.A. and AGL2010-20801-C02-01 to E.C.P.). We thank the "Servicio de Apoyo a la Investigación" of the University of Murcia for their assistance with cell culture and gene expression analysis and the "Centro Oceanográfico de Murcia, Instituto Español de Oceanografía" for their assistance with fish care.

7. Appendix I

The recipes of the buffers that appeared in table 3.

0.01M PBS pH 7.4			
Reagent	Mw (g/mol)	Concentration	Amount of reagent (g/l)
Na$_2$HPO$_4$	141.96	9 mM	1.335
NaH$_2$PO$_4$	156.01	2 mM	0.414
NaCl	58.44	0.15 M	8.775

PBT pH 7.4		
Reagent	Concentration	Amount of reagent
PBS	0.01 M	1 l
BSA	0.01%	0.1 g/l
Triton-X100	0.2%	2 ml/l

0.01 M Coons pH 7.4			
Reagent	Mw (g/mol)	Concentration	Amount of reagent (g/l)
Veronal	206.177	0.01 M	2.06
NaCl	58.44	0.1 M	8.5

CBT, pH 7.4		
Reagent	Concentration	Amount of reagent
Coons buffer	0.01 M	1 l
BSA	0.01%	0.1 g/l
Triton-X100	0.2%	2 ml/l

8. References

Abad, M. E.; García-Ayala, A.; Lozano, M. T. & Agulleiro, B. (1992). Somatostatin 14- and somatostatin 25-like peptides in pancreatic endocrine cells of *Sparus aurata* (teleost): a light and electron microscopic immunocytochemical study. *General and Comparative Endocrinology*, Vol.86, No.3, (June 1992), pp. 445-452, ISSN 0016-6480.

Ausubel, F. M.; Brent, R.; Kingston, R.; Seidman, J. G.; Smith, J. A. & Struhl, K. (Eds.). (1995). *Current Protocols in Molecular Biology*, Vol.13. Wiley, ISBN 9780471142720, New York.

Banati, R.B.; Schubert, P.; Rothe, G.; Gehrmann, J.; Rudolphi, K.; Valet, G. & Kreutzberg, G. W. (1994). Modulation of intracellular formation of reactive oxygen intermediates in peritoneal macrophages and microglia/brain macrophages by propentofylline. *Journal of Cerebral Blood Flow and Metabolism*, Vol.14, No.1 (January 1994), pp. 145-149, ISSN 0271-678X.

Besseau, L. & Faliex, E. (1994). Resorption of unemitted gametes in *Lithognathus mormyrus* (Sparidae, Teleostei): a possible synergic action of somatic and immune cells. *Cell &Tissue Research*, Vol.276, No.1 (April 1994), pp. 123–132, ISSN 0302-766X.

Cabas, I.; Chaves-Pozo, E.; García Alcazar, A.; Meseguer, J.; Mulero, V. & García-Ayala, A. (2011). Dietary intake of 17alpha-ethinylestradiol promotes leukocytes infiltration in the gonad of the hermaphrodite gilthead seabream. *Molecular Immunology*, Vol.48, No.15-16, (September 2011), pp. 2079-2086, ISSN 0161-5890.

Chaves-Pozo, E.; Pelegrín, P.; Mulero, V.; Meseguer, J. & García-Ayala, A. (2003). A role for acidophilic granulocytes in the testis of the gilthead seabream (*Sparus aurata* L., Teleostei). *Journal of Endocrinology*, Vol.179, No.2, (November 2003), pp. 165-174, ISSN 0022-0795.

Chaves-Pozo, E.; Pelegrín, P.; García-Castillo, J.; García-Ayala, A.; Mulero, V. & Meseguer, J. (2004a). Acidophilic granulocytes of the marine fish gilthead seabream (*Sparus aurata* L.) produce interleukin-1beta following infection with *Vibrio anguillarum*. *Cell and Tissue Research*, Vol.316, No.2, (May 2004), pp. 189-195, ISSN 0302-766X.

Chaves-Pozo, E.; Mulero, V.; Meseguer, J. & García-Ayala, A. (2004b). Flow cytometry based techniques to study testicular acidophilic granulocytes from the protandrous fish gilthead seabream (*Sparus aurata* L.). *Biological Procedures Online*, Vol.6, (June 2004), pp. 129-136, ISSN 1480-9222.

Chaves-Pozo, E.; Mulero, V., Meseguer, J. & García-Ayala, A. (2005a). An overview of cell renewal in the testis throughout the reproductive cycle of a seasonal breeding teleost, the gilthead seabream (*Sparus aurata* L). *Biology of Reproduction*, Vol.72, No.3, (March 2005), pp. 593-601, ISSN 0006-3363.

Chaves-Pozo, E.; Mulero, V.; Meseguer, J. & García-Ayala, A. (2005b). Professional phagocytic granulocytes of the bony fish gilthead seabream display functional adaptation to testicular microenvironment. *Journal of Leukocyte Biology*, Vol.78, No 2 (August 2005), pp. 345-351, ISSN 0741-5400.

Chaves-Pozo, E.; Muñoz, P.; López-Muñoz, A.; Pelegrín, P.; García-Ayala, A.; Mulero, V. & Meseguer, J. (2005c). Early innate immune response and redistribution of inflammatory cells in the bony fish gilthead seabream experimentally infected with *Vibrio anguillarum*. *Cell and Tissue Research*, Vol.320, No.1 (April 2005), pp. 61-68, ISSN 0302-766X.

Chaves-Pozo, E.; Liarte, S.; Vargas-Chacoff, L.; García-López, A.; Mulero, V.; Meseguer, J.; Mancera, J. M. & García-Ayala, A. (2007). 17Beta-estradiol triggers postspawning in spermatogenically active gilthead seabream (*Sparus aurata* L.) males. *Biology of Reproduction*, Vol.76, No.1, (January 2007), pp. 142-148, ISSN 0006-3363.

Chaves-Pozo, E.; Liarte, S.; Fernández-Alacid, L.; Abellán, E.; Meseguer, J.; Mulero, V. & García-Ayala, A. (2008a). Pattern of expression of immune-relevant genes in the gonad of a teleost, the gilthead seabream (*Sparus aurata* L.). *Molecular Immunology*, Vol.45, No.10, (May 2008), pp. 2998-3011, ISSN 0161-5890.

Chaves-Pozo, E.; Castillo-Briceño, P.; García-Alcázar, A.; Meseguer, J.; Mulero, V. & García-Ayala, A. (2008b). A role for matrix metalloproteinases in granulocyte infiltration and testicular remodelation in a seasonal breeding teleost. *Mollecular Immunology*, Vol.45, No.10, (May 2008), pp. 2820-2830, ISSN 0161-5890.

Chaves-Pozo, E.; Arjona, F. J.; García-López, A.; García-Alcázar, A.; Meseguer, J. & García-Ayala, A. (2008c). Sex steroids and metabolic parameter levels in a seasonal breeding fish (*Sparus aurata* L.). *General and Comparative Endocrinology*, Vol.156, No.3 (May 2008), pp. 531-536, ISSN 0016-6480.

Chaves-Pozo, E.; Liarte, S.; Mulero, I.; Abellán, E.; Meseguer, J. & García-Ayala, A. (2009). Early presence of immune cells in the developing gonad of the gilthead seabream (*Sparus aurata* Linnaeus, 1758). *The Journal of Reproduction and Development*, Vol.55, No.4, (August, 2009), pp. 440-445, ISSN 0916-8818.

Esteban, M. A.; Mulero, V.; Muñoz, J. & Meseguer, J. (1998). Methodological aspects of assessing phagocytosis of *Vibrio anguillarum* by leucocytes of gilthead seabream (*Sparus aurata* L.) by flow cytometry and electron microscopy. *Cell and Tissue Research*, Vol.293, No.1, (July 1998), pp. 133-141, ISSN 0302-766X.

Corrales, J.; Mulero, I.; Mulero, V. & Noga, E. J. (2010). Detection of antimicrobial peptides related to piscidin 4 in important aquacultured fish. *Developmental and Comparative Immunology*, Vol.34, No.3, (March 2010), pp. 331-343, ISSN 1879-0089.

García-Ayala, A.; Villaplana, M.; García-Hernández, M. P.; Chaves-Pozo, E. & Agulleiro, B. (2003). FSH-, LH-, and TSH-expressing cells during development of *Sparus aurata* L. (Teleostei). An immunocytochemical study. *General and Comparative Endocrinology*, Vol.134, No.1 (October 2003), pp. 72-79, ISSN 1095-6840.

García-Castillo, J.; Chaves-Pozo, E.; Olivares, P.; Pelegrín, P.; Meseguer, J. & Mulero, V. (2004). The tumor necrosis factor alpha of the bony fish seabream exhibits the *in vivo* proinflammatory and proliferative activities of its mammalian counterparts, yet it functions in a species-specific manner. *Cellular and Molecular Life Sciences*, Vol.61, No.11, (June 2004), pp. 1331-1340, ISSN 1420-682X.

Hopwood, D. (1977). Fixation and fixatives, In: *Theory and practice of histological techniques*, J.D. Bancroft & A. Stevens (Eds.), 21-42, Churchill Livingstone Inc, ISBN 0-443-03559-8, New York, EEUU.

Khan, U. W. & Rai, U. (2007). Differential effects of histamine on Leydig cell and testicular macrophage activities in wall lizards: precise role of H1/H2 receptor subtypes. *Journal of Endocrinology*, vol.194, No.2, (August 2007), pp. 441-448, ISSN 0022-0795.

Kilemade, M.; Lyons-Alcántara, M.; Rose, T.; Fitzgerald, R. & Mothersill, C. (2002). Rainbow trout primary epidermal cell proliferation as an indicator of aquatic toxicity: an *in vitro/in vivo* exposure comparison. *Aquatic Toxicology*, Vol.60, No.1-2, (October 2002), pp. 43-59, ISSN 0166-445X.

Kohler, G. & Milstein, C. (1975). Continuous cultures of fused cells secreting antibody of predefined specificity. *Nature*, Vol.256, No.5517, (August 1975), pp. 495-497, ISSN 0028-0836.

Kubben, F. J.; Peeters-Haesevoets, A.; Engels, L. G.; Baeten, C. G.; Schutte, B.; Arends, J. W.; Stockbrugger, R. W. & Blijham, G. H. (1994). Proliferating cell nuclear antigen (PCNA): a new marker to study human colonic cell proliferation. *Gut*, Vol.35, No.4, (April 1994), pp. 530-535, ISSN 0017-5749.

Leber, T. M. & Balkwill, F. R. (1997). Zymography: a single-step staining method for quantitation of proteolytic activity on substrate gels. *Analytical Biochemistry*, Vol.249, No.1, (June 1997), pp. 24-28, ISSN 0003-2697.

Leenaars, M. & Hendriksen, C. F. M. (2005). Critical Steps in the Production of Polyclonal and Monoclonal Antibodies: Evaluation and Recommendations. *ILAR Journal*, Vol.46, No.3, pp. 269-279, ISSN 1084-2020.

Leenaars, P.; Hendriksen, C.; Leeuw, W.; Carat, F.; Delahaut, P.; Fischer, R.; Halder, M.; Hanly, W.; Hartinger, J.; Hau, J.; Lindblad, E.; Nicklas, W.; Outschoorn, I. & Stewart-Tull, D. (1999). The Production of Polyclonal Antibodies in Laboratory Animals. In: *ECVAM Workshop Report, ATLA*, Vol.27, pp. 79-102.

Liarte, S.; Chaves-Pozo, E.; García-Alcázar, A.; Mulero, V.; Meseguer, J. & García-Ayala, A. (2007). Testicular involution prior to sex change in gilthead seabream is characterized by a decrease in DMRT1 gene expression and by massive leukocyte infiltration. *Reproductive Biology and Endocrinology*, Vol.5, (June 2007), pp. 20-35, ISSN 1477-7827.

Lipman, N. S.; Jackson, L. R.; Trudel, L. J. & Weis-García, F. (2005). Monoclonal versus polyclonal antibodies: distinguishing characteristics, applications, and information resources. *ILAR Journal, Vol.*46, pp, 258-268, ISSN 1084-2020.

Maga, G. & Hubscher, U. (2003). Proliferating cell nuclear antigen (PCNA): a dancer with many partners. *Journal of Cell Science*, Vol.116, No.15, (August 2003), pp. 3051-3060, ISSN 0021-9533.

Meijide, F. J.; Lo Nostro, F. L. & Guerrero, G. A. (2005). Gonadal development and sex differentiation in the cichlid fish *Cichlasoma dimerus* (Teleostei, Perciformes): a light- and electron-microscopic study. *Journal of Morphology*, Vol.264, No.2, (May 2005), pp. 191-210, ISSN 0362-2525.

Micale, V.; Perdichizzi, F. & Santangelo, G. (1987). The gonadal cycle of captive white bream *Diplodus sargus* (L.). *Journal of Fish Biology*, Vol.31, pp, 435-440, ISSN 0022-1112.

Miltenyi, S.; Muller, W.; Weichel, W. & Radbruch, A. (1990). High gradient magnetic cell separation with MACS. *Cytometry*, Vol.11, No.2, pp. 231-238, ISSN 0196-4763.

Modig, C.; Raldua, D.; Cerda, J. & Olsson, P. E. (2008). Analysis of vitelline envelope synthesis and composition during early oocyte development in gilthead seabream (*Sparus aurata*). *Molecular Reproduction and Development*, Vol.75, No.8, (August 2008, pp. 1351-1360, ISSN 1098-2795.

Morgado, I.; Santos, C. R.; Jacinto, R. & Power, D. M. (2007). Regulation of transthyretin by thyroid hormones in fish. *General and Comparative Endocrinology*, Vol.152, No.2-3, (June-July 2007), pp. 189-197, ISSN 0016-6480.

Mulero, I.; Sepulcre, M.P.; Roca, F. J.; Meseguer, J.; García-Ayala, A. & Mulero, V. (2008). Characterization of macrophages from the bony fish gilthead seabream using an antibody against the macrophage colony-stimulating factor receptor.

Developemental and Comparative Immunology, Vol.32, No.10, (April 2008), pp. 1151-1159, ISSN 0145-305X.

Parvinen, M.; Soder, O.; Mali, P.; Froysa, B. & Rtizen, E. M. (1991). In vitro stimulation of stage-specific deoxyribonucleic acid synthesis in rat seminiferous tubule segments in interleukin-1 alpha. Endocrinology, Vol.129, No.3, (September 1991), pp. 1614-1620, ISSN 0013-7227.

Pelegrín, P.; Chaves-Pozo, E.; Mulero, V. & Meseguer, J. (2004). Production and mechanism of secretion of interleukin-1beta from the marine fish gilthead seabream. Developmental and Comparative Immunology, Vol.28, No.3, (March 2004), pp. 229-237, ISSN 0145-305X.

Picchietti, S.; Abelli, L.; Buonocore, F.; Randelli, E.; Fausto, A. M.; Scapigliati, G. & Mazzini, M. (2006). Immunoglobulin protein and gene transcripts in sea bream (Sparus aurata L.) oocytes. Fish and Shellfish Immunology, Vol.20, No.3, (March 2006), pp. 398-404, ISSN 1050-4648.

Picchietti, S.; Mazzini, M.; Taddei, A. R.; Renna, R.; Fausto, A. M.; Mulero, V.; Carnevali, O.; Cresci, A. & Abelli, L. (2007). Effects of administration of probiotic strains on GALT of larval gilthead seabream: Immunohistochemical and ultrastructural studies. Fish and Shellfish Immunology, Vol.22, No.1-2, (January-February 2007), pp. 57-67, ISSN 1050-4648.

Pinto, P. I.; Estevao, M. D.; Redruello, B.; Socorro, S. M.; Canario, A. V. & Power, D. M. (2009). Immunohistochemical detection of estrogen receptors in fish scales. General and Comparative Endocrinology, Vol.160, No.1, (January 2009), pp. 19-29, ISSN 1095-6840.

Pirone, A.; Lenzi, C.; Marroni, P.; Betti, L.; Mascia, G.; Giannaccini, G.; Lucacchini, A. & Fabiani, O. (2008). Neuropeptide Y in the brain and retina of the adult teleost gilthead seabream (Sparus aurata L.). Anatomy, Histology and Embryology, Vol. 37, No.3, (June 2008), pp. 231-240, ISSN 1439-0264.

Pöllänen, P.; Soder, O. & Parvinen, M. (1989). Interleukin-1 alpha stimulation of spermatogonial proliferation in vitro. Reproduction, Fertility and Development, Vol.1, No.1, (January 1989), pp. 85-87, ISSN 1031-3613.

Radaelli, G.; Patruno, M.; Maccatrozzo, L. & Funkenstein, B. (2003). Expression and cellular localization of insulin-like growth factor-II protein and mRNA in Sparus aurata during development. Journal of Endocrinology, Vol.178, No.2, (August 2003), pp. 285-299, ISSN 0022-0795.

Radaelli, G.; Patruno, M.; Rowlerson, A.; Maccatrozzo, L. & Funkenstein, B. (2005). Cellular localisation of insulin-like growth factor binding protein-2 (IGFBP-2) during development of the marine fish, Sparus aurata. Cell and Tissue Research, Vol. 319, No.1 (January 2005), pp. 121-131, ISSN 0302-766X.

Ruigrok, V. J.; Levisson, M.; Eppink, M. H.; Smidt, H. & van der Oost, J. (2011). Alternative affinity tools: more attractive than antibodies? Biochemistry Journal, Vol.436, No.1, (May 2011), pp. 1-13, ISSN 1470-8728.

Santos, C. R.; Ingleton, P. M.; Cavaco, J. E.; Kelly, P. A.; Edery, M. & Power, D. M. (2001). Cloning, characterization, and tissue distribution of prolactin receptor in the sea bream (Sparus aurata). General and Comparative Endocrinolology, Vol. 121, No. 1, (January 2001), pp. 32-47, ISSN 0016-6480.

Schulz, R. W.; de Franca, L. R.; Lareyre, J. J.; Le Gac, F.; Chiarini-García, H.; Nobrega, R. H. & Miura, T. (2010). Spermatogenesis in fish. *General and Comparative Endocrinology*, Vol.165, No.3, (February 2010), pp. 390-411, ISSN 1095-6840.

Scott, A. P. & Sumpter, J.P. (1989). Seasonal variations in testicular germ cell stages and in plasma concentrations of sex steroids in male rainbow trout (*Salmo gairdneri*) maturing at 2 years old. *General and Comparative Endocrinology*, Vol.73, No1, (January 1989), pp. 46-58, ISSN 1095-6840.

Sepulcre, M. P.; Pelegrín, P.; Mulero, V. & Meseguer, J. (2002). Characterisation of gilthead seabream acidophilic granulocytes by a monoclonal antibody unequivocally points to their involvement in fish phagocytic response. *Cell and Tissue Research*, Vol.308, No.1, (April 2002), pp. 97-102, ISSN 0302-766X.

Sepulcre, M. P.; López-Muñoz, A.; Angosto, D.; García-Alcazar, A.; Meseguer J. & Mulero, V. (2011). TLR agonists extend the functional lifespan of professional phagocytic granulocytes in the bony fish gilthead seabream and direct precursor differentiation towards the production of granulocytes. *Molecular Immunology*, Vol.48, No.6-7, (March 2011), pp. 846-859, ISSN 1872-9142.

Shapiro, H. M. (2005). *Practical Flow Cytometry*, John Wiley & Sons, Inc, ISBN 0-471-41125-6, Hoboken, EEUU.

Sternberger, L. A. (1986). *Immunocytochemistry* (Edited by 3rd). Wiley, ISBN 0471867217 9780471867210, New York, EEUU.

Syed, V.; Stéphan, J. P.; Gérard, N.; Legrand, A.; Parvinen, M.; Bardin, C. W. & Jégou, B. (1995). Residual bodies activate Sertoli cell interleukin-1 alpha (IL-1 alpha) release, which triggers IL-6 production by an autocrine mechanism, through the lipoxygenase pathway. *Endocrinology* Vol.136, No.7, (July 1995), pp. 3070-3078, ISSN 0013-7227.

Villaplana, M.; García-Ayala, A.; García-Hernández, M. P. & Agulleiro, B. (1997). Ontogeny of immunoreactive somatolactin cells in the pituitary of gilthead sea bream (*Sparus aurata* L., Teleostei). *Anatomy and Embryology*, Vol.196, No.3, (September 1997), pp. 227-34, ISSN 0340-2061.

Yamashita, S. (2007). Heat-induced antigen retrieval: mechanisms and application to histochemistry. *Progress in Histochemistry and Cytochemistry*, Vol.41, No.3, (December 2007), pp. 141-200, ISSN 0079-6336.

Yokoyama, W. M. (2001). Production of Monoclonal Antibodies. In: *Current Protocols in Cell Biology*, K. Chambers (Ed.), 16.1.116.1.17, Wiley, ISBN 9780471143031.

Zapata, A. G.; Chibá, A. & Varas, A. (1996). Cells and tissues of the immune system of fish. In: *The Fish Immune System: Organism, Pathogen, and Environment*, G. Iwama; T. Nakanishi, (Eds.), 1-62, Academic Press, ISBN 01-235-04392, San Diego, EEUU.

7

Immunocytochemical Approaches to the Identification of Membrane Topology of the Na$^+$/Cl$^-$-Dependent Neurotransmitter Transporters

Chiharu Sogawa, Norio Sogawa and Shigeo Kitayama
Okayama University
Japan

1. Introduction

Transporters, located in the cell surface plasma membrane and intracellular organelle membrane, compose a major class of integral membrane proteins, as distinguished with their specialized functions as channels and receptors. They are divided into two groups of gene family; one is the solute carrier (SLC) super family and the other is the ATP-binding cassette (ABC) gene family (Giacomini & Sugiyama, 2006). The transporters for neurotransmitters in the plasma membrane, which belong to the SLC6 gene family, terminate synaptic neurotransmission by Na$^+$/Cl$^-$-dependent uptake of released neurotransmitters into the neuronal and/or glial cells (Iversen, 1971). They are responsible for the reuptake not only of neurotransmitters, monoamines (dopamine (DA), noradrenaline (NA), and serotonin (5-HT)) and amino acids (γ-aminobutyric acid (GABA) and glycine), but also of neuromodulators and/or osmolytes (proline, taurine, betaine, and creatine). Cloning of their cDNA has facilitated the understanding of the primary structure, gene expression, and their roles in neuronal functions (Amara & Kuhar, 1993). Cellular and molecular aspects of these transporters are needed for the clarification of their physiological and pathological relevance, since it seems likely that alterations of their structure, function and expression produce their anatomical and functional divergence, resulting in an involvement of a number of neurological and psychiatric disorders.

Elucidating the molecular structure of these transporters should be a first step to clarify their functional mechanisms, such as transport processes, regulations and consequent phenomena. Particularly, information of the membrane topology is substantially important for clarifying their structure-function relationship. Among the Na$^+$/Cl$^-$-dependent neurotransmitter transporters, this chapter focuses on the monoamine neurotransmitter transporters such as those for DA (DAT), NA (NET), and 5-HT (SERT). These monoamine transporters are of particular interest, since they are a target of drugs of abuse and/or antidepressants, and are involved in various neuronal disorders (Gether et al., 2006). Therefore, understanding of the expression and function of these transporters could promise clues to clarify their pathophysiological significance and thereby to develop medications for treatments of neuronal disorders mentioned above.

Cloning of the GABA transporter (GAT) cDNA and subsequently the NET cDNA have led to the identification of the gene family, such as SLC6 (Amara & Kuhar, 1993). Hydropathy analysis[1] of the primary structure of these transporters showed a common model for their membrane topology; the twelve hydrophobic regions that consist of α-helical transmembrane domains (TMDs) interrupted by alternating intra- and extracellular loops, one large putatively extracellular loop, and intracellularly localization of N- and C-termini (Fig. 1A). Based on this information, initial studies have examined the membrane topology of the transporters using various biochemical techniques including, for example, an immunological technique with specific antibodies against hydrophilic regions predicted as intra- and extracellular loops, or antibodies against known epitope tags incorporated at hydrophilic regions of the neurotransmitter transporters in the transiently or stably expressing cell lines. Recent success of the X-ray crystallography of the bacterial homologue of the Na^+/Cl^--dependent neurotransmitter transporter LeuT confirmed the proposed model and extended the structural understanding (Yamashita et al., 2005) (Fig. 1B). However, it is still difficult to understand fully the structure of mammalian plasma membrane neurotransmitter transporters. In addition to more studies on X-ray crystallography of bacterial homologues, we are needed to address directly to the mammalian transporter proteins to determine their membrane topology.

We have been investigating the expression and function of DAT and NET isoforms produced by alternative RNA splicing (Kitayama & Dohi, 2003). Recently, we found novel variants of human DAT (hDAT) and NET, skipping the region encoded by exon 6 (Sogawa et al., 2010). A hydropathy analysis of the variant designated hDATΔEX6 revealed 11 putative TMDs, suggesting a membrane topology different from that of full-length (FL) hDAT. To explore the unique structure of this variant, an immunocytochemical analysis with confocal microscopy was performed using two specific antibodies, one recognizing the second extracellular region (anti-hDAT-EL2 antibody) and the other recognizing the intracellular C-terminus (anti-hDAT-Ct antibody) in the tranfected cells treated with (permeabilized) or without (non-permeabilized) surfactant Triton. Immunoreactivity to the anti-hDAT-Ct antibody was only observed in FL hDAT-expressing cells treated with Triton, in contrast to the detection even in the untreated cells expressing hDATΔEX6, strongly suggesting the C-terminus of hDATΔEX6 to be located extracellulary. This information of the membrane topology is substantially important for clarifying the structure-function relationship. Since changes in membrane topology could influence the expression and function of the transporter at large, our observations provide clue to understand the functional modifications and expressional alterations of the DAT/NET splice variants.

In this chapter, we summarize the recent progress in our understanding of the contribution of the immunocytochemical approaches to determining a membrane topology of the plasma membrane neurotransmitter transporters. We discuss a usefulness of specific antibodies against the epitopes located in the extra- and/or intra-cellular region of plasma membrane protein, that led us to identify the unique structure of the DAT/NET splice variants. We also discuss the additional molecular biological and protein engineering techniques in

[1] An index that progressively evaluates the hydrophilicity and hydrophobicity of a protein along its amino acid sequence, thereby predicts higher structure of proteins. For this purpose, a hydropathy scale has been composed wherein the hydrophilic and hydrophobic properties of each of the 10-20 amino acid side-chains is taken into consideration.

comparison of their benefits and problems. In addition, we summarized basic methods concerning cell preparations, gene transfection techniques into mammalian cell lines, and fixation and permeabilization of cell cultures.

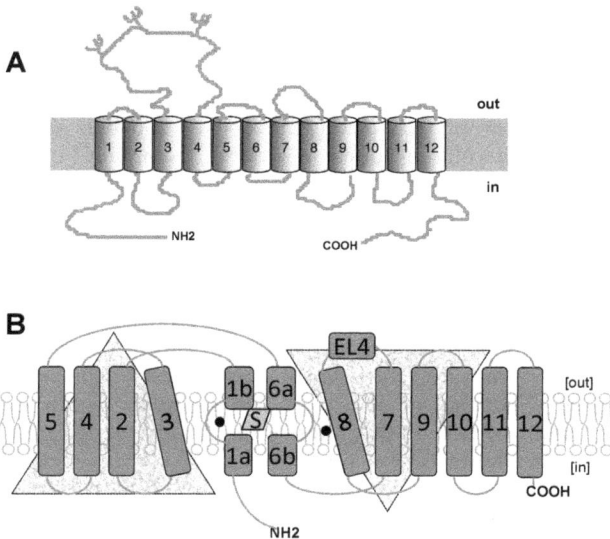

Fig. 1. Schematic presentation of the topological model of the neurotransmitter transporter (A) and LeuT (B). (A) Transmembrane helices are shown as cylinders and are numbered 1-12. Helical features are based on the hydropathy analysis of deduced amino acid sequence of the cloned transporter cDNA. The C and N termini are intracellular. (B) Model of LeuT in the plasma membrane based on its crystal structure. S: substrate leucine, closed circle: Na+.

2. General consideration of the neurotransmitter transporter

2.1 Cloning of cDNAs

Cloning of their cDNA has facilitated the understanding of the primary structure, gene expression, and their roles in neuronal functions. The GABA transporter-1 (GAT-1) cDNA was cloned from rat brain as a first one of the gene family by means of protein purification (Guastella et al., 1990). Hydropathy analysis of the deduced protein suggested multiple TMDs that did not display homology to any previously identified proteins as assessed by database search, indicating that GAT-1 appeared to be a member of a previously uncharacterized family of transporter.

Soon after the cloning of GAT-1, human NET cDNA was cloned by direct expression of pools of clones from a human SK-N-SH cell cDNA library in COS-1 cells and screening for transfectants expressing NET using radiolabelled noradrenaline (NA) analogue m-iodobenzylguanidine (Pacholczyk et al., 1991). The predicted protein sequence of NET demonstrated significant amino acid identity with GAT-1, identifying a new gene family for neurotransmitter transporter proteins.

This conclusion was confirmed by the subsequent cloning of other neurotransmitter transporters, such as DAT and SERT, in which degenerative oligonucleotides based on conservative amino acid sequences between GAT-1 and NET were used to amplify cDNA fragments from brain mRNA to screen cDNA library and clone them. Hydropathy analysis of their primary structures again suggested a membrane topological model for the proteins belonging to the Na^+/Cl^- -dependent transporter family (Fig. 1A). All members of this family share the twelve hydrophobic regions that consist of α-helical TMDs interrupted by alternating intra- and extracellular loops, one large putatively extracellular loop, and intracellularly localization of N- and C-termini (Amara & Kuhar, 1993).

2.2 Functional roles

The major function of the neurotransmitter transporters located on the cell surface plasma membrane is to terminate synaptic neurotransmission by Na^+/Cl^--dependent uptake of released neurotransmitters into the neuronal and/or glial cells (Iversen, 1971). Transport of substrate such as neurotransmitter is driven by Na^+ gradient as energy source created by Na^+,K^+-ATPase, thereby being usually from extracellular to intracellular side. This intracellular accumulation of neurotransmitters acts not only to maintain synaptic clearance of the released neurotransmitters but also to regulate the storage and consequently release of neurotransmitters in the synaptic vesicles in concert with the vesicular neurotransmitter transporters located on the synaptic vesicle membrane. Therefore, they regulate spatiotemporal components of the neurotransmission. This has been supported by analysis of the neurotransmitter transporter knockout mice. The first evidence came from DAT knockout mice that were characterized by a dopamine-deficient but hyperdopaminergic state as a result of reduced dopamine clearance and reduced vesicular storage of dopamine (Giros et al., 1996; Jones et al., 1998). Loss of a transport function could thus cause severe disease or lethality, for instance, in the case of loss-of-function DAT mutants in infantile parkinsonism-dystonia in humans (Kurian et al., 2009).

According to the transport process mentioned above, changes in driving force such as Na^- gradient might cause reversal of transport. The discovery of reverse transport has been brought about by the study of the pharmacology of sympathomimetic amines including amphetamine, a psychostimulant drug abused widely, and now probably all the neurotransmitter transporters are considered to possess bidirectional transport (Sitte & Freishmuth, 2010). Another feature of the neurotransmitter transporter is channel-like activity, showing coupled and uncoupled currents (Gerstbrein & Sitte, 2006). These are important issues to consider roles of the neurotransmitter transporter in physiology and pathology, however, out of scope in this chapter. For details see those references.

3. Biochemical approaches to the membrane topology

A number of approaches have been designed to determine the topological arrangement of membrane spanning segments in protein subunits, and allowed to reveal an insight into the structure of membrane proteins. Fig. 2 summarizes such strategies, which include:

1. Development of site-directed antibodies for use in immunocytochemistry (Fig. 2A)
2. Detection of glycosylation sites inserted into hydrophilic domains (Fig. 2B)
3. Determination of the orientation of a reporter domain linked to a series of C-terminal truncation using a protease protection assays (Fig. 2C)
4. Chemical modification of specific amino acid residues using membrane-permeant and -impermeant reagents (Fig. 2D)
5. Immunocytochemical detection of the epitope-tag inserted into hydrophilic domains or fused to truncated (intact) C-terminal (Fig. 2E)

Each approach has its own benefit but also limitation, because many of the approaches require modification of the protein being studied, which often results in changes of the functional properties due to perturbation of the structure of the protein (Green et al., 2001). The neurotransmitter transporter protein is not tolerant of the alterations in primary sequence at conserved domain, such as the insertion of reporter epitope or the mutagenesis of several amino acid residues. Therefore, it appears to allow minor modification, such as substitution of single amino acid residue with cysteine, for example. Since substitution with cysteine appears to be very tolerated, it is useful for evaluating the membrane topology to determine the intra- and extracellular orientation of substituted cysteine residues using membrane-permeant and -impermeant sulfhydryl-reactive agents. A residue which is water-accessible from the cytoplasmic side but not from the extracellular side of the membrane should not react at an appreciate rate with extracellularly applied MTSET or MTSES (impermeant reagent), but might react with MTSEA (permeant reagent). In this way, topology can be assessed under conditions in which protein function and structure remain unaltered (Javitch, 1998).

All members of the Na$^+$/Cl$^-$dependent neurotransmitter transporter gene family share the twelve hydrophobic regions that might consist of α-helical transmembrane domains (TMDs) interrupted by alternating intra- and extracellular loops, one large putatively extracellular loop, and intracellularly localization of N- and C-termini. Based on this information, initial studies examined the membrane topology of the transporters by various biochemical techniques. For example, study with immunological technique using specific antibodies against hydrophilic regions predicted as intra- and extracellular loops or antibodies against known epitope tags incorporated at hydrophilic regions of the neurotransmitter transporters transiently or stably expressed in the cell lines, strongly confirm the topological model, as described below.

3.1 Site-directed antibodies

Mabjeesh and Kanner have reported the first evidence for the membrane topology of GAT using antibodies raised against synthetic peptides corresponding to several regions of the rat brain GAT-1 (Mabjeesh & Kanner, 1992). According to the model based on the hydropathy analysis, these 4 antibodies against amino and carboxy termini and predicted 3rd and 4th intracellular loops recognized the intact transporter on Western blots. Interestingly, GAT protein digested partially at amino and carboxy termini by protease revealed transport activity when reconstituted in liposomes, suggesting that these regions are not essential for transport function (Mabjeesh and Kanner, 1992). These studies did not address directly to explore the membrane topology of the transporter, however, provided useful information for the subsequent topological studies (Fig. 2A).

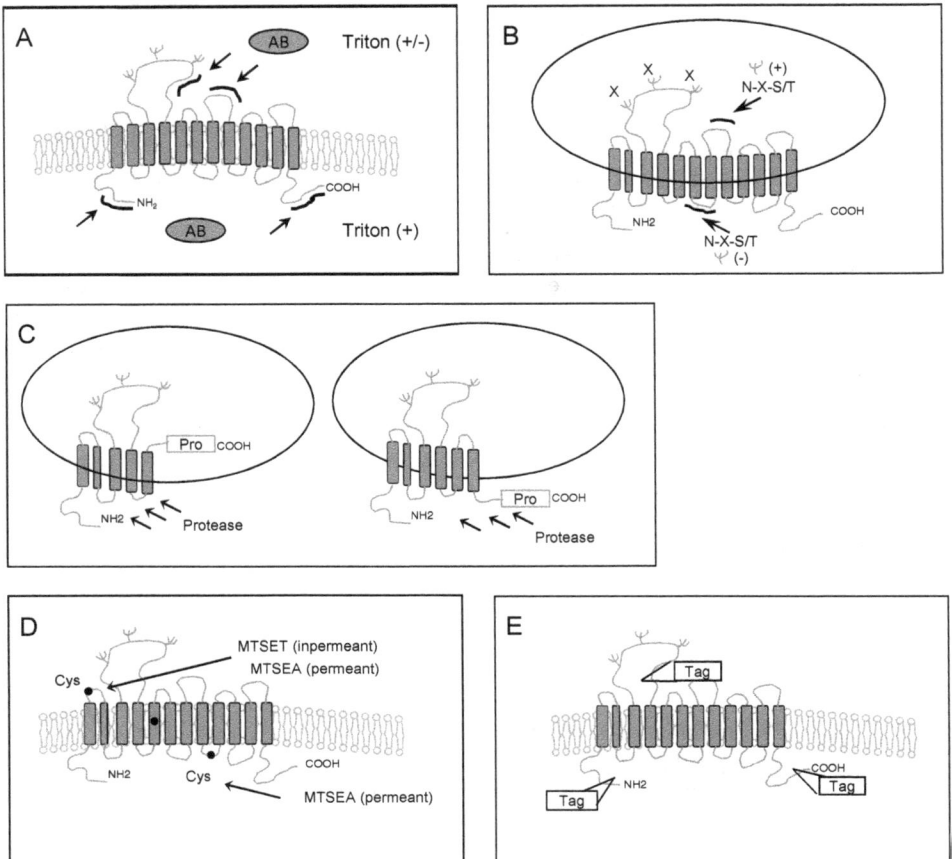

Fig. 2. Schematic presentations of the strategies for identifying membrane topology. (A) Site-directed antibodies were developed for use in immunocytochemistry. (B) Detection of glycosylation sites (N-X-S/T) inserted into hydrophilic domains was performed after deletion of endogenous sites (X). (C) The orientation of a reporter domain linked to a series of C-terminal truncation was determined using a protease protection assays. (D) Chemical modification of specific amino acid residues (Cys) was performed using membrane-permeant and -inpermeant reagents. (E) The epitope-tag was inserted into hydrophilic domains, or fused to truncated (intact) C-terminal, and the immunocytochemical detection was performed.

Antibodies have been raised against synthetic peptides derived from the predicted primary sequence of the human NET at several hydrophilic regions (Melikian et al., 1994). Immunocytochemical study with one antibody raised against a putative intracellular loop (N430, hNET 430-444) demonstrated that it detected hNET expression in a stably transfected cell line (LLC-NET) only in the presence of detergent, suggesting the predicted membrane topology. Bruss et al. investigated the membrane topology of human NET using polyclonal antibodies raised against peptides corresponding to hydrophilic sequences; N- and C-terminal peptides of the hNET, as well as against peptides from the putative extracellular loops between TMDs 3 and 4 and TMDs 7 and 8 (Bruss et al., 1995). They confirmed the topological model mentioned above.

Site-directed antibodies to the human DAT have been raised against its hydrophilic regions (Vaghan et al., 1993). Using these antibodies, the authors identified ligand binding domains of DAT with the photoaffinity compounds and limited proteolysis (Vaughan, 1995; Vaughan & Kuhar, 1996). The data experimentally verified an aspect of theoretical model of the transporter. Hersch et al. have raised monoclonal antibodies against the predicted N-terminus and second extracellular loop of DAT, and determined the subcellular location of DAT by immunoperoxidase and immunogold electron microscopy (Hersch et al., 1997). Immunogold labeling the intracellular and extracellular epitopes was found in the complementary positions that would be predicted based on molecular models of DAT in which the N-terminus is intracellular and the large third extramembranous loop is extracellular.

3.2 N-glycosylation scanning mutagenesis

Kanner's group further examined the membrane topology of GAT-1 by N-glycosylation scanning mutagenesis (Bennett & Kanner, 1997). Insertion of glycosylation sites into hydrophilic regions predicted as linker loop was tolerated well, although the approach is not straightforward in accordance with earlier studies using this approach on transporters (Fig. 2B). Overall, the results supported the theoretical 12 TMD model. However, they indicated that the loop connecting putative TMD 2 and 3, which was predicted to be located intracellularly, could be glycosylated *in vivo*. In addition, studies with permeant and impermeant methane sulfonate reagent suggested that Cys74, located in the hydrophilic loop connecting TMD 1 and 2, was intracellular rather than extracellular. Based on these results, they proposed a model in which the topology deviated from the theoretical one in the amino-terminal third of the transporter; the highly conserved TMD1 did not form a conventional TMD.

Olivares et al. proposed a similar topological model for GLYT1 using same approach (Olivares et al., 1997). The data supported a rearrangement of the first third of the protein, and consistent with Kanner's proposal that hydrophobic domain 1 seems not to span the membrane, and the loop connecting hydrophobic domain 2 and 3, formally believed to be intracellular, appears to be extracellularly located.

However, the subsequent studies on the membrane topology of SERT have argued against this alternative topology (Chen et al., 1998). They examined the SERT membrane topology by measuring the reactivity of selected lysine and cysteine with extracellular reagents. The cysteine-specific biotinylation reagent N-biotinylaminoethylmethanethiosulfonate (MTSEA-biotin) labeled wild-type SERT but not a mutant in which Cys-109, predicted to lie in the first external loop, was replaced with alanine. All of the mutants tested were active and therefore likely to be folded correctly. Therefore, these results support the original transmembrane topology. Recent findings of the X-ray crystallography of the bacterial homologue of the neurotransmitter transporter confirmed the original model (Yamashita et al., 2005), and this is discussed in the later section.

3.3 Protease protection assay

Clark performed different approach to clarify the membrane topology of GAT-1 (Clark, 1997). She generated a series of C-terminal truncations to which a prolactin epitope was fused. Following expression of transporter-prolactin chimeras in *Xenopus* oocytes, protease protection assays were performed to determine the transmembrane orientation (Fig. 2C). The data indicated that N- and C-termini were cytosolic and hydrophobic domains spanned the membrane in a manner consistent with the predicted hydropathy model.

Furthermore, the author showed that residues in the loops connecting hydrophobic domains 3 and 4 (predicted EL2), and those 7 and 8 (EL4) are accessible to protease in the cytoplasm, suggesting the presence of pore loop structures which extend into the membrane from the extracellular face (Clark, 1997). However, as mentioned above, the neurotransmitter transporter protein is not tolerant of the alterations in primary sequence at conserved domain, the insertion of reporter epitope such as prolactin might perturb the structure of the protein like the case for N-gycosylation scanning mutagenesis. Again, the findings of LeuT structure did not support pore loop structure (Yamashita et al., 2005).

3.4 Chemical modification

Of particular importance for investigating the membrane topology is to maintain functional properties intact under experimental conditions. As already mentioned, application of the substituted cysteine accessibility method (SCAM) can indirectly probe structure and conformational dynamics in the neurotransmitter transporters (Javitch, 1998) (Fig. 2D). The human DAT has 13 cysteines. Ferrer and Javitch have sought to identify those cysteine residuesthe modification of which affects cocaine analogue binding and to determine the topology of these reactive cysteines by mutating each of the cysteines, one at a time and in various combinations (Ferrer & Javitch, 1998). They demonstrated that Cys-90 and Cys-306 appear to be extracellular, and Cys-135 and Cys-342 appear to be intracellular, each of these residues is predicted to be in extramembranous loops, supporting the hypothetical membrane topology of the neurotransmitter transporter. More importantly, the binding of cocaine increases the rate of reaction of MTSEA and MTS ethyltrimethylammonium with the extracellular Cys-90 and therefore acts by inducing a conformational change. Usefulness of SCAM to investigate structure-function relationship of the neurotransmitter transporters led to the subsequent success in this field.

3.5 Epitope-tagging

Insertion of epitope-tag into hydrophilic region often perturbs structure of the neurotransmitter transporter proteins. However, a hemagglutinin epitope (HA) tag has been successfully introduced into the second extracellular loop (EL2) of DAT (Sorkina et al., 2006) (Fig. 2E). They confirmed that all their attempts to introduce epitopes into other parts of EL2 or into EL6 produced mutants that did not efficiently exit the endoplasmic reticulum. Ciliax et al. demonstrated an antibody to the C-terminal part of EL2 that recognizes the human DAT (Ciliax et al., 1995). However, Sorkina et al. could not find conditions under which the antibody binds DAT in living cells, and they suggested that this part of the loop is not assessable under physiological conditions (Sorkina et al., 2006).

4. X-ray crystallography of LeuT

The structure of *Aquifex aeolicus* leucine transporter (LeuT$_{Aa}$), a bacterial homologue of mammalian Na$^+$/Cl$^-$-dependent neurotransmitter transporter, has been solved recently (Yamashita et al., 2005). Although the overall sequence identity between LeuT and its mammalian counterparts is low (20-25%), several regions within transmembrane segments 1, 3, 6, and 8 displays ~50% conservation. Examination of the structure of the crystallized LeuT protein revealed such structure as a "occluded form " with substrate leucine and Na$^+$. There is an unexpected structural repeat in the first ten TMDs that relates TMD1-5 with TMD6-10 around a pseudo-twofold axis of symmetry located in the membrane plane (Fig. 1B). The pseudo-repeats are oriented antiparallel to one another with the two central TMDs, 1 and 6, which are unwound near the substrate and Na$^+$ binding sites, located halfway across the lipid bilayer. This unwinding structure is important for understanding the structure-function relationship, because it exposes helix dipoles to maintain carbonyl oxygen and amide nitrogen for substrate binding and Na$^+$ coordination.

Subsequent studies added further structural models by indicating an occluded form with tricyclic antidepressant (TCA) (Singh et al., 2007; Zhou et al., 2007) and an outward-facing form with competitive inhibitor tryptophan (Singh et al., 2008). The structure of TCA-LeuT complex reveals a TCA binding pocket, called "extracellular vestibule", composed of 7 amino acid residues, which are located approximately 11 Angstrom above the substrate leucine binding site, and accommodated with EL4. Comparison of the original LeuT structure and TCA-LeuT complex reveals a hypothetical extracellular gate in a closed conformation formed by the conserved R30 in TMD 1 and D404 in TMD 10 that is formed when two coordinating water molecules are displaced. A second salt bridge observed in the original LeuT structure formed intracellularly between R5 in N-terminus and D369 in TMD 8, which is proposed to play a role in intracellular gating (Yamashita et al., 2005). Inward-facing conformation of LeuT has not been identified yet, that helps to understand fully the alternative access model of the transporter.

There are limits to the evaluation of mammalian neurotransmitter transporters by what a distantly related bacterial homologue can provide. EL2 as well as N- and C-termini are considerably shorter in LeuT, and none of the intracellular loops harbor consensus sequences for phosphorylation. Furthermore, it seems unlikely to be evaluated using electrophysiological techniques required to characterize channel activities of the neurotransmitter transporters, as mentioned in the section 2.2 (Gerstbrein & Sitte, 2006).

5. Immunocytochemical approaches to the DAT splice variants

We have been investigating the expression and function of DAT and NET isoforms produced by alternative RNA splicing (Kitayama & Dohi, 2003). Recently we found novel splice variants of human DAT and NET designated hDATΔEX6 and hNETΔEX6, skipping the region encoded by exon 6 (Sogawa et al., 2010). Hydropathy analysis of the variants revealed putative 11 TMDs, and an immunocytochemical analysis with confocal microscopy suggested a membrane topology different from that of full-length (FL) hDAT, such as that the C-terminus could be located extracellulary. In this section, we summarize these findings, and discuss further possibility of future works.

5.1 C-terminal orientation of the splice variants

C-terminal of the neurotransmitter transporters is no doubt intracellular. A hydropathy analysis of hDATΔEX6 revealed 11 putative transmembrane domains (TMDs), suggesting a membrane topology different from that of the original full-length (FL) hDAT. If TMDs are inserted behind the truncation, the C-terminus could be located extracellulary. To explore this possibility, we performed an immunocytochemical analysis with confocal microscopy using two specific antibodies, one recognizing the second extracellular region (anti-hDAT-EL2 antibody) and the other recognizing the intracellular C-terminus (anti-hDAT-Ct antibody). Among the MDCK cells stably expressing FL hDAT, immunoreactivity to the anti-hDAT-Ct antibody was only observed in those cells treated with Triton (Fig. 3A). Among the MDCK cells expressing hDATΔEX6, however, immunoreactivity to the anti-hDAT-Ct antibody was detected even in the untreated cells (Fig. 3B). Control experiments using the anti-hDAT-EL2 antibody showed immunoreactivity in the Triton-treated and untreated cells expressing both FL hDAT (Fig. 3A) and hDATΔEX6 (Fig. 3B). These results strongly suggest the C-terminus of hDAT ΔEX6 to be located extracellulary.

Deletion of the region encoded by exon 6 in the hDATΔEX6 variant is believed to affect the orientation of those TMD regions that follow. The present findings, suggesting the C-terminal region to be located extracellularly, did not indicate the exact membrane topology of each TMD including the 7-12th TMDs. A previous study suggested an alternative membrane topology consisting of TMD in the EL2 region corresponding to the involvement of N-terminal regions TM1 and TM2, as mentioned in the previous section. A predicted membrane topology of hDATΔEX6 simply reflected the sequence of hydrophobic regions considered as TMDs. Therefore, further study is needed to determine the precise membrane topology of the hDATΔEX6 variant.

Since C-terminal region is well known to participate an important role in the expression and function of neurotransmitter transporters, a putative extracellular location of C-terminus in the hDATΔEX6 variant suggests a differential feature of its roles. Previous studies with DAT mutants have suggested several possibilities of the C-terminus-dependent mechanisms underlying DAT regulation; (1) an elimination of phosphorylation in the C-terminal region by PKC (Holton et al., 2005), (2) a loss of interaction with proteins such as PICK1 in the PDZ domain at the C-terminal end (Torres et al., 2001), and (3) an unknown mechanism independent of interaction with the PDZ protein at the C-terminus (Bjerggaard et al., 2004). Studies on C-terminal splice variants of hNET also documented a critical contribution of the hNET C-terminus to transporter trafficking, stability, and function (Sogawa et al., 2007). These explanations seem unlikely in the case of hDATΔEX6, since the present findings suggested the C-terminus of hDAT ΔEX6 to be located extracellulary. At present, it is unknown how trafficking of hDATΔEX6 to the plasma membrane is regulated, and further study is needed to clarify this.

5.2 Oligomerization

We also found that hDATΔEX6 had a dominant negative effect on FL hDAT, possibly through the formation of heterooligomeric complexes, as suggested by the results of the immunoprecipitation assays (Sogawa et al., 2010). A cell surface biotinylation assay with an

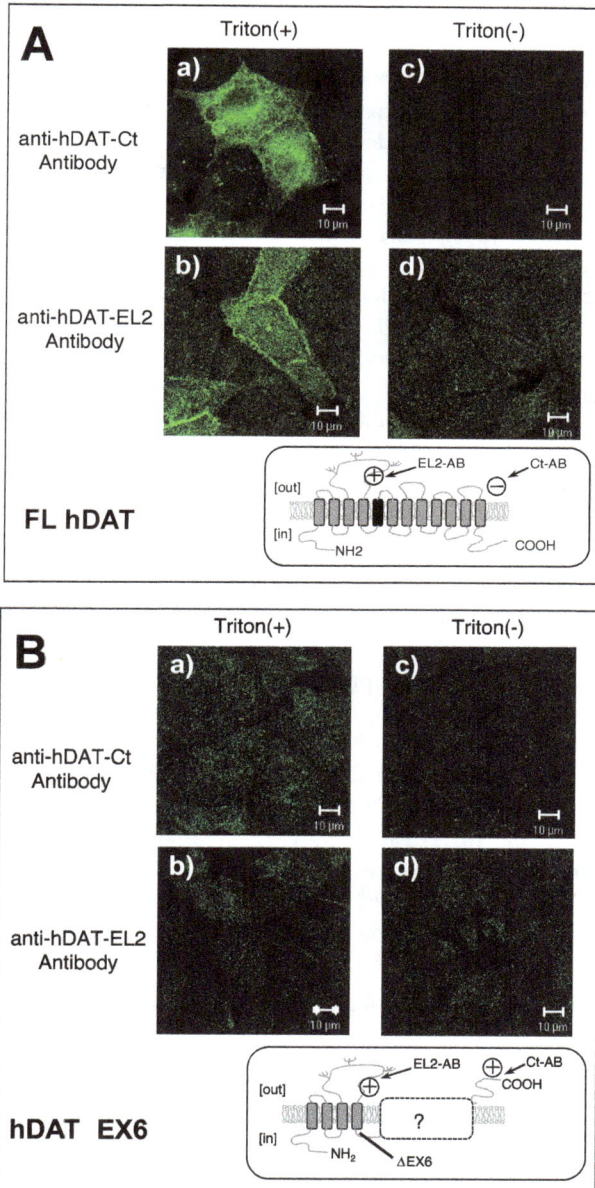

Fig. 3. Cell surface expression of the DAT splice variants determined immunocytochemically with two antibodies against different epitopes in stably transfected MDCK cells. Cells stably expressing FL hDAT (A) and hDAT Δ EX6 (B) were treated with (a, b) or without (c, d) Triton, and examined for immunocytochemistry with two different antibodies. From Sogawa et al., 2010 (doi: 10.1371/jounal.pone.0011945.g007).

immunoblot analysis using the anti-hDAT-EL2 antibody demonstrated a reduction in the expression of the 80-85kDa hDAT at the cell surface, on co-expression with hDATΔEX6. An immunoblot analysis using the anti-HA antibody to the HA-tagged FL hDAT demonstrated that co-expression of hDATΔEX6 decreased the cell surface expression of the 80-85kDa mature hDAT, in consistent with the decreased V_{max} of [^{3}H]DA uptake. Interaction among the isoforms was further examined by conducting immunopreciptation assays using tagged forms of hDAT, such as HA-hDAT and N-terminally His-tagged hDATΔEX6 (His-hDATΔEX6). The proteins precipitated with the anti-His antibody were subjected to an immunoblot analysis using the anti-hDAT-EL2 antibody and anti-HA antibody. Isolation of His-hDATΔEX6 with the anti-His antibody allowed the detection of HA-hDAT using the anti-HA antibody when the two proteins were expressed simultaneously, indicating that FL hDAT and hDATΔEX6 form heterooligomeric complexes. It seems unlikely to be a nonspecific aggregation with other membrane proteins including neurotransmitter transporters, since an immunoprecipitation with anti-His antibody did not detect HA-tagged mouse glutamate transporter GLT-1 in COS-7 cells co-transfected with His-hDATΔEX6, in association with the lack of an effect on [^{3}H]glutamate uptake through mGLT-1 (Sogawa et al., 2010).

There is a growing body of evidence that oligomerization is necessary for the cell surface expression of neurotransmitter transporters including DAT (Torres et al., 2003). The structure of LeuT suggested dimerization through interaction at the 9th and 12th TMDs (Yamashita et al., 2005). However, it seems unlikely that FL hDAT and hDATΔEX6 interact in these regions, since according to the predicted membrane topology of hDATΔEX6, the C-terminal region is located extracellulary, preventing direct interaction. Torres et al. found that the C-terminal of DAT was not essential for oligomerization, and that a small fragment comprising the first two TMDs inhibited the wild-type transporter function but not when the leucine repeat motif present in the 2nd TMD was mutated (Torres et al., 2003). However, it is unclear whether FL hDAT and hDATΔEX6 associate in the same way, since immunoprecipitation assays do not reveal modes of interaction and no information is available about the membrane topology of hDATΔEX6 except the C-terminus. Therefore, further study is needed to determine the mechanism by which hDATΔEX6 forms a heterooligomeric complex with FL hDAT.

It might be probable that hDATΔEX6 interacts with FL hDAT at the plasma membrane to produce a dominant negative effect on the activity of FL hDAT, since a part of the hDATΔEX6 protein was observed in the plasma membrane (Sogawa et al., 2010). However, there is evidence that the functional unit of the transporter is a monomer, though isoforms or different transporters such as NET and SERT consist of heterodimers (Kocabas et al., 2003). Further study is needed to clarify this possibility.

6. Protocol

The followings are methods related to our report described in the previous section 5. We summarize the points necessary to perform the immunocytochemical approach to the membrane topology of DAT as a typical model of the neurotransmitter transporters.

Immunocytochemical Approaches to the Identification of Membrane Topology of the Na⁺/Cl⁻-Dependent Neurotransmitter Transporters

153

6.1 Cell preparation and transfection

6.1.1 Cell culture

COS-7 cells and MDCK cells (RIKEN Cell Bank, Tsukuba, Japan) were maintained in Dulbecco's modified Eagle's medium (DMEM) supplemented with 10% fetal bovine serum, 100 units/ml penicillin-G, 100 µg/ml streptomycin and 2.5 µg/ml fungizone at 37 °C under 5% CO_2/95% air.

6.1.2 Vector construction

Cloning of full-length cDNAs of human DAT variants, FL hDAT and hDATΔEX6, were performed by RT-PCR at three discrete and overlapping regions in first strand cDNA pool synthesized from white blood cells total RNA. The products were isolated from agarose gel, digested with KpnI, BanHI, NgoMIV or XhoI, and subcloned into pcDNA3 at KpnI/XhoI digestion site. The isolated each clone was analyzed by restriction enzyme digestion and nucleotide sequencing[2].

Human NET variant cDNAs (FL hNET and hNETΔEX6) were isolated from human cDNA library made from SK-N-SH cells by PCR, and cloned into pcDNA3 at EcoRI/XhoI digestion site (Kitayama et al., 2001). mGLT-1 cDNA (AK134609) in plasmid vector pFLC1 was obtained from RIKEN Mouse FANTOM FLS through KK DNAFORM, and a BamHI fragment containing ORF was subcloned into pcDNA3, as described (Sogawa et al., 2010).

6.1.3 Transfection

COS-7 cells at subconfluence were harvested and transfected with pcDNA3 alone or with pcDNA3 containing hDAT, hNET, or mGLT cDNA and/or variant cDNA by electroporation or using FuGENE6 according to the manufacturer's directions (Roche Diagnostics, Mannheim, Germany). Electroporated cells were diluted in culture medium, plated in 24-well tissue culture plates and cultured for 2-3 days. For immunological analyses, COS-7 cells at subconfluence in 60-mm diameter Petri dishes (for Western blotting) or Falcon BIOCOAT® Cellware rat tail collagene, type I 4-well culture slides (Becton Dickinson Labware, Bedford, MA) (for immunocytochemistry) were transfected with cDNAs using FuGENE6.

6.1.4 Cloning of cell lines

To generate stably transfected MDCK cells, we transfected pcDNA3 harboring FL hDAT or hDATΔEX6 using FuGENE6 into MDCK cells and selected in 600 µg/ml Geneticin (G418). Individual cells were used to generate clonal lines. Multiple lines tested positive for immunocytochemistry using rabbit anti-hDAT polyclonal antibody. MDCK cells stably expressing FL hDAT and hDATΔEX6 were passaged in DMEM culture media containing 10% fetal bovine serum, 100 U/ml penicillin, 100 µg/ml streptomycin, 2.5 µg/ml fungizone, and 600 µg/ml G418.

---

[2] Request for these clones and related materials should be addressed to Dr. Shigeo Kitayama,

6.2 Immunocytochemistry and confocal microscopy

6.2.1 Fixation

For immunological analyses, MDCK cells were diluted in culture medium, plated in Falcon BIOCOAT® Cellware rat tail collagene, type I 4-well culture slides and cultured for 2 days. Cells were initially rinsed twice with Ca^{2+}- and Mg^{2+}-containing PBS, and then fixed in 4% paraformaldehyde (PFA) for 5 min at room temperature. After three washes with PBS, cells were treated with (permeabilized) or without (non-permeabilized) 0.25% Triton X-100 for 5 min at room temperature. If methanol or ethanol fixative was used, every cell was permeabilized. We choose 4% PFA fixation, because we need distinguish the permeabilized cells from non-permeabilized cells.

6.2.2 Immunostaining

After permeabilization, the slides were washed with PBS and incubated in blocking solution (2% goat serum) for 30 min at room temperature. Cells were incubated with rabbit anti-hDAT polyclonal antibody (anti-hDAT-Ct antibody against NH_2-CEKDRELVDRGEVRQFTLRHWL, Chemicon, AB1766, 1:500; anti-hDAT-EL2 antibody against NH_2-CHLHQSHGIDDLGPPRWQ, Chemicon, AB5802, 1:250)) for overnight at 4 °C, followed by incubation with FITC-conjugated anti-rabbit secondary antibody (Sigma-Aldrich Corporation, St. Louis, MO). Cells were then washed three times with PBS, and the filter with cells was excised from its support and mounted on a slide glass with Perma Fluor® aqueous mounting medium (Thermo Shandon, Pittsburgh, PA, USA).

6.2.3 Confocal microscopy

After a final wash for immunostaining of MDCK cells grown on Falcon BIOCOAT® culture slides (Becton Dickinson Labware, Bedford, MA) performed as above, the cells were covered by coverslip with mounting medium. Immunofluorescent images were generated using a Zeiss laser scanning confocal microscope (LSM510).

7. Conclusion

The plasma membrane neurotransmitter transporters, such as members of the SLC6 gene family, act to terminate neurotransmission by Na^+/Cl^- -dependent uptake of released neurotransmitters, thereby maintaining their synaptic clearance, and fine-tuning synaptic transmission. Common structure of these transporters predicted by hydropathy analysis showed twelve hydrophobic transmembrane domains with intracellular amino- and carboxy-termini. Biochemical and immunocytochemical studies supported this structure one by one. Recent findings of the X-ray crystallography of a bacterial homologue, LeuT, confirmed such structure as a "occluded form " with substrate leucine and Na^+, and the subsequent studies added further occluded form with antidepressant and out-ward facing form with competitive inhibitor tryptophan. Based on these observations, combinatory approaches might provide further evidence supporting the membrane topology of the neurotransmitter transporter. These strategies will develop the future studies on the structure-function relationship of the neurotransmitter transporters that promises clues to develop new targets for therapeutics of transporter-associated disorders.

8. References

Amara, S.G. & Kuhar, M.J. (1993). Neurotransmitter transporters: Recent progress. *Annual Review of Neuroscience*, Vol.16, (March 1993), pp. 73-93, ISSN 0147-006X.

Bennett, E.R. & Kanner, B.I. (1997). The membrane topology of GAT-1, a (Na⁺ + Cl⁻)-coupled γ-aminobutyric acid transporter from rat brain. *The Journal of Biological Chemistry*, Vol. 272, No.2, (January 1997), pp. 1203-1210, ISSN 0021-924X.

Bjerggaard, C., Fog, J.U., Hastrup, H., Madsen, K., Loland, C.J., Javitch, J.A. & Gether, U. (2004). Surface targeting of the dopamine transporter involves discrete epitopes in the distal C terminus but does not require canonical PDZ domain interactions. *The Journal of Neuroscience*, Vol.24, No.31, (August 2004), pp. 7024-7036, ISSN 0270-6474.

Bruss, M., Hammermann, R., Brimijoin, S. & Bonisch, H. (1995). Antipeptide antibodies confirm the topology of the human norepinephrine transporter. *The Journal of Biological Chemistry*, Vol.270, No.16, (April 1995), pp. 9197-9201, ISSN 0021-924X.

Chen, J.G., Liu-Chen, S. & Rudnick, G. (1998). Determination of external loop topology in the serotonin transporter by site-directed chemical labeling. *The Journal of Biological Chemistry*, Vol.273, No.20, (May 1998), pp. 12675-12681, ISSN 0021-924X.

Ciliax, B.J., Heilman, C., Demchyshyn, L.L., Pristupa, Z.B., Ince, E., Hersch, S.M., Niznik, H.B. & Levey, A.I. (1995). The dopamine transporter: immunochemical characterization and localization in brain. *The Journal of Neuroscience*, Vol.15, No.3 Pt1, (March 1995), pp. 1714-1723, ISSN 0270-6474.

Clark, J.A. (1997). Analysis of the transmembrane topology and membrane assembly of the GAT-1 γ-aminobutyric acid transporter. *The Journal of Biological Chemistry*, Vol.272, No.23, (Jun 1997), pp. 14695-14704, ISSN 0021-924X.

Ferrer, J. & Javitch, J. (1998). Cocaine alters the accessibility of endogenous cysteines in putative extracellular and intracellular loops of the human dopamine transporter. *Proceedings of the National Academy of Sciences of the United States of America*, Vol.95, No.16, (August 1998), pp. 9238-9243, ISSN 0027-8424.

Gerstbrein, K. & Sitte, H.H. (2006). Currents in neurotransmitter transporters, In: *Neurotransmitter Transporters (Handbook of Experimental Pharmacology)*, Vol.175, Sitte, H.H. & Freissmuth, M., pp. 95-111, Springer, ISBN 13: 9783540297833, Berlin.

Gether, U., Andersen, P.H., Larsson, O.M. & Schousboe, A. (2006). Neurotransmitter transporters: molecular function of important drug targets. *Trends in Pharmacological Sciences*, Vol.27, No.7, (July 2006), pp. 375-383, ISSN 0165-6147.

Giacomini, K.M. & Sugiyama, Y. (2006). Membrane transporters and drug response. In: *Goodman & Gilman's The Pharmacological Basis of Therapeutics*, Brunton, L.L., Lazo, J.S. & Parker, R.L. pp. 41-70, McGraw-Hill, ISBN 9780071624428, New York.

Giros, B., Jaber, M., Jones, S.R., Wightman, R.M. & Caron, M.G. (1996). Hyperlocomotion and indifference to cocaine and amphetamine in mice lacking the dopamine transporter. *Nature*, Vol.379, No.6566, (Februaly 1996), pp. 606–612, ISSN 0028-0836.

Green, N., Fang, H., Kalies, K.U. & Canfield, V. (2001). Determining the topology of an integral membrane protein. *Current Protocols in Cell Biology*, Chapter 5, Unit 5.2, (May 2001), ISSN 1934-2500.

Guastella, J., Nelson, N., Nelson, H., Czyzyk, L., Keynan, S., Miedel, M.C., Davidson, N., Lester, H.A. & Kanner, B.I. (1990). Cloning and expression of a rat brain GABA transporter. *Science*, Vol.249, No.4974, (September 1990), pp. 1303-1306, ISSN 1095-9203.

Hersch, S.M., Yi, H., Heilman, C.J., Edwards, R.H. & Levey, A.I. (1997). Subcellular localization and molecular topology of the dopamine transporter in the striatum and substantia nigra. *The Journal of Comparative Neurology*, Vol.388, No.2, (November 1997), pp. 211-227, ISSN 1096-9861.

Holton, K.L., Loder, M.K. & Melikian, H.E. (2005). Nonclassical, distinct endocytic signals dictate constitutive and PKC-regulated neurotransmitter transporter internalization. *Nature Neuroscience*, Vol. 8, No. 7, (July 2005), pp. 881-888, ISSN 1097-6256.

Iversen, L.L. (1971). Role of transmitter uptake mechanisms in synaptic neurotransmission. *British Journal of Pharmacology*, Vol.41, No.4, (April 1971), pp. 571-591, ISSN 1476-5381.

Javitch, J. (1998). Probing structure of neurotransmitter transporters by substituted-cysteine accessibility method. *Methods in Enzymology*, Vol.296, pp. 331-346, ISSN 0076-687.

Jones, S.R., Gainetdinov, R.R., Jaber, M., Giros, B., Wightman, R.M. & Caron, M.G. (1998). Profound neuronal plasticity in response to inactivation of the dopamine transporter. *Proceedings of the National Academy of Sciences of the United States of America*, Vol.95, No.7, (March 1998), pp. 4029-4034, ISSN 0027-8424.

Kitayama, S. & Dohi, T. (2003). Norepinephrine transporter splice variants and their interaction with substrates and blockers. *European Journal of Pharmacology*, Vol.479, (August 2003), pp. 65-70, ISSN 0014-2999.

Kitayama, S., Morita, K. & Dohi, T. (2001). Functional characterization of the splice variants of human norepinephrine transporter. *Neuroscience Letters*, Vol.312, No.2, (October 2001), pp. 108-112, ISSN 0304-3940.

Kocabas, A.M., Rudnick, G. & Kilic, F. (2003). Functional consequences of homo- but no: hetero-oligomerization between transporters for the biogenic amine neurotransmitters. *Journal of Neurochemistry*, Vol.85, No.6, (Jun 2003), pp. 1513-1520, ISSN 0022-3042.

Kurian, M.A., Zhen, J., Cheng, S.Y., Li, Y., Mordekar, S.R., Jardine, P., Morgan, N.V., Meyer, E., Tee, L., Pasha, S., Wassmer, E., Heales, S.J.R., Gissen, P., Reith, M.E.A. & Maher, E.R. (2009). Homozygous loss-of-function mutations in the gene encoding the dopamine transporter are associated with infantile parkinsonism-dystonia. *Journal of Clinical Investigation*, Vol.119, No.6, (Jun 2009), pp. 1595-1603, ISSN 0021-9738.

Mabjeesh, N.J. & Kanner, B.I. (1992). Neither amino nor carboxy termini are required for function of the sodium- and chloride-coupled γ-aminobutyric acid transporter from rat brain. *The Journal of Biological Chemistry*, Vol.267, No.4, (February 1992), pp. 2563-2568, ISSN 0021-924X.

Melikian, H.E., McDonald, J.K., Gu, H., Rudnick, G., Moore, K.R. & Blakely, R.D. (1994). Human norepinephrine transporter. Biosynthetic studies using a site-directed polyclonal antibody. *The Journal of Biological Chemistry*, Vol.269, No.16, (April 1994), pp. 12290-12297, ISSN 0021-924X.

Olivares, L., Aragon, C., Gimenez, C. & Zafrai, F. (1997). Analysis of the transmembrane topology of the glycine transporter GLYT1. *The Journal of Biological Chemistry*, Vol.272, No.2, (January 1997), pp. 1211-1217, ISSN 0021-924X.

Pacholczyk, T., Blakely, R.D. & Amara, S.G. (1991). Expression cloning of a cocaine- and antidepressant-sensitive human noradrenaline transporter. *Nature*, Vol.350, No.6316, (March 1991), pp. 350-354, ISSN 0028-0836.

Singh, S.K., Yamashita, A. & Gouaux, E. (2007) Antidepressant binding site in a bacterial homologue of neurotransmitter transporters. *Nature*, Vol.448, No.7156, (August 2007), pp. 952-956, ISSN 0028-0836.

Singh, S.K., Piscitelli, C.I., Yamashita, A. & Gouaux, E. (2008). A competitive inhibitor traps LeuT in an open-to-out conformation. *Science*, Vol.322, No.5908, (December 2008), pp. 1655-1661, ISSN 0028-0836.

Sitte, H.H. & Freissmuth, M. (2010). The reverse operation of Na⁺/Cl⁻-coupled neurotransmitter transporters - why amphetamines take two to tango. *Journal of Neurochemistry*, Vol.112, No.2, (January 2010), pp. 340-355, ISSN 0022-3042.

Sogawa, C., Kumagai, K., Sogawa, N., Morita, K., Dohi, T. & Kitayama, S. (2007). C-terminal region regulates the functional expression of human noradrenaline transporter splice variants. *The Biochemical Journal*, Vol.401, No.1, pp. 185-195, ISSN 0264-6021.

Sogawa, C., Mitsuhata, C., Kumagai-Morioka, K., Sogawa, N., Ohyama, K., Morita, K., Kozai, K., Dohi, T. & Kitayama, S. (2010). Expression and function of variants of human catecholamine transporters lacking the fifth transmembrane region encoded by exon 6. *PLoS One*, Vol.5, No.8, (August 2010), e11945, ISSN 1932-6203.

Sorkina, T., Miranda, M., Dionne, K.R., Hoover, B.R., Zahniser, N.R. & Sorkin, A. (2006). RNA interference screen reveals an essential role of Nedd4-2 in dopamine transporter ubiquitination and endocytosis. *The Journal of Neuroscience*, Vol.26, No.31, (August 2006), pp. 8195-8205, ISSN 0270-6474.

Torres G.E., Yao, W.D., Mohn, A.R., Quan, H., Kim, K.M., Levey, A.L., Staudinger, J. & Caron, M.G. (2001). Functional interaction between monoamine plasma membrane transporters and the synaptic PDZ domain-containing protein PICK1. *Neuron*, Vol.30, No.1, (April 2001), pp. 121-134, ISSN 0896-6273.

Torres, G., Carneiro, A., Seamans, K., Fiorentini, C., Sweeney, A, Yao, W.D. & Caron, M.G. (2003). Oligomerization and trafficking of the human dopamine transporter. Mutational analysis identifies critical domains important for the functional expression of the transporter. *The Journal of Biological Chemistry*, Vol.278, No.4, (January 2003), pp. 2731-2739, ISSN 0021-924X.

Vaughan, R.A., Uhl, G.R. & Kuhar, M.J. (1993). Recognition of dopamine transporters by antipeptide antibodies. *Molecular and Cell Neuroscience*, Vol.4, No.2, (April 1993), pp. 209-215, ISSN 1044-7431.

Vaughan, R.A. (1995). Photoaffinity-labeled ligand dinding domains on dopamine transporters identified by peptide mapping. *Molecular Pharmacology*, Vol.47, No.5, pp. 956-964, ISSN 0026-895X.

Vaughan, R.A. & Kuhar, M.J. (1996). Dopamine transporter ligand binding domains. Structural and functional properties revealed by limited proteolysis. *The Journal of*

Biological Chemistry, Vol.271, No.35, (August 1996), pp. 21672-21680, ISSN 0021-924X.

Yamashita, A., Singh, S.K., Kawate, T., Jin, Y. & Gouaux, E. (2005). Crystal structure of a bacterial homologue of Na⁺/Cl⁻ dependent neurotransmitter transporters. *Nature*, Vol.437, No.7056, (September 2005), pp. 215-223, ISSN 0028-0836.

Zhou, Z., Zhen, J., Karpowich, N.K., Law, C.J., Reith, M.E. & Wang, D.N. (2007). LeuT-desipramine structure reveals how antidepressants block neurotransmitter reuptake. *Science*, Vol.317, No.5843, (September 2007), pp. 1390-1393, ISSN 1095-9203.

Part 2

Specific Applications of Immunocytochemistry

8

Spermiomics: A New Term Describing the Global Survey of the Overall Sperm Function by the Combined Utilization of Immunocytochemistry, Metabolomics, Proteomics and Other Classical Analytical Techniques

Joan E. Rodríguez-Gil

Dept. Animal Medicine & Surgery, Autonomous University of Barcelona
Spain

1. Introduction

Mammalian sperm function is a very complex discipline, involving a great number of functional pathways. This complexity is the result of an extreme specialization in the mechanisms involved in the control of sperm function. We have to remember that sperm is designed with the ultimate role of transmitting male genetic information to the next generation. To this purpose, mammalian sperm firstly undergoes a very complex process of formation, which is initiated at the time of entry to meiosis of the testicular spermatogonia and finished during the epididymal maturation. Subsequently, mature mammalian sperm enter into the female genital tract during the ejaculation. This process initiates a new set of very complicated processes of functional changes, known as capacitation. Capacitation has to be very finely regulated, since there is a close chronological relationship between the time that sperm cells are present in the oviduct and ovulation. This allows sperm cells to reach their optimal functional status at the time of oocyte penetration. However, we must remind that the evolutionary reproductive strategy chosen by each species is different. This has as a direct consequence that the regulatory mechanisms that sperm utilize to modulate their function during their lifespan will be very different among species. These differences will be dependent on the specific schedule of events followed by each species from ejaculation to oocyte fecundation. All of this complexity implies that the experimental procedures that investigators have utilized to know sperm function have shown only a partial picture of it. This includes immunocytochemistry, one of the most important tools in the study of mammalian sperm function. Taking into account these points, it would be necessary to make a brief review of the phenomena during all of these processes in order to get a better comprehension of the techniques by which these phenomena are studied.

1.1 A Brief review on sperm formation and its subsequent epididymal maturation

Sperm formation and subsequent epididymal maturation in mammals is a very long and complex process that, in the majority of species, lasts for about two months, with relatively few variations among species (see Clermont, 1972; Wistuba et al., 2007). The formation of a spermatozoon that can display a full ability to reach oocyte fertilization is a process that is commonly divided in three chronological steps; spermatogenesis, spermiogenesis and sperm maturation.

The first step in the mammalian sperm formation is spermatogenesis. This phase is carried out into testes and consists in the formation of haploid germ cells, named spermatocytes, from a basal line of diploid germinal cells named spermatogonia. For this purpose, spermatogonia undergo a meiotic process in which each spermatogonia yield four separate, haploid spermatocytes (Wolgemuth et al., 1995). But meiosis is not only a process in which a diploid cell produces haploid descendants. Meiosis also hosts a process of DNA recombination yielding four haploid cells with distinct, individual changes in their genome (Yauk et al., 2003). The control of spermatogenesis is not well known, basically due to its complexity, although it is well known the control role played by Leydig cells, which acts through secretion of steroid hormones that are introduced inside the tubular stroma through specific androgen binding proteins, as well as the role of apoptosis in the control of spermatocytes formation (Sharpe et al., 1990; Sharpe et al., 1992; Giampietri et al., 2005). All of these processes involve the expression of many separate proteins, as well as many functional changes in these proteins, including variations in their phosphorylation levels (see as example Maekawa e al., 2002) that can be analyzed through proteomics and genomics.

The second step in the mammalian sperm formation is the spermiogenesis. This process is also carried out into the testes. In fact, spermiogenesis in closely linked to spermatogenesis, both being the subsequent step and being underwent in the same placement that spermatogenesis, namely the spermatic tubulae. In the spermiogenesis, the spermatocytes obtained after the meiotic phase undergo a morphological and functional transformation that converts these cells in the direct precursors of spermatozoa (Sharpe, 1994). These precursors, named spermatids, undergo many changes, including the substitution of histones by protamines, the elimination of practically all of the cytoplasm, the transformation of the Golgi apparatus in the acrosome and a pronounced shape change from a rounded one to the typical, elongated shape of the mammalian sperm, including the formation of the tail and the mitochondrial sheath (Tesari et al., 1998). Again, the control of this process is complex and poor understood, although it is well known the key role that Sertoli cells are playing here. Notwithstanding, and in a similar manner than that indicated for the first step of the spermatogenic process, spermiogenesis also involves the expression of many separate proteins, as well as many functional changes in these proteins, including variations in their phosphorylation levels (Hecht, 1995) that can be again analyzed through proteomics and genomics.

The third and final step in the mammalian sperm formation is the epididymal maturation. As its name indicates, this process takes place into the epididymis a long, tubular structure annexed to the testes (Robaire & Hinton, 2002). Spermatozoa formed after the finalization of the spermiogenesis are released from the spermatic tubules to their lumen, which communicates with the epididymis. Afterwards, spermatids travel through the entire epididymis during a time lapse of about fourteen days. During this period, spermatids undergo several very important modifications which allow these cells to reach their fertilization ability (Jones, 1999). Several of these modifications include total elimination of cytoplasmic remnants in the form of cytoplasmic droplets, final condensation of sperm nucleus and release of sperm proteins to the epididymal lumen, the pehomenon that is concomitant with the uptake of extracellular, epididymal proteins by sperm (Lasserre et al., 2001). As in the other phases of sperm formation, the regulation of all of these changes is not well known and, again, the genomic and proteomic techniques will be basic to the complete understanding of this process.

1.2 Changes in sperm function after ejaculation: capacitation

The final result of all of the above described processes is a mature spermatozoon, which is stored into the final zone of the epididymis until ejaculation or its destruction and subsequent reabsortion. However, the mature spermatozoon that is stored in the final segment of the epididymis lacks neither motility nor a real fertilizing ability in mammals (Vijayaraghavan et al., 1996). These properties are only reached after ejaculation. In this way, motility is stimulated after the contact of spermatozoa with seminal plasma and female genital tract fluids during and immediately after ejaculation, through several signaling compounds that are found by sperm cells after ejaculation such as prostaglandins, estrogens or even neurotransmitters like dopamine (Lindholmer, 1974). On the other hand, the fully fertilizing ability of mammalian spermatozoa is only reached after undergoing a progressive process named capacitation. The capacitation is progressively undertaken during the journey of spermatozoa through the entire female genital tract. In this way, spermatozoa that reached oviduct and waiting there the pass of the ovulated oocytes have undergone several considerable functional changes that allow them to be able for oocyte penetration (Chang, 1984). In the last years there has been a considerable amount of literature devoted to this point and capacitation implies important changes in sperm function such as a complete transformation of the cell membrane fluidity, activation of acrosome components, changes in the tyrosine phosphorylation and hence in the protein activity, of a great array of sperm proteins, displacements of calcium intracellular fluxes or the adoption of an specific motility pattern named hyperactivated motility (Bedford, 1983; Yanagimachi, 1994a; Visconti et al., 1998). Notwithstanding, although these changes are well described in many mammalian species, there are many unknown aspects regarding the regulation of sperm capacitation yet and, at this moment, this is one of the most studied aspects of sperm function.

Thus, as this brief summary has tried to highlight, mammalian sperm physiology is a very long and complex process, not fully studied and understood in any means. The importance of the study of these processes is very great if we want to optimize the strategies by which we can optimize the reproductive indexes in all of the mammalian species, including man. For this purpose, the utilization of tools like genomics and proteomics, specially adapted to the peculiar characteristics of sperm function will be basic in the understanding of the overall sperm function.

2. Common instrumental approaches to the study of mammalian sperm function

Until now, mammalian sperm function has been studied by using techniques that are focused in individual and very particular aspects of this overall function. As an example, the achievement of capacitation has been linked to a myriad of processes, such as increase in overall protein tyrosine phosphorylation, structural changes of cell membrane, activation of protein kinase A or processing of proacrosin to acrosin (Bedford, 1983; Langlais & Roberts, 1985; Yanagimachi, 1994a; Visconti et al., 1998; Lefièvre et al., 2005). These studies have been carried out separately, and the global interpretation of data can be inferred only after a long and often arduous process of data integration and comparison of results among the published literature. Additionally, if we consider the considerable differences among species, the interpretation of all the collected data are even more difficult, delaying thus a comprehensible interpretation of the overall mammalian sperm function. Despite these problems, the utilized techniques have yielded a great deal of useful information, which has allowed investigators to have a reasonable knowledge of the overall sperm function. Basically, the common instrumental approaches utilized until now here have been of three separate types, namely enzymatic analyses, determination of metabolite levels and detection of protein expression and location.

2.1 Enzymatic analyses

This chapter is not devoted to the detailed description of these techniques, although they are important, since enzymatic analyses allow investigators to determine the exact, precise activity of an individual enzyme in a specific point of the sperm lifespan, evaluating thus putative variations of these activities during the entire life of the cell. Concomitantly, the enzymatic analyses also allow investigators to determine the catalytic properties of a specific sperm enzyme, aiding thus to the identification of the exact isozyme that is present in the sperm. In this way, enzymatic analyses are important in order to reflect sperm function in a particular situation.

There are many examples of determination of enzymatic activities in sperm ejaculates, from regulatory enzymes of sperm energy metabolism, like hexokinases (Medrano et al., 2006a), to proteins related to acrosome activity, like acrosin (Cui et cal., 2000). In all cases, these techniques are all based in spectrophotometric techniques, in which the specific activity of the studied enzyme is utilized to form (or destroy) a final substrate with optical activity, such as NAD$^+$ (Passoneau & Lowry, 1993). This serves to detect the rhythm of variation of this optically active substrate during a controlled time lapse in standard conditions of total volume and temperature. After this, the specific activity of the studied proteins can be expressed as specific activities of the analyzed enzymes or results can be utilized to determine the enzyme kinetics by using of diagrams such as the Lineweaver/Burke representation, after determination of enzyme activity in the presence of increasing concentrations of the specific substrate (see Fernández-Novell et al., 2006 as an example of both expressions of hexokinase activity in boar sperm). Although these techniques have been also widely utilized in any tissue of cell sample, the determination of enzymatic activities in mammalian sperm can present a specific variant that is almost uniquely developed for these cells. This variant is due to the very high proportion of non-soluble structures together with the extraordinarily low amount of cytoplasm that presents mammalian sperm when compared to other eukaryotic cells. This peculiarity implies that many sperm proteins are distributed among the soluble and the no-soluble fraction of sperm extracts after their homogenization and centrifugation under following standard protocols (see Medrano et al., 2006a; Medrano et al, 2006b as examples). This distribution and its putative changes following overall changes in the sperm function can contribute in a relevant form in the control of the specific activities of these enzymes, being thus an additional mechanism of control of the enzyme activity that is almost specific for sperm. In this way, determination of a specific enzymatic activity in mammalian sperm would not be uniquely directed towards the analysis of supernatants obtained after homogenization/centrifugation of samples, but also towards the corresponding pellets, that would have resuspended in known quantities of the corresponding homogenization buffer (see Fernández-Novell et al., 2004 as an example of the analysis in both supernatants and pellets of homogenized samples of hexokinase activity in boar sperm).

2.2 Metabolic analyses: metabolite levels

The control of the energy levels is a very important point in order to maintain sperm function during their complete time life. This control is not only regulated by changes in the activity of position of the implied enzymes, but also through variations in the intracellular levels of metabolites involved in this process, such as ATP (Hammersted & Lardy, 1983). As in the case of the techniques devoted to the analysis of enzymatic activities, this chapter is not devoted to describe the techniques developed to determine intracellular levels of metabolites, and they are described elsewhere in the literature. Summarizing, the great majority of metabolite analysis are based in spectrophotometric techniques, similar to that developed for evaluating enzyme activities. However, in case of metabolites, the developed techniques are more frequently developed in an endpoint basis, in which all of the metabolite present in the sample is completely degraded by the addition of the corresponding, specific enzymes and other substrates. This leads to the obtainment of the maximal levels of optically active derived substance obtained from the degradation of the studied metabolite. Treatment of samples usually required as a first step the deproteinization of samples. This is done either through acid precipitation with substances as perchloric acid, as for determining ATP (Lambrecht & Transtschold, 1984) of basic precipitation with substances like sodium hydroxide, as for determining fructose 2,6-bisphosphate (Gómez-Foix et al., 1991). This implies that all of the protein content of sperm samples will be precipitated and, in these conditions, is not possible to evaluate a putative fractioning of metabolites between the soluble and the no-soluble fraction of sperm homogenates treated as per the determination of enzyme activities. In case of polymeric metabolites, such as glycogen, samples must be pretreated in order to degrade the polymer into their monomers by using the appropriate degrading enzymes. For instance, determination of intracellular sperm glycogen levels requires a previous incubation of α-amyloglucosydase, in order to release all of the glucose contained in the polymer (Ballester et al., 2000; Palomo et al., 2003). Then, this released glucose is determined through a standard, spectrophotometric technique (Ballester et al., 2000; Palomo et al., 2003).

2.3 Protein expression analysis: Western blot

The determination of the catalytic activity of an enzyme is not the unique form to study the functional role of this protein. In fact, there are many proteins that have not enzymatic properties. In this manner, these no-enzyme proteins can not be studied through techniques involving enzymatic analyses. Thus, another valid approximation to determine the role of a specific protein in the overall sperm function is to evaluate other characteristics that can control its activity. The two most important mechanisms involving control of protein activity are modulating both the total protein content and the phosphorylation levels on tyrosine, serine and threonine residues (Isen et al., 2006). This is especially important in mature mammalian spermatozoa, since these cells can not control the total content of a protein through modulation of its gene expression (Watd & Coffey, 1991). In this manner, molecular biology techniques involving studies of gene expression are not relevant in spermatozoa. This leads to that the most important techniques to determine both total content and phosphorylation levels of a sperm protein involved the utilization of specific antibodies. The most common of the techniques utilized to determine both the total content and phosphorylation levels of a specific protein is the Western blotting of transferred samples after being subjected to an electrophoresis onto an SDS-polyacrylamide support (SDS-PAGE). Again, this chapter is not devoted to the description of this technique, although it is necessary to indicate that Western blotting is currently one of the most widely utilized techniques in the study of mammalian sperm function, existing a very high amount of articles published in which this technique has been utilized to analyze proteins like hexose transporters (Rigau et al., 2002; Sancho et al., 2007), protein kinases (Vijayaraghavan et al., 1997; Breitbart & Naor, 1999) or even nucleoproteins (Flores et al., 2008; Flores et al., 2011). The usefulness of Western blotting is evident, although it is a technique that can cause many troubles to the unwary. One of the most important troubles that can be present in this technique is the lack of specificity of the antibody utilized to detect the studied protein. This is a problem that can have no easy solution, and, in this way, Western blotting requires in an unavoidable manner the presence of the adequate negative and positive controls in order to assure that the detected bands corresponds without doubts with the studied protein.

2.4 Protein location analysis: immunocytochemistry

Sperm protein function can be also studied through another technique involving the utilization of specific antibodies. In this way, sperm proteins can modify their activity also through changes in their specific location inside the sperm structure. These location changes can be studied through immunocytochemistry, which detect not only these changes but also other aspects like variations in the intensity of protein phosphorylation following functional changes (see 41 as an example). The immunocytochemistry has been also a widely utilized technique, which can be described in a very great number of articles devoted to sperm function (see Baccetti et al., 1988; Sutovsky et al., 2001; Albarracín et al., 2004 as examples). However, and in a similar for to that indicated for the Western blotting, the immunocytochemistry is not a totally straightforward technique, existing the possibility to fall into mistakes with a non easy resolution. In this way, troubles related with the specific fixative technique utilized can be of importance. For example, fixation of samples with ethanol rendered a different location of several hexose transporters in boar sperm when compared with similar samples fixed with paraformaldehyde (Rigau et al, 2002; Sancho et al., 2007; Bucci et al., 2010). In this way, there are several applications of immunocytochemistry that are completely valid for spermatozoa. In this manner, one of the most utilized is that centered in the location of submembrane proteins, like hexose transporters or proteins linked to other sperm structures, such as peri-mitochondrial actin and mitofusin-2. The most feasible immunocytochemistry of these proteins is carried out as follows:

Spermatozoa are washed three times with PBS. After the third washing, samples are centrifuged at 600 g for 10 min at 25°C, and the subsequent cellular pellet is resuspended in a 4% (w/v) paraformaldehyde solution in PBS for 15 min at 25°C. Fixed cells are centrifuged again at 600 g for 3 min at 25°C, and the supernatants are discarded. The cellular pellet is resuspended in 500 μL of PBS. This cell suspension is subsequently seeded into gelatin-coated slides of 76 mm x 26 mm of surface. Slides are afterwards covered with a 0.2% (v/v) Triton X-100 solution in PBS for 30 min. This step is absolutely necessary in spermatozoa, since the structure of these does not allow antibodies for an easy entering into the cell. Thus, spermatozoa need to be permeabilized in order to allow antibodies for their penetration inside the sperm cell structure. Slides are then washing with PBS and further incubated in a blocking solution containing 1% (w/v) bovine serum albumin (BSA) for 30 min at 4°C. After a new washing with PBS, slides are then incubated at 4°C with the appropriated primary antibody solution. The dilution and the incubation time will be variable, depending on the specific, utilized antibody. Notwithstanding, as a general rule, the incubation with the majority of the utilized primary antibodies will be carried out overnight, with an antibody dilution in the range of 1/100-to-1/500 (v/v) in PBS. After the incubation with the primary antibody, the slides are washed again with PBS and subsequent incubated with the appropriate, fluorochrome-conjugated secondary antibody. Again, the dilution and the incubation time will be dependent on the specific secondary antibody that will be utilized, although as a general rule a time incubation of 1h is enough. Moreover, the secondary antibody solution is mixed with another solution of the nuclear stain Hoechst 33258 at a final concentration of 1 μg/mL. The addition of the nuclear staining is needed in order to perform and adequate identification of the whole sperm structure, facilitating thus the precise location of the obtained markings. Afterwards, slides are thoroughly washed for 4 times with PBS and mounted in the DABCO solution. This solution is composed by 50% (v/v) glycerol and 25 mg/mL 1,4 diazabicyle [2,2,2] octane in water. Mounted slides are stored at 4°C in the dark. In these storing conditions, fluorescence can be optimally maintained for a maximum of 15 days. Figure 1 shows the final results obtained with this technique, after the observation of samples through a laser confocal microscope.

An interesting variation of the fixation technique consists in the application of cryoconservation techniques. This technique is specially indicated in the study of nuclear sperm proteins, which are difficult to detect in whole cells. Effectively, mature spermatozoa are cells with a very low volume and this study is usually carried out in whole cells. This is not a problem for the great majority of sperm proteins, since the small volume of sperm allows antibodies for a good penetration into the cell after its detergent-caused permeabilization. However, there are proteins that are not accessible for antibodies in whole spermatozoa. The most important of these proteins are those linked to the DNA in the nuclear structures, such as protamines and histones. The only manner in which antibodies against these proteins can contact with their specific proteins is after sectioning of sperm. However, this is practically impossible in samples treated with a standard fixation/inclusion procedure, since the width of the slice that investigators can obtain render a very small percentage of good sperm sections. Cryofixation and further cryosection allows investigators to obtain better results, as shown in (Flores et al., 2008; Flores et al., 2011) and Figure 2, and the procedure is the following: The procedure will start by the washing of sperm samples three times with PBS and subsequent fixation with 500 μL of a 2% (w/v) paraformaldehyde solution in PBS for 15 min at 25°C. Fixed samples will be centrifuged at 600 g for 3 minutes, and the supernatants will be discarded. The cellular pellet

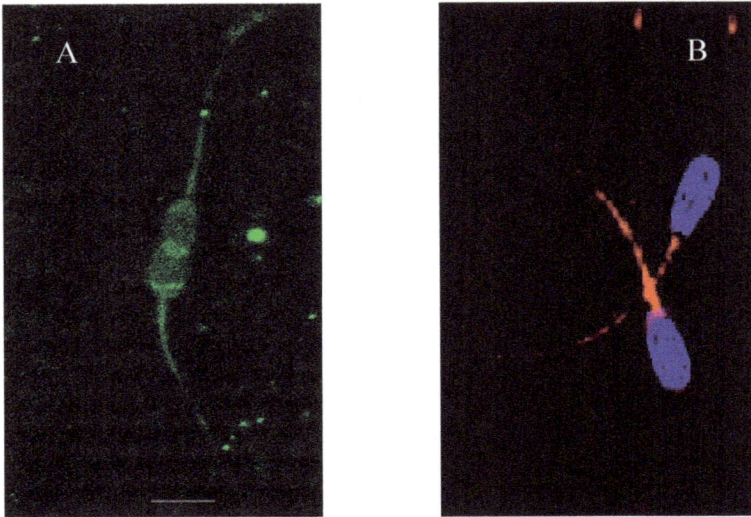

Fig. 1. Immunocytochemistry of actin (A) and mitofusin-2 (B) performed in boar spermatozoa. Figures show the location of both proteins after the utilization of an Alexa 488-conjugated donkey anti-goat secondary antibody for actin and an Alexa 647-conjugated goat anti-rabbit secondary antibody for mitofusin-2. The nuclear stain with Hoechst 33258 is also evident in (B). Bars indicate a real size of 7 μm. Figures are taken from photographs made for the work published in Flores et al. (2010).

will be then resuspended in 500 μL of PBS and centrifuged again at 600 g for 3 min. Supernatants will be again discarded, and the pellets obtained will be embedded in 40 μL of the any cryo-inclusion medium, like the OCT1 (Leica Instruments; Wetzlar, Germany). Samples will be immediately frozen with liquid N_2 and stored until their processing at -80°C. When stated, the included samples will be sectioned in slices of 1 mm of thickness by using a cryostat. Sections will be subsequently placed onto gelatin-coated slides (76 mm x 26 mm). Immediately, the slides will be covered with a PBS solution containing 0.1 (v/v) commercial Hoechst 33258 solution (Boehringer Mannheim). This stain will allow for the determination of an exact co-localization between the signal obtained with the specific antibody and the sperm nuclear DNA in case of the study of the interaction between DNA and a specific nuclear protein. Incubation with Hoechst 33258 will be maintained for 15 min at 38.5°C, preventing any light source from reaching the slides. Afterwards, samples will follow the standard immunocytochemistry procedure.

Although fixation can be a serious trouble in the immunocytochemistry procedure, it is not the only one that can be found. In fact, the worst troubles could be linked with the specificity of the antibodies utilized to detect a protein. This trouble, that is similar to that described for the Western blotting, can cause many problems of identification and interpretation and, similarly to that indicated for Western blotting, immunocytochemistry requires in an unavoidable manner the presence of the adequate negative and positive controls in order to assure that the detected marks correspond without doubts with the studied protein.

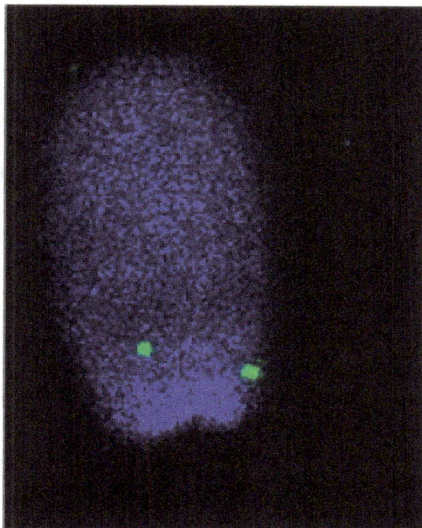

Fig. 2. Detection of histone H1 in the head of freshly obtained boar spermatozoa following the cryofixation and further cryosection technique. Green spots show the location of histone H1 in the head of boar sperm. Nuclear staining has been carried out after the utilization of the Hoechst 33258 stain. The image has been taken from Flores et al. (2011).

3. Morphological and functional characteristics that mediates application of technical global approaches in mammalian sperm

As described above, the study of mammalian sperm has been carried out by applying analytical techniques that are common to all of the other cell types. However, mammalian sperm is a very specific cell, with many typical characteristics that difficult not only the application of the classical techniques, but also the interpretation of the obtained results. Taking into account this, a succinct description of these characteristics is needed in order to a correct interpretation of the new techniques applied to the study of mammalian sperm biology. For this purpose, we have classified these sperm particularities in morphological aspects and functional characteristics.

3.1 Morphological characteristics of mammalian sperm

The mammalian sperm is a cell with a very characteristic shape, which can not be confused with any other cell. This is due to the fact that these cells are specifically designed to reach their ultimate goal, the penetration of an oocyte. This characteristic morphology also implies that the whole cellular structure of sperm is totally different to the other cells. These differences include aspects such as the practical absence of cytoplasm and cytoplasmic organelles like the Golgi apparatus, ribosomes, lysosomes or endoplasmic reticulum, the presence of a haploid, highly compacted nucleus and the existence of specialized organelles that are specialized products from classical cell structures. In this last group the most important are the acrosome, which is a derived of the Golgi apparatus, and the mitochondrial sheath, which is a specialized structure formed with very tightly bound mitochondria (see Eddy, 1988). Taking into account these structures, the mammalian sperm is morphologically structured in four structures; the head, the neck, the midpiece and the tail (see Figure 2).

The head is the apical section of the sperm cell. It contains four structures, the acrosome, the post-acrosomal dense plate, the sub-acrosomal space and the nucleus (Figure 2). Thus, this structure contains all of the genetic information that the male contribute to the future embryo, as well as the machinery that allows sperm to penetrate into the oocyte.

The acrosome is a vesicle located in the apical extreme of the head, covering the nucleus. This structure contains an amorphous material that corresponds with an enzymatic cocktail consisting in lythic enzymes designed to disaggregate the external oocyte structures during the sperm penetration. The acrosome is confined with a double cell membrane that is independent from the main cellular membrane, having a specific lipo-proteic structure derived from the Golgi apparatus (Eddy, 1988).

The post-acrosomal dense plate is an homogeneous plate composed with a electro-dense and fibrous material in which the inner acrosomal membrane is tightly fixed. Its location is at the distal apex of the acrosome and its function is the fixation of the acrosome until the acrosome reaction.

The sub-acrosomal space is the free space between the acrosome and the nucleus. It is especially wide under the apical zone of the acrosome, whereas it is practically absent in the lateral areas of the head between the acrosome and the nucleus. It is composed by cytoplasm and acts as a protective area for sudden changes of the external conditions.

Finally, the nucleus is the greatest and most important structure of the head. It is composed by DNA linked to specific nucleoproteins. The most important of these proteins is protamine, which can be present in two forms, the protamine 1 and the protamine 2 (Balhorn, 2007). The DNA/protamine structure is very tight, due to the fact that protamines are linked in the inner groove of the DNA helix (Biegeleisen, 2006). This originates a hypercondensed structure, which impedes any possibility of DNA expression. However, although protamines are the main proteins in the nucleus, there is also a significant proportion of histones (as much as the 15% of the total nucleoprotein content, see Wykes and Krawetz, 2003; O'Brien et al., 2005), indicating thus that a small proportion of the sperm nuclear structure would be similar to that of the somatic, eukaryotic cells. On the other hand, this strongly suggests that the sperm nucleus structure is heterogeneous, with main protamine-rich domains and concrete histone-rich domains, with an unknown function at this moment (Flores et al., 2011).

The sperm head connects in its distal apex with the neck or connecting piece. This small structure is, however, very important, since it has two key roles. The first role is to transmit the kinetic energy produced in the tail to the head, causing thus a progressive movement of the sperm. The second role is to transmit to the oocyte the centriolum that will be required to make the first cellular division after the fussion between both germ cells (Eddy, 1988). The neck contains then two types of structures. The first structures are all linked to the role of transmitting movement of the head. These structures, namely the basal plate, the laminar bodies, the capitulum and the segmented columns, are composed by fibrous, electrodense material, and they are organized in a complex and rigid manner, enabling thus the movement transmission (Figure 3). The second main structure is the centriolum, as described above. It is similar to that any other centriolum observed in any somatic cell. Finally, the apical apex of the axoneme that will form the entire sperm tail is embedded into the distal zone of the neck (Figure 3), facilitating thus the transmission of the movement originated in the tail to the other sperm areas.

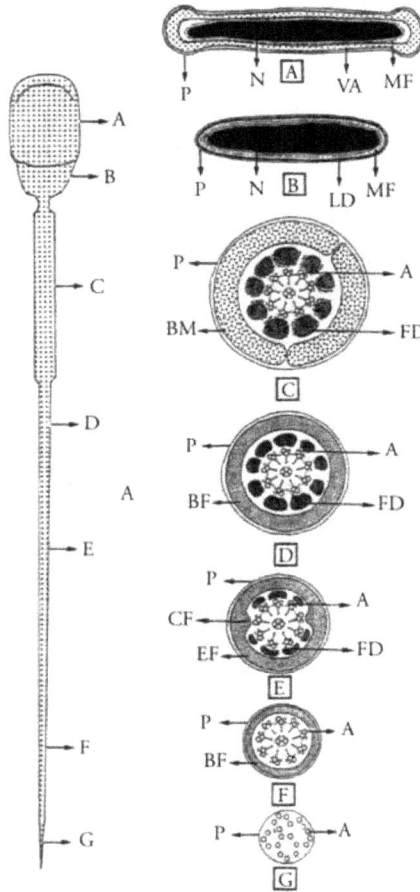

Fig. 3. Schematic representation of several transversal sections of a mammalian sperm indicating the present structures. A: Head, acrosomal area. B: Head, post-acrosomal area. C: Midpiece. D: Tail, proximal area. E: Tail, medial area. F: Tail, distal area. G: Tail, terminal area. MF: Perinuclear fiberous material. N: Nucleus. P: Cytoplasm. VA: Acrosome. LD: Postacrosomal dense plate. A: Axoneme. BM: Mitochondrial sheath. FD: Dense fibers. CF: Columns of the fibrous sheath. EF: Ribs of the fibrous sheath. Taken with permission from Bonet et al. (2000).

The structure that is placed immediately after the neck is the midpiece. This structure, in fact, corresponds to the apical zone of the tail, but it is characterized by the presence of a mitochondrial sheath that covers the inner tail structure (Eddy, 1988 and Figure 3). The mitochondrial sheath is composed by 150-200 mitochondria disposed in a helicoidal belt. These mitochondria are in a tight contact among them, although they are not fused together. This mitochondrial sheath finishes in its distal apex in a dense, annular structure known as the Hensen's ring (Figure 3). At this point, the midpiece finishes, giving place to the tail. It is not well known the exact role that this mitochondrial sheath plays in mammalian sperm function. A briefly discussion on this point will be made below. As an advance, we can say that several authors have suggested that the main role is the obtainment of energy for maintaining the sperm movement. However, recent data are not in agreement with this hypothesis. It is probable that other roles, such as the maintenance of a correct redox environment or the regulation of the capacitation could be more important than the suggested fuelling role.

The mitochondrial sheath of the midpiece involves the complete structure that will be continued in the tail. This structure is formed by a central axoneme with a structure similar to any other eukaryotic axonemes and a series of fibrous components that completely covers the axoneme (Eddy, 1988 and Figure 3). Of course, the axoneme is the main responsible for generating the sperm movement. However, the components that cover the axoneme are also very important, since they are responsible for transforming the axoneme movement in optimal for the sperm progressivity. The structures that cover the axoneme are basically a fibrous sheath and a variable number, in dependence of species, of longitudinal columns, which are a continuation of the fibrous columns that are present in the midpiece (Eddy, 1988 and Figure 3). As indicate above, these structures confers to the sperm movement its progressivity and planarity into the space. As a result, any defect in these structures will have disastrous consequences for the sperm movement. The final section of the tail, named terminal piece, is characterized by the lack of these complementary structures, leaving thus only the axoneme as the final sperm structure (Eddy, 1988 and Figure 3).

3.2 Functional characteristics of mammalian sperm

Mammalian sperm have a myriad of specific functional characteristics, which are derived from their enormous specialization. A thorough study of these functional characteristics will need then the writing of a complete book, and here is not the place to make this. In this way, we will only made a brief introduction into one of the most intriguing questions regarding sperm function, the control of the energy metabolism and its relationship with the maintenance of motility.

Obviously, the control mechanisms of sperm energy management play an essential role. This is because of the fact that practically all of the reactions that maintain the functional status of the cell (control of tyrosine phosphorylation levels, maintenance of the membrane proteins glycosylation, etc.) need significant energy consumption. Thus, the optimal function of all of these mechanisms will depend, to a great extent, on a correct functioning of control mechanisms modulating sperm energy management. Unfortunately, and despite the great amount of knowledge that many investigators have accumulated in the past 15 years, there are several commonplaces regarding sperm energy metabolism, which, in fact, obstruct an optimal, practical application of this knowledge. Thus, everybody knows that the spermatozoon is a totally strict, glycolytic cell. Of course, this is an unquestionable fact. However, the adoption of this assertion, without any doubt, can lead to the opinion that spermatozoa are almost exclusively glycolytic so, then, they have practically no other modulator system to manage their energy levels (Mann 1975). On the other hand, if the spermatozoon is an exclusively glycolytic cell, what is the role of sperm mitochondria and the associated Krebs cycle? In this respect, it is noteworthy that many investigators indicate as an absolute fact that the energy obtained through the Krebs cycle is, in all conditions, absolutely necessary for the maintenance of sperm motility in all species (Nevo et al. 1970; Ford and Harrison 1985; Halangk et al. 1985; Folgero et al. 1993; Ruíz-Pesini et al. 1998), despite the same investigators maintaining an absolute pre-eminence of glycolysis to obtain sperm energy, without noticing the energy contradiction that the simultaneous assumption of both principles implies. These contradictions highlight the complexity of the question, which has to be approached with an open mind. Only in this manner some valid and general conclusions with practical applicability on sperm conservation should be attained.

Once monosaccharides have been uptaken, transformed and phosphorylated to obtain glucvose 6-phosphate (G 6-P), they undertake the appropriate metabolic pathway designed for energy synthesis. If we preclude any possible anabolic pathway (glycogen synthesis, pentose phosphate cycle, etc.; see Urner and Sakkas 1999; Ballester et al. 2000), the utilization of G 6-P to obtain ATPs starts with the glycolytic pathway, which culminates in the obtainment of pyruvate. Afterwards, the obtained pyruvate can be sent either to the formation of extracellular lactate or to its introduction into the mitochondrial Krebs cycle. This last step is controlled by the lactate dehydrogenase (LDH) activity, and it is noteworthy that the further metabolization of pyruvate/lactate through the Krebs cycle yields a great amount of ATPs, as well as important levels of reducing potential in the form of nicotinamide adenine dinucleotide hydrogenase (NADH). This NADH is of great importance not only in the maintenance of anabolic pathways, but also in the control of intracellular redox and pH levels. The equilibrium between sugar metabolization through simple glycolysis or through glycolysis plus the Krebs cycle depends on a great number of factors, such as the O_2 pressure, the pH, the intracellular levels of ATP and the action of several intracellular signalling factors, like nitric oxide (Stryer 1995). All of these factors allow for a very fine regulation in order to maintain the appropriate ATP intracellular levels and, thus, the required sperm energy levels at each point of its life-time. If we analyse the published data regarding the glycolysis/Krebs cycle equilibrium in mammalian sperm, we can observe the existence of clearly contradictory results, as indicated above. Thus, there is a general consensus about the fact that practically all mammalian sperm from fresh ejaculates has very high glycolytic activity (it can comprise more than 95% of the formed ATP in boar sperm from fresh ejaculates; see Marín et al. 2003). This very high glycolytic rhythm is one of the main factors that preclude the obtainment of a stable, estequiometric equilibrium in this pathway (Hammersted and Lardy 1983). Nonetheless, several authors have described that the energy obtained from the Krebs cycle is absolutely necessary to maintain sperm motility in species such as bull, despite the fact that the energy obtained from this pathway is very small (Nevo et al. 1970; Ford and Harrison 1985; Halangk et al. 1985; Folgero et al. 1993; Ruíz-Pesini et al. 1998). To complicate this question further, other authors have described that, in fact, the energy from the Krebs cycle is not necessary to maintain motility in other species like mice (Mukai and Okuno 2004). How can all of these contradictory results have a global meaning?

As in other points, we can only speculate about the exact meaning of the data shown above. In spite of this, some points can be discussed in order to clarify this fundamental aspect of sperm energy management. As a first question, we must consider the hypothesis of the existence of separate metabolic phenotypes in mammalian sperm (Rodríguez-Gil, 2006). If we consider this hypothesis, we can also suggest that each metabolic phenotype will have a separate equilibrium between the catabolic and the anabolic pathways. These different equilibriums will lead to concomitant differences in the preponderance of the principal energy obtaining pathways, whether anaerobic (glycolysis) or aerobic (Krebs cycle). In this sense and concerning fresh ejaculates, dog sperm, which has a very active glycogen metabolism concomitantly to high intracellular G 6-P and other hexose 6-phosphate levels, shows a mitochondrial oxidative activity higher than boar spermatozoa, which have a much less active glycogen metabolism and practically absent G 6-P levels (Rigau et al. 2002; Marín et al. 2003; Medrano et al. 2006a). This difference implies that the equilibrium between glycolysis- and Krebs cycle-obtained ATPs will be very separate between both species and as a consequence, the specific regulation of both pathways would also differ. Following this comparison, the existence of species in which ATPs from Krebs cycle would be absolutely necessary to maintain motility is easily assumable whereas ATPs of a similar origin would be of a much lesser importance in other mammals. Another possible explanation for the observed discrepancies can be related to the fact that authors have ignored that mammalian sperm is a cell with a very active and complex life-cycle, with enormous changes in its specific function during its life-time. Thus, we can only bear in mind that sperm motility in sperm from fresh ejaculates in all species is very different to that from the hyperactivated sperm subjected to capacitation (Yanagimachi 1994b). Therefore, if we assume that capacitated sperm has energy requirements far greater than those from recently ejaculated cells in order to maintain hyperactivated motility and all of the energy consuming processes linked to acrosome reaction and oocyte penetration, it is logical to assume that the energy-obtainment mechanisms will change during capacitation to obtain a higher energy-production rhythm. This implies that a direct comparison of the energy obtained from glycolysis alone and the Krebs cycle in freshly obtained and capacitated sperm from the same ejaculate will show very great differences between each other. A clear example of this point is boar spermatozoa, where the proportion of energy that is obtained from the Krebs cycle in cells from fresh ejaculates is very low, less than 5% of the total generated energy (Marín et al. 2003). However, other experiments show that the attainment of "in vitro" capacitation induces a constant increase in the mitochondrial activity of these cells, which was measured through specific staining and analysis of O_2 consumption (Ramió-Lluch et al., 2011). These data indicate that mammalian sperm has the ability to equilibrate its energy-obtainment systems depending on its necessities. These necessities will surely be modulated by factors such as variations in the intracellular levels of ATP and other related nucleotides (ADP and AMP), since these levels tend to be maintained within very narrow limits, at least those of ATP, in separate species subjected to different study conditions (Rigau et al. 2002; Medrano et al. 2006a). This ability causes the direct comparison of the results obtained from separate laboratories to be very difficult, since the small differences that each laboratory will introduce into its precise methodology can result in great effects on the specific energy equilibrium. This can also to be a partial explanation of the contradictions shown in the bibliography that have greatly hampered a global comprehension of the mammalian sperm energy-levels management.

4. Technical approaches to the global study of whole cell function: metabolomics and proteomic arrays

4.1 Metabolomics

In the last years, a series of novel techniques have been developed in order to obtain a more global vision of the overall cell function. In this sense, specialties such as metabolomics have been applied with success in many cellular systems. Metabolomics has been defined as the study of the chemical processes in a cell involving the processing of metabolites as a whole. This discipline allows investigators a more comprehensive knowledge of all the processes related to the obtainment of energy in a whole cell. This is especially important in mammalian sperm, in which the metabolization of energy substrates is not only related to the obtainment of the energy, but also with the regulation of other cellular processes, like phospho-dephosphorylation of proteins not directly related with the energy regulation (Urner & Sakkas, 2003). Thus, metabolomics is a very important tool in the interpretation of the regulation of mammalian sperm function.

However, the application of metabolomics in mammalian sperm requires several particularities, linked not only to the specific characteristics of sperm energy metabolism, but also to the specific morphological characteristics of sperm cells. It must be reminded that mammalian sperm present morphological characteristics that difficult in a great measure their processing for the analysis of any molecular mechanisms. One of the most important of these characteristics, as it has been indicated above in this chapter, is the very great percentage of non-soluble structures integrating the whole cell. Some of these structures are the nucleus, the axoneme and even the mitochondrial sheath, which are mostly composed by very tightly components (hyper stabilized DNA, axoneme proteins, fibrous sheath and longitudinal columns proteins, etc, see Klaus & Hunnicutt, 2006) These proteins are very difficult to isolate and even separate from other components of the sperm structure. This greatly difficult the correct processing of all of the molecules related with these sperm structures which, in turn, impedes in many occasions to reach to correct interpretations of the data yielded after this processing. Furthermore, the correct processing of semen samples would be different depending on the structure or the specific functional aspect that was studied. As an example, the study of membrane-linked functions such as uptake of metabolites will require a sample treatment in which membrane components will be detached from the rest of sperm structures, usually combining the homogenization of samples with the addition of a detergent in the homogenization buffer, and, in some cases, the study will further require a subsequent enrichment of the homogenized samples through ultracentrifugation. On the contrary, the study of some of sperm nuclear function will require a much harder sample treatment, in order to release nuclear components that are very tightly bounded among them, such as protamines. Other important question that has limited the use of metabolomic approximations to the study of the sperm function is the high economical cost needed for these studies. This cost is basically due to that metabolomics requires the utilization of sophisticated equipment, such as mass-spectrophotometric analyzers and the work of expert personnel in the computerized analysis of data obtained from the rhythm of variations of isotopic levels in treated samples. This implies that these works are not accessible to investigators and in the majority of times the co-operation among several interdisciplinary groups is mandatory. As a consequence, there are very few manuscript published until now in which metabolomics has been applied to the study of sperm functionality, and the following explanation will be based in the work published in Marín et al. (2003).

The most important question arising from a metabolomic study is to determine the fate of an specific substrate when it is processed by cell and the evaluation of putative changes in the rhythm of pathway in which this substrate is processed when cell function changes. Centering on the sperm studies, they have been centered in determining the fate of both glucose and lactate as an energy source of freshly obtained boar sperm (Marín et al., 2003). Taking into account this purpose, metabolomic studies was carried out through the combination of three separate techniques. The first technique was a conventional analysis of intracellular levels of the perhaps most important intermediate glucose metabolite, G 6-P. The second technique was the determination of the rhythm of glucose oxidation, a marker of the Krebs cycle, through incubation of sperm with the randomly [14C] radioactive substrates [U-14C] glucose and [U-14C] lactate and the subsequent analysis of the $^{14}CO_2$ formation. The third technique was the mass isotopomer analysis by gas chromatography/mass spectrometry (GC/MS) of the intermediate metabolites originated after the incubation of boar sperm with the no-radioactive substrate [1,2-13C2]glucose (Marín et al, 2003).

4.1.1 Determination of glucose 6-phosphate intracellular levels

As commented above, the determination of the intracellular levels of G 6-P is carried out through a spectrophotometric technique, described in Michal (1984). Briefly, it consists in two successive steps. The first step is the homogenization of samples. The second step is the incubation of samples with the appropriate substrates to obtain an optically active product, and its determination by spectrophotometry.

Regarding the homogenization of samples (Fillat et al., 1992; Marín et al., 2003), this procedure must separate low-molecular weight, soluble metabolites such as sugars from all of the other cell components with a greater molecular weight, like nucleic acids or proteins. For this purpose, samples will be homogenized in the presence of acid, in order to precipitate all of the high-molecular weight components. The most common acid utilized for this purpose is an aqueous solution of perchloric acid ($HClO_4$) at a concentration of 10% (v:v). Homogenization can be carried out by using different systems, like sonication or mechanical rupture, either through a manual technique or an automatized one. However, in this precise technique, the homogenization system is not as crucial as that in other techniques devoted to the obtainment of other cell components, such as proteins, since the existence of a very acidic environment will cause the precipitation of the majority of the cell components by itself. One vital point to consider, however, is that the homogenization has to be carried out at 4°C, since greater temperatures can cause a rapid destruction of the soluble metabolites that we want to determine. Thus, the $HClO_4$ solution has to be stored and utilized at 4°C, and the homogenization technique has to be performed at this temperature (Rigau et al., 2002; Medrano et al., 2006a). The $HClO_4$ volume that is added to the samples varies depending on the cell concentration, although it is usually in a range between 200 μL and 500μL. After homogenization, samples will be subjected to centrifugation at 29,000 g for 5 minutes at 4°C, and clear supernatants will be collected, carefully avoiding a possible mechanical remixing with the obtained pellet (Fillat et al., 1992; Marín et al., 2003).

At this moment, low-molecular weight metabolites like G 6-P will be contained in the clear supernatant. Notwithstanding, they are solubilized in a very acidic medium, in which their structural stability is very short. To avoid this, samples have to be immediately neutralized after their obtainment. For this purpose, the obtained supernatans will be added with a low volume (i.e., 5 µL) of any sort of liquid pH indicator, which suffers a color change depending on the medium pH. Immediately afterwards, we will add to the samples very low volumes of a concentred solution of a strong basic solution. The most common solution is an aqueous one of 5M K_2CO_3, which will yield a no-soluble $KClO_4$ precipitate when it reacts with the excess of the $HClO_4$ in the medium (Fillat et al., 1992; Marín et al., 2003). The addition of the K_2CO_3 must be performed very slowly and in very small quantities (i.e., volumes of about 5 µL), and, after the addition of one of these small volumes, samples will be mixed and the color of samples will be observed. When samples color will indicate that they have reached a pH of about 7, we will stop the addition of K_2CO_3 and the exact volume of the added basic solution will be annotated in order to determine the exact, final volume of the sample (Fillat et al., 1992; Marín et al., 2003). Finally, the neutralized will be subjected to another centrifugation at 29,000 g for 15 minutes at 4°C in order to eliminate the produced $KClO_4$ pellets, and supernatants will be immediately subjected to the G 6-P determination. It is noteworthy that the stability of the obtained, neutralized samples is not very great. In fact, samples can be utilized only for a few hours after their obtainment, and they must be placed in this time at 4°C. These samples can not be conserved frozen for a long time also, and they should be processed as longer as 2-3 days after their obtainment.

The most usually spectrophotometric technique utilized for determining G 6-P levels is that based in the ability of G 6-P to be oxidized by the enzyme glucose 6-phosphate dehydrogenase (G 6-P DH, see Michal, 1984). This reaction will be yield in the presence of NAD^+, which will be reduced to NADH. The NADH is an optically active substance at a wavelength of 340 nm, and, in this way, the absorbance change at 340 nm will be a direct result of the presence of G 6-P in the sample. The technique, based in that published in Michal (1984), is performed as follows:

An aliquot of 300 µL will be mixed with 225 µL of a reaction mixture of an adjusted buffer (Ph 7.4) containing 0.1M Tris, 1M Cl_2Mg and 20 µg/µL NAD^+. The mixture will be incubated for 2 minutes at 37°C. Afterwards, a first optical lecture will be made to obtain the initial absorbance of the sample ($\mathring{A}_0$). After this lecture, the mixture will be added with 15 µL of a solution of G 6-P DH with a specific activity of 30 U/mL. This will be mixed and further incubated for 10 minutes at 37°C. After this time, a final optical lecture will be made to obtain the final absorbance of the sample ($\mathring{A}_1$). The G 6-P levels of the sample will be obtained then through the difference between $\mathring{A}_1$ and $\mathring{A}_0$, after applying the logical corrections with the appropriate G 6-P standards and the correction for the intrinsic absorbance of the sample. Finally, G 6-P levels will be normalized through the determination of the total protein content of samples. This will be determined in the pellets obtained after the homogenization in the presence of $HClO_4$. These pellets will be resuspended in 400 µL of 1M K_2CO_3 and heated at 60°c until the resuspension of pellets. The total protein content of these resuspended pellets will be determined through the Bradford method (Bradford, 1976), by using a commercial kit, in our case from BioRad (Hercules, CA).

4.1.2 Determination of the rhythm of formation of $^{14}CO_2$ formation from [U-^{14}C]-marked substrates

In this technique, spermatozoa until a final volume of 250 μL will be incubated in a standard Krebs-Ringer-Henseleit medium added with the [U-^{14}C] marked substrate (Rodríguez-Gil et al., 1991; Marín et al., 2003). In the sperm metabolomics works published until now, the utilized substrate is [U-^{14}C]-glucose at a final concentration of 10 mM (Marín et al., 2003). In all cases the specific radioactivity will be of 5000 cpm/μmol substrate. The incubation will be carried out in Eppendorf tubes in which a small piece of a standard filter paper totally soaked in β-phenylethylamine will be cased into the tap of the tube. The β-phenylethylamine is a compound that will trap all of the CO_2 that will be released by the cells. It is essential to avoid that the sperm cell suspension will contact this piece of paper Eppendorf tubes will be closes as tightly as possible and then cells will be subjected to incubation during 60 minutes at 37°C in very gentle shaking. Again, this shaking will be enough gentle to avoid the contact between filter papers and cell suspensions. After this time, cells will be killed by the addition of 350 μL of 10 % (v:v) $HClO_4$. The acid will also release all of the CO_2 that will be accumulated into the cells, including the $^{14}CO_2$ obtained after the cellular processing of the ^{14}C-marked utilized substrates. Samples will be further incubated for 30 additional minutes at 37°C in gentle shaking, which will allow for a complete release of the intracellular CO_2. Afterwards, the β-phenylethylamine-soaked filter papers of the tubes will be gently extracted, avoiding again any contact with the liquid of the tubes. These papers will be lent to air dry into a laminar flux chamber. Once dried, the radioactivity presented in the filter papers will be determined by using a liquid scintillation counter previous immersion of papers in a vial containing 2,5-diphenyl oxazol. Afterwards, the rhythm of substrate oxidation by cells will be easily calculated taking into account the proportion of radioactive counts found in the papers when comparing with the total radioactive counts placed in the incubation medium.

4.1.3 Determination of the rhythm of formation of intermediate metabolites through gas chromatography/mass spectrometry

The utilization of [U-^{14}C]-marked substrates only allows investigators to determine the final fate of the substrates after their metabolization by the cells. However, this technique does not allow investigators to known what are the exact metabolism pathways by which sperm are able to metabolize these substrates. This will impede a global survey of the metabolic pathways utilized by spermatozoa to utilize these substrates through the analysis with [U-^{14}C] marked substrates. However, this problem can be overcome if investigators are able to utilize marked substrates but not in a uniform manner, but with substrates in which marking is linked to one specific carbonil radical. Moreover, if the utilized marked substrate is not radioactive, all of the problems that are inherent to their utilization will be eliminated and, in this way, the handling of experiments will be much easier. This is possible if investigators can utilize ^{13}C-marked substrates, which can be analyzed through GC/MS. Centering on the results published regarding sugar utilization pathways by boar spermatozoa (Marín et al., 2003), the chosen marked substrate was the no-radioactive [1,2-$^{13}C_2$]-glucose. This specific marking allows investigators to elucidate the proportion of substrate that is metabolized by boar sperm cell through their entry into de glycolytic pathway, the accumulation of this marked glucose into glycogen and even the pass of marked glucose through the other possible metabolic pathways, namely pentose phosphate cycle and Krebs cycle after the glycolytic pathway. Likewise, this specifically marked substrate is able to determine the percentage of glucose that can be derived to the synthesis of fatty acids. Finally, a putative reconversion of glucose into gluconeogenesis, which will be linked with an indirect pathway for glycogen synthesis, can also be determined following the fate of the [1,2-^{13}C]-glucose. The Figure 4 shows how the [1,2-^{13}C2]-glucose can be tracked for all their putative metabolic pathways in the cell. This Figure is excerpted from Marín et al. (2003).

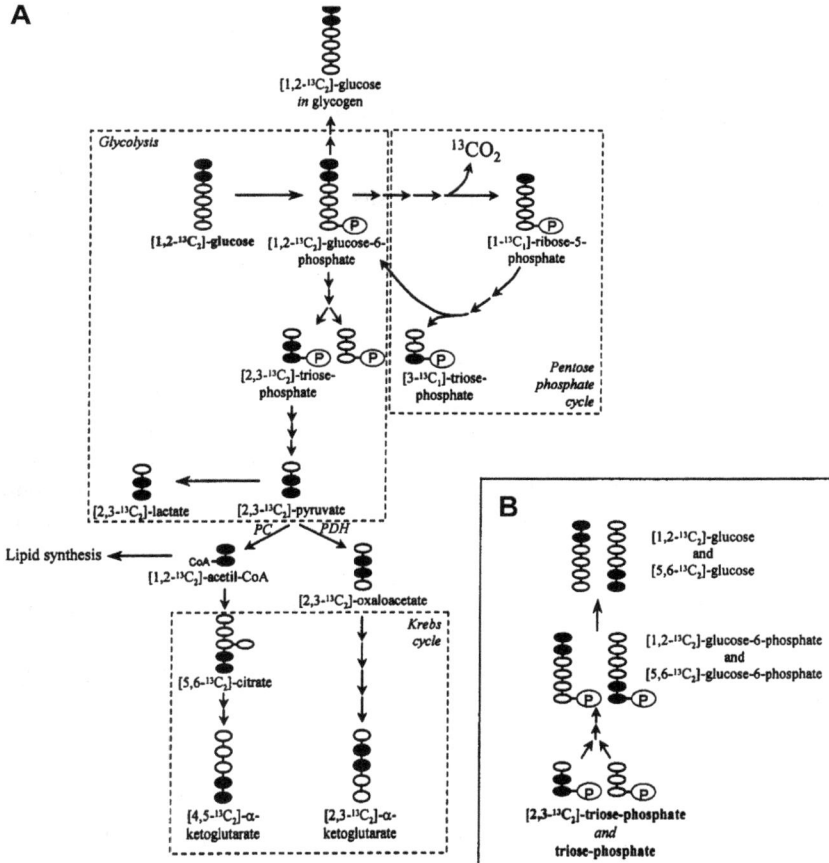

Fig. 4. Expected mass isotopomers in lactate, ribose, newly synthesized fatty acids, and glutamate from incubations with [1,2-^{13}C2]glucose (A) and in newly synthesized glucose from gluconeogenesis (B). When [1,2-^{13}C2]glucose enters into the cell, it is converted into glucose-6-phosphate (G6P), which can undergo glycogen synthesis, and enter the glycolytic pathway or the pentose phosphate cycle (PPC). From glycolysis, two triose-phosphate molecules are formed, one of them with two ^{13}C and the other one without ^{13}C. Both can then form pyruvate, and therefore 50% [2,3-^{13}C2]lactate. Pyruvate can also enter lipid synthesis, forming molecules with a paired number of ^{13}C atoms, or the Krebs cycle, obtaining two different labelling distributions in α-ketoglutarate (which is in equilibrium

with medium glutamate) depending on the enzyme used to enter the Krebs cycle: pyruvate carboxylase (PC) or pyruvate dehydrogenase (PDH). When G6P enters the PPC, one ^{13}C is lost in CO_2 formation, giving ribose-5-phosphate with only one labelled atom. This molecule can also enter the non-oxidative pentose phosphate pathway forming triose-phosphate molecules with only one ^{13}C, and all subsequent products labelled in one atom. Furthermore, when products from glycolysis of [1,2-[13]C2]glucose undergo gluconeogenesis, two different labelling patterns are expected in glucose isotopomers: [1,2-[13]C2]glucose, which is the initial isotope, and [5,6-[13]C2]glucose, formed as a result of the isotopic equilibrium between the labelled and unlabeled triose-phosphates.

The application of the GC/MS requires a previous preparation of the samples that will be different, depending upon the isotopes that investigators are looking for. In the case of glucose metabolites, the preparation of samples will start with the extraction from semen samples of the low-molecular weight. This will be done through the homogenization of samples until obtaining a perchloric acid-pH adjusted supernatant likewise to that described for the determination of G 6-P levels (Fillat et al., 1992). Once obtained this extract, samples will be treated as described in Tserng et al. (1984); Lee et al. (1996) and Kurland et al. (2000) For this purpose samples will be taken to dryness and they will be subsequently resuspended in 400 μL of 0.5 % (w:v) hydroxylamine hydrochloride in pyridine. The mixture will be heated to 100 "C for 1 h and will be then evaporated to dryness under nitrogen. To this residue will be added 100 pL of pyridine and 20 pL of acetic anhydride. The solvent and excess reagent will be removed by evaporation under nitrogen, and the obtained pellet will be dissolved in 20-50 pL of pyridine for the GC/MS analysis. The GC/MS induces the chemical ionization of the components of the sample. This ionization actives molecules fot the GC/MS detectors, and the activated-^{13}C-marked compounds can be fractionated and further detected through the mass spectrophotmetry system. As indicated in Kurland et al. (2000) for the determination of glucose derivatives, the GC conditions will be of a 6 feet X 2-mm, inner diameter glass column packed with 3% SP2340 on 100/120 Supelcoport. The helium flow rate will be of 20 mL/min at a column temperature of 235°C. The retention time for glucose derivative will be of 2.4 min, and chemical ionization will be performed by using methane as the carrier gas. This treatment will create active, detectable forms in the range 0f 327-336 m/z. This range will include ^{13}C-marked glucose and all of their derivatives (Marín et al., 2003).The molar ratios of the isotopomers included in this range will be calculated from the ion intensities using a weighted multiple linear regression analysis (Hammersted & Lardy, 1983). The distribution of isotopomeric species will be expressed as molar fractions of the total glucose concentration. The percent content of [^{13}C]glucose will be calculated as the weighted average ^{13}C content of the isotopomers.

4.2 Proteomic arrays

Whereas metabolomics allows investigators to elucidate the pathway/s by which a substance is utilized by a cell, arrays allows investigators to determine the global status of a specific cell function through estimation of either the specific amount of the proteins involved in this function or the phosphorylation/glycosylation levels of these proteins. In this way, investigators can have a global survey of the exact situation of a specific molecular mechanism in the moment of the analysis. There are, of course, other types of arrays than those devoted to the study of proteins. In this sense, microarrays are widely utilized for the study of global genic expression and location, and they are the basis of the named genomics. However, genomics is not a very useful tool in the study of mature sperm function, since these cells do not have gene expression (Ward & Coffey, 1991), and their response to both external and internal stimuli are totally based in changes in the degradation/loss, location and post-translational structure, without any possibility to arrange newly synthesized proteins from their nuclear machinery. Thus, the study of a local sperm cell function mechanism has to be approached by using protein-devoted miniarrays, following the discipline named as proteomics.

There are in the literature many examples of proteomic studies in many cells and tissues. However, sperm proteomics is practically absent. At this moment, we can only detected in the literature one article, regarding changes in serine, tyrosine and threonine phosphorylation levels of a wide arrays of proteins involved in the control of the global sperm function, like protein kinases and phosphatases in both dog and boar mature sperm subjected to incubation with either glucose or fructose (Fernández-Novell et al., 2011). This article, however, can be a good basis for describing the application of the proteomic arrays in sperm as follows.

First of all, it is noteworthy that proteomic arrays are commercial products manufactured and commercialized by several different commercial firms. The precise technique to manufacture these arrays is, of course, subjected to patent protection and, in this manner, this is not possible to offer a detailed explanation of the technique utilized for arrays manufacture. Taking into account this limitation, a miniarray consists in an inert basis. This basis is usually a small square of a material similar to that utilized in the protein transferences in the Western blotting techniques, although other materials can be also utilized. Manufacturers placed in very concrete points of these arrays a known amount of a specific antibody against one protein of interest, in a manner in which the putative antigen-antibody reaction that investigators will be look for will located in a small point onto the inert basis. Thus, the basis can be full by a great number of small points in which specific antibodies for one protein can be placed (see Figure 5).

Once the studied proteins have been immobilized onto the array through their linking to their specific antibody, investigators can study these proteins through two different ways. The first way is the analysis of the amount of each protein. This is done through a direct developing of the performed antigen-antibody reaction through a system equal to that carried out in the Western blotting analysis with the transferred samples. In fact, as indicated above, the miniarray is similar to a transferred membrane of a Western blotting analysis and, thus, it can be analyzed in the same manner. The second way is the analysis of post-translational modifications of the proteins immobilized on the arrays after the antigen-antibody reaction. For this purpose, the miniarrays on which the selected proteins have been fixed are subsequent incubated with a specific protein from a concrete post-translational modification mechanism. The most common post-translational mechanisms that control protein function are the changes in the phosphorylation levels of these proteins (Isen et al., 2006). Protein phosphorylation can be only be made on serine, tyrosine or threonine residues (Isen et al., 2006), and specific antibodies against protein serine, tyrosine and threonine phosphorylation have been developed by several commercial firms. Thus, membranes are incubated in the presence of one of these specific antibodies in order to analyze the amount of serine, tyrosine or threonine phosphorylation of all of the proteins previously immobilized there. Afterwards, results are developed in the same way that as for the Western blotting technique (Burnette, 1981), since, in fact, both systems are very similar.

Taking into account all of these information, the miniarrays techniques that has been published regarding mature sperm function have been carried out under the following procedure (Fernández-Novell et al., 2011).

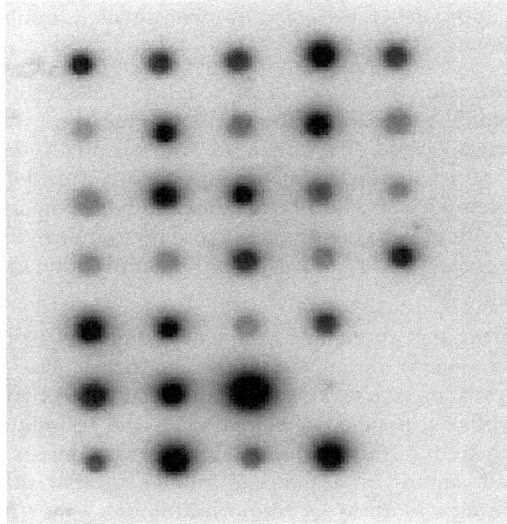

Fig. 5. Aspect of a proteomic miniarray in which positive spots indicate the presence of a specific antigen-antibody reaction between the specific antibody placed onto the inert base and the antigen present in the sample placed on the array. This image has been taken from Fernández-Novell et al. (2011).

Sperm samples are homogenised in 1mL of an ice-cold extraction solution comprising a 15mM Tris/HCl buffer (pH 7.5) plus 120mM NaCl, 25mM KCl, 2mM EGTA, 2mM EDTA, 0.1mM DTT, 0.5% Triton X-100, 10 mg/mL leupeptin, 0.5mM PMSF and 1mM Na_2VO_4 (extraction solution). Of these buffer components, EGTA and EDTA are ion chelants that can act as protein phosphatases inhibitors. Another protein phosphatise inhibitor is Na_2VO_4, thus avoiding artifactual changes in the phosphorylation levels of the studied proteins. The detergent Triton X-100 facilitates sperm homogenization through cell membrane lysis, whereas leupeptin and PMSF are known protease inhibitors (Roche Applied Science, 2004), avoiding thus the artifactual decrease of the sample protein content. Afterwards, homogenised samples are left for 30 min at 4°C and then centrifuged at 10,000g for 15 min at 4°C. Supernatants are taken and then used to test the degree of tyrosine, serine and threonine phosphorylation of selected proteins included in the chosen arrays, which is purchased to a commercial firm that followed a custom array of tested proteins. The analysis is performed following the standard protocol provided by the manufacturer. Briefly, each sample is diluted in 2mL of an extraction solution supplied by the commercial firm and containing 1% dry milk to reach a final protein concentration of 2 mg/mL.

Simultaneously, the supplied array membranes containing the specific antibodies are placed in standard 60mmx15mm suspension culture dishes and incubated with a blocking solution containing 150mM NaCl, 25mM Tris and 0.05% (v/v) Tween-20 (TBST; pH 7.5) plus 5%(w/v) dry milk and left for 1 h at room temperature under slow shaking. Membranes are then incubated with the samples for 2 h at room temperature under slow shaking. After incubation, membranes are washed 3 times, 15 minutes each with TBST. Afterwards, the membranes are incubated with 10 mg/mL of an HRP-conjugated antibody against serine, tyrosine or threonine phosphorylated proteins diluted in TBST for 2 h at room temperature under slow shaking. Finally, samples are washed 3 times, 15 minutes each with TBST and then incubated with a peroxidase substrate and exposed to a commercial X-ray film. Concomitantly, the total protein content of the supernatants is determined by the Bradford method (Bradford, 1976) using a commercial kit from BioRad. The intensity of the spots obtained is quantified using any specific software for image analysis of blots and arrays, in which background has to be previously made uniform for all of the arrays analysed. The values obtained for the control samples (i.e. those incubated in the absence of sugars) have to be transformed in order to obtain a basal arbitrary value of 100, from which the intensity values for the other samples will be calculated. Furthermore, two types of negative control must be applied. In one, one or two arrays must be incubated with a randomly chosen sample but without further incubation with the primary antibody. In the other negative control, one or two arrays have to be incubated with the antibodies but without samples. Statistical differences between groups can be checked by using the Student–Neumann–Keuls test or any of their derivatives. However, since the analytical technique has an intrinsic subjective component, investigators must consider that true differences would be only considered as those for which a $P0.05$ and a percentage difference above 20% were detected.

5. Conclusion

The study of mammalian sperm biology is a very open field with many unanswered questions at this moment. This is mainly due to the fact that mammalian sperm have a very complex life, which is translated in the existence of many, concomitant and complex molecular mechanisms that control the whole sperm function. The overall study of this complexity is not completely possible with the classical analytical tools, based on the analysis of concrete, punctual functional aspects. This is also true for immunocytochemistry, which has been mainly devoted to the location of specific proteins in the study of sperm function. In this way, only the application of integrated analytical systems, such as the metabolomics or the miniarray studies can achieve a better and deeper knowledge of the mature mammalian sperm functionality. In this book, we proposed that the coordinated utilization of these integrated analytical systems was named "spermiomics". In this manner, spermiomics would be the best tool for future investigations of the overall, mature mammalian sperm function.

6. References

Albarracín JL, Fernández-Novell JM, Ballester J, Rauch MC, Quintero-Moreno A, Peña A, Mogas T, Rigau T, Yañez A, Guinovart JJ, Slebe JC, Concha II, Rodríguez-Gil JE

(2004). Gluconeogenesis-linked glycogen metabolism is important in the achievement of in vitro capacitation of dog spermatozoa in a medium without glucose. *Biol Reprod* 71, 1437-1445.

Baccetti B, Burrini AG, Collodel G, Magnano AR, Piomboni P, Renieru T (1988). Immunocytochemistry and sperm pathology. *J Submicrosc Cytol Pathol* 20, 209-24.

Balhorn R (2007). The protamine family of sperm nuclear proteins. *Genome Biol* 8:227, 1-8.

Ballester J, Fernández-Novell JM, Rutllant J, García-Rocha M, Palomo MJ, Mogas T, Peña A. Rigau T, Guinovart JJ, Rodríguez-Gil JE (2000). Evidence for a functional glycogen metabolism in mature mammalian spermatozoa. *Mol Reprod Develop* 56, 207-219.

Bedford JM (1983). Significance of the need for sperm capacitation before fertilization ir eutherian mammals. *Biol Reprod* 28, 108-120.

Biegeleisen K (2006). The probable structure of protamine-DNA complex. *J Theor Biol* 241 533-540.

Bonet S, Briz M, Pinart E, Sancho S, García-Gil N, Badia E (2000). *Morphology of boar spermatozoa.* Insitut d'Estudis Catalans, ISBN 84-7283-533-2, Barcelona (Spain).

Bradford MM (1976). A rapid and sensitive method for the quantification of microgram quantities of protein following the principle of protein-dye binding. *Anal Biochem* 112, 195-203.

Breitbart H, Naor Z (1999). Protein kinases in mammalian sperm capacitation and the acrosome reaction. *Rev Reprod* 4, 151–159.

Bucci D, Isani G, Spinaci M, Tamanini C, Mari G, Zambelli D, Galeati G (2010). Comparative immunolocalization of GLUTs 1, 2, 3 and 5 in boar, stallion and dog spermatozoa. *Reprod Domest Anim* 45, 315-322.

Burnette WN (1981). "Western blotting": electrophoretic transfer of proteins from sodium dodecyl sulfate-polyacrylamide gels to unmodified nitrocellulose and radiographic detection with antibody and radioiodinated protein A. *J Anal Biochem* 112, 195–203.

Chang MC (1984). The meaning of sperm capacitation: a historical perspective. *J Androl* 5, 45-50.

Clermont Y (1972). Kinetics of spermatogenesis in mammals: seminiferous epithelium cycle and spermatogonial renewal. *Physiol Rev* 52, 198-236.

Cui YH, Zhao RL, Wang Q, Zhang ZY (2000). Determination of sperm acrosin activity for evaluation of male fertility. *Asian J Androl*, 229-23.

Eddy EM (1984). The Spermatozoon. In: *The physiology of reproduction*, Knobil E, Neill JD, eds, 27-68. Raven Press, New York (USA), ISBN 0-88167-281-5.

Fernández-Novell JM, Ballester J, Medrano A, Otaegui PJ, Rigau T, Guinovart JJ, Rodríguez-Gil JE (2004). The presence of a high-Km hexokinase activity in dog, but not in boar, sperm. *FEBS Lett* 570, 211-216.

Fernández-Novell JM, Ballester J, Altirriba J, Ramió-Lluch L, Barberà A, Gomis R, Guinovart JJ, Rodríguez-Gil JE (2011). Glucose and fructose as functional modulators of overall dog, but not boar sperm function. *Reprod Fertil Develop* 23, 468-480.

Fillat C, Rodríguez-Gil JE, Guinovart JJ (1992). Molybdate and tungstate act like vanadate on glucose metabolism in isolated hepatocytes. *Biochem J* 282, 659-663.

Flores E, Cifuentes D, Fernández-Novell JM, Medrano A, Bonet S, Briz MD, Pinart E, Peña A, Rigau T, Rodríguez-Gil JE (2008). Freeze-thawing induces alterations in the protamine-1/DNA overall structure in boar sperm. *Theriogenology* 69, 1083-1094.

Flores E, Fernández-Novell JM, Peña A, Rigau T, Rodríguez-Gil JE (2010). Cryopreservation-induced alterations in boar spermatozoa mitocondrial function are related to changes in the expresión and location of midpiece mitofusin-2 and actin network. *Theriogenology* 74, 354-363.

Flores E, Ramió-Lluch L, Bucci D, Fernández-Novell JM, Peña A, Rodríguez-Gil JE (2011). Freezing-thawing induces alterations in histone H1-DNA binding and the breaking of protamine-DNA disulfide bonds in boar sperm. *Theriogenology* In Press, doi 10.1016/j.theriogenology.2011.05.039.

Folgero T, Bertheussen K, Lindal S, Torbergsen T, Oian P (1993). Mitochondrial disease and reduced sperm motility. *Human Reprod* 8, 1863–1868.

Ford WC, Harrison A (1985). The presence of glucose increases the lethal effect of alpha-chlorohydrin on ram and boar spermatozoa in vitro. *J Reprod Fertil* 73, 197–206.

Giampietri C, Petrungaro S, Coluccia P, D'Alessio A, Starace D, Riccioli A, Padula F, Palombi F, Ziparo E, Filippini A, De Cesaris P (2005). Germ cell apoptosis control during spermatogenesis. *Contraception* 72, 298-302.

Gómez-Foix AM, Rodríguez-Gil JE, Guinovart JJ, Bosch F (1991). Prostaglandins E_2 and $F_{2\alpha}$ increase fructose 2,6-bisphosphate levels in isolated hepatocyes. *Biochem J* 274, 309-312.

Halangk W, Bohneback R, Kunz W (1985). Interdependence of mitochondrial ATP production and ATP utilization in intact spermatozoa. *Biochim Biophys Acta* 808, 316–322.

Hammersted RH, Lardy HA (1983). The effects of substrate cycling on the ATP yield of sperm glycolysis. *J Biol Chem* 258, 8759–8768.

Hecht NB (1995). The making of a spermatozoon: a molecular perspective. *Dev Genetics* 16, 95-103.

Isen JV, Blagoev B, Gnad F, Macek B, Kumar C, Mortensen P, Mann M (2006). Global, in vivo, and site-specific phosphorylation dynamics in signaling networks. *Cell* 127, 635–48.

Jones R (1999). To store or mature spermatozoa? The primary role of the epididymis. *Int J Androl* 22, 57–67.

Klaus A, Hunnicutt G (2006). Ultrastructural Features of Mammalian Sperm: Applications of Cold Field-Emission Scanning Electron Microscopy. *Microsc Microanal* 12, 232-233.

Kurland IJ, Alcivar A, Bassilian S, Lee WNP (2000). Loss of [^{13}C]glycerol carbon via the pentose cycle. Implications for gluconeogenesis measurement by mass isotoper distribution analysis.*J Biol Chem* 275, 36787-36793.

Lambrecht M, Transtschold D (1984). ATP determination with hexokinase and glucose 6-phosphate dehydrogenase. In: *Methods of enzymatic analysis.* Bergmeyer HU, ed., 543-551. Verlag Chemie, Wheinheim (Germany), ISBN 3-527-26046-3.

Langlais J, Roberts KD (1985). A molecular membrane model of sperm capacitation and the acrosome reaction of mammalian spermatozoa. Gamete Res 12, 183–224.

Lasserre A, Barrozo R, Tezón, JG, Miranda PV, Vazquez-Levin MH (2001). Human epididymal proteins and sperm function during fertilization: un update. *Biol Res* 34, 165-178.

Lee WP, Edmond J, Bassilian S, Morrow J (1996). Mass isotopomer study of glutamine oxidation and synthesis in primary culture of astrocytes. *Dev Neurosci* 18, 469-477.

Lefièvre L, Jha KN, De Lamirande E, Visconti PE, Gagnon C (2002). Activation of protein kinase A during human sperm capacitation and acrosome reaction. *J Androl* 23, 709-16.

Lindholmer CH (1974). The importance of seminal plasma for human sperm motility. *Biol Reprod* 10, 533-542.

Maekawa M, Toyama Y, Yasuda M, Yagi T, Yuasa S (2002). Fyn tyrosine kinase in sertoli cells is involved in mouse spermatogenesis. *Biol Reprod* 66, 211–221.

Mann T (1975). Biochemistry of semen. In: *Handbook of Physiology*. Greep RO, Astwood EB, eds., 321-347. American Physiology Society, Washington DC (USA), ISBN 0-7216-3182-7.

Marin S, Chiang K, Bassilian S, Lee WNP, Boros LG, Fernández-Novell JM, Centelles JJ, Medrano A, Rodriguez-Gil JE, Cascante M (2003). Metabolic strategy of boar spermatozoa revealed by a metabolomic characterization. *FEBS Lett* 554, 342-346.

Medrano A, García-Gil N, Ramió L, Rivera MM, Fernández-Novell JM, Ramírez A, Peña A, Briz MD, Pinart E, Concha II, Bonet S, Rigau T, Rodríguez-Gil JE (2006a). Hexose-specificity of hexokinase and ADP-dependence of pyruvate kinase play important roles in the control of monosaccharide utilization in freshly diluted boar spermatozoa. *Mol Reprod Develop* 73, 1179-1194.

Medrano A, Fernández-Novell JM, Ramió L, Alvarez J, Goldberg E, Rivera M, Guinovart JJ, Rigau T, Rodríguez-Gil JE (2006b). Utilization of citrate and lactate through a lactate dehydrogenase and ATP-regulated pathway in boar spermatozoa. *Mol Reprod Develop* 73, 369-378.

Michal G. (1984) Glucose 6-phosphate. In: *Methods of Enzymatic Analysis*, Bergmeyer HU, ed., 191-197. Verlag Chemie, Weinheim (Germany), ISBN 3-527-26046-3.

Mukai C, Okuno M (2004). Glycolysis plays a major role for adenosine triphosphate supplementation in mouse sperm flagellar movement. *Biol Reprod* 71, 540–547.

Nevo AC, Polge C, Frederick G (1970). Aerobic and anaerobic metabolism of boar spermatozoa in relation to their motility. *J Reprod Fertil* 22, 109–118.

O'Brien J, Zini A (2006). Sperm DNA integrity and male infertility. *Urology* 65, 16-22.

Palomo MJ, Fernández-Novell JM, Peña A, Guinovart JJ, Rigau T, Rodríguez-Gil JE (2003). Specific location of hexose-induced dog sperm glycogen synthesis. *Mol Reprod Develop* 64, 349-359.

Passonneau JV, Lowry OH (1993). *Enzymatic analysis: a practical guide*. Totowa NJ, ed., 85-110. Humana Press, Doorwerth (The Netherlands), ISBN 0-89603-238- 8.

Ramió-Lluch L, Fernández-Novell JM, Peña A, Colás C, Cebrián-Pérez JA, Muiño-Blanco T, Ramírez A, Concha II, Rigau T, Rodríguez-Gil JE (2011). "In vitro" capacitation and acrosome reaction are concomitant with specific changes in mitochondrial activity in boar sperm: evidence for a nucleated mitochondrial activation and for the existence of a capacitation-sensitive subpopulational structure. *Reprod Domest Anim* 46, 664-673.

Ramírez AR, Castro MA, Angulo C, Ramió L, Rivera MM, Torres M, Rigau T, Rodríguez-Gil JE, Concha II (2009). The presence and function of dopamine type-2 receptors in boar sperm: a possible role for dopamine in viability, capacitation and modulation of sperm motility. *Biol Reprod* 280, 753-761.

Rigau T, Rivera M, Palomo MJ, Mogas T, Ballester J, Peña A, Otaegui PJ, Guinovart JJ, Rodríguez-Gil JE (2002). Differential effects of glucose and fructose on energy metabolism in dog spermatozoa. *Reproduction* 123, 579-591.

Rodríguez-Gil JE (2006). Mammalian sperm energy resources management and survival during conservation in refrigeration. *Reprod Dom Anim* 41 (Suppl. 2), 11–20.

Robaire B, Hinton BT (2002). *The epididymis: from molecules to clinical practice. A comprehensive survey of the efferent ducts, the epididymis, and the vas deferens.* Kluwer Academic/Plenum Publishers, ISBN 0-306-46684-8, Norwell, Dordrecht (Great Britain/The Netherlands).

Roche Applied Science (2004). *The complete guide for protease inhibition.* Roche Biomedical Guides. Roche Diagnostics, retrieved from www.roche-applied-science.com/proteaseinhibitor.

Rodríguez-Gil JE, Gómez-Foix AM, Fillat C, Bosch F, Guinovart JJ (1991). Activation by vanadate of glycolysis in hepatocytes from diabetic rats. *Diabetes* 40, 1355-1359.

Ruíz-Pesini E, Díez C, Lapena AC, Pérez-Martos A, Montoya J, Alvarez E, Arenas J, López-Pérez M (1998). Correlation of sperm motility with mitochondrial enzymatic activities. *Clin Chem* 44, 1616–1620.

Sancho S, Casas I, Ekwall H, Saravia F, Rodriguez-Martinez H, Rodriguez-Gil JE, Flores E, Pinart E, Briz M, Garcia-Gil N, Bassols J, Pruneda A, Bussalleu E, Yeste M., Bonet S (2007). Effects of cryopreservation on semen quality and the expression of sperm membrane hexose transporters in the spermatozoa of Iberian pigs. *Reproduction* 134, 111-121.

Sharpe RM, Maddocks S, Kerr JB (1990). Cell-cell interactions in the control of spermatogenesis as studied using Leydig cell destruction and testosterone replacement. *Am J Anat* 188, 3-20.

Sharpe RM, Maddocks S, Millar M, Kerr JB, Saunders PT, McKinell C (1992). Testosterone and spermatogenesis. Identification of stage-specific, androgen-regulated proteins secreted by adult rat seminiferous tubules. *J Androl* 13, 172-84.

Sharpe RM (1994). Regulation of spermatogenesis. In: *The physiology of reproduction.* Knobil E, Neil JD, eds., 1363-1434. Raven Press, ISBN 0-88167-281-5, New York (USA).

Stryer L (1995). Citric acid cycle. In: *Biochemistry.* Stryer L, ed., 509–525. Freeman Co., ISBN 0-7167-3051-0, New York (USA).

Sutovsky P, Moreno R, Ramalho-Santos J, Dominko T, Thompson WE, G Schatten. A putative, ubiquitin-dependent mechanism for the recognition and elimination of defective spermatozoa in the mammalian epididymis. *J Cell Sci* 114: 1665-1675, 2001.

Tesarik J, Sousa M, Greco E, Mendoza C (1998). Spermatids as gametes: indications and limitations *Human Reprod* 13, 89-107.

Tserng KY, Gilfilan CA, Kalhan SC (1984). Determination of carbon-13 labeled lactate in blood by gas chromatography/ mass spectrometry. *Anal Chem* 56, 517-523.

Urner F, Sakkas D. (2003). Protein phosphorylation in mammal spermatozoa. *Reproduction* 125, 17–26.

Vijayaraghavan S, Stephens DT, Trautman K, Smith GD, Khatra B, Da Cruz e Silva EF, Greengard P (1996). Sperm motility development in the epididymis is associated with decreased glycogen synthase kinase-3 and protein phosphatase 1 activity. *Biol Reprod* 54, 709-18.

Vijayaraghavan S, Trautmann K D, Goueli S A, Carr DW (1997). A tyrosine phosphorylated 55-kilodalton motility-associated bovine sperm protein is regulated by cyclic adenosine 30,50 monophosphate and calcium. *Biol Reprod* 56, 1450–1457.

Visconti PE, Galantino-Homer H, Moore GD, Bailey JL, Ning TX, Fornes M, Kopf GS (1998). The molecular basis of sperm capacitation. *J Androl* 19, 242–8.

Ward WS, Coffey DS (1991). DNA packaging and organization in mammalian spermatozoa: comparison with somatic cells". *Biol Reprod* 44, 569–74.

Wistuba J, Stukenborg JB, Luetjens CM (2007). Mammalian spermatogenesis. *Funct Develop Embriol* 1, 99-117.

Wolgemuth DJ, Rhee K, Wu S, Ravnik SE (1995). Genetic control of mitosis, meiosis and cellular differentiation during mammalian spermatogenesis. *Reprod Fertil Develop* 7 669 – 683.

Wykes SM, Krawetz A (2003). The structural organization of sperm chromatin. *J Biol Chem* 278, 29471-29474.

Yanagimachi R (1994a). Mammalian fertilization. In: *The Physiology of Reproduction*. Knobil E. Neill JD, eds., 135-185. Raven Press, ISBN   0-88167-281-5, New York (USA).

Yanagimachi R (1994b). Hyperactivation of Spermatozoa. In: *The Physiology of Reproduction*. Knobil E, Neill JD, eds., 152-154. Raven Press, ISBN 0-88167-281-5, New York (USA).

Yauk CL, Bois PRJ, Jeffreys AJ (2003). High-resolution sperm typing of meiotic recombination in the mouse MHC Eβ gene. *EMBO J* 22, 1389 – 1397.

9

Application of Immunocytochemistry to Sputum Cells to Investigate Molecular Mechanisms of Airway Inflammation

Kittipong Maneechotesuwan and Adisak Wongkajornsilp
Faculty of Medicine Siriraj Hospital,
Mahidol University
Thailand

1. Introduction

The definition of *sputum* is not equivalent to *mucus*, but sometime clinicians use both interchangeably. The definition includes the pathological secretion dispelled by cilia and expectorated with coughing. It maintains airway hydration and traps particulates, bacteria, and viruses.

The contents of sputum comprise damaged ciliated epithelium and inflammatory cells. Airway mucus possesses antioxidant, antiprotease, and antimicrobial activities. Its volume is increased during chronic airway inflammation. The sputum composition can be altered by the underlying disease and its severity. The expansion of sputum neutrophil is variably in severe asthma or in chronic obstructive pulmonary disease (COPD).

Cough and ciliary clearance greatly depend on the viscosity of the secretion to the ciliary surface (Voynow & Rubin 2009). Surface tension and surfactant interactions can overpower surface forces. Tenacity or adhesivity is the greatest determinant for the efficiency of cough to eliminate secretion (Voynow & Rubin 2009).

Sputum analysis is a non-invasive approach to dissect underlying pathophysiology of inflammatory airway diseases. The cellular and biochemical constituent of sputum correlates well with both bronchial wash (BW) and bronchoalveolar lavage (BAL), but to a lesser degree with bronchial biopsies (Fahy, et al. 1995, Pizzichini, et al. 1998, Maestrelli, et al. 1995). These observations imply the differences in the luminal and mucosal phase of airway inflammation. The fraction of neutrophils decreases from central (20-30%) to peripheral airways (<2%) while the opposite is true for macrophages (Rankin, et al. 1992). Based on this observation, sputum represents the more proximal airway, whereas BW and BAL represent the more peripheral airways. Successive sputum collection after a single sputum induction exhibited a stepwise neutrophil decrement and macrophage increment (Holz, et al. 1998a, Richter, et al. 1999) reflecting the sampling of more distal airway from later collection. The standardization of sputum induction is required to reduce inter- and intra-subject variability.

2. Sputum induction

The sputum induction technique allows the noninvasive collection of the airway content and provides an opportunity to identify biomarkers of airway inflammation in several conditions (e.g., asthma and COPD). It is superior to spontaneous sputum expectoration in the higher quantity of collected secretion from the lower airways. Sputum induction requires a high degree of cooperation from the patient. The procedure should be performed in a quiet environment and conducted by an experienced technician under the supervision of an experienced physician.

Ultrasonic neubulizers are recommended for sputum induction since other nebulizers dc not provide the sufficient output of saline aerosol. The output of nebulizers should be accurately tested. A spirometry provides real-time assessment of the baseline airway caliber and promptly alerts for the excessive bronchoconstriction during the saline inductior (Paggiaro, et al. 2002). A spirometer is superior to a peak flow meter since it provides greater sensitivity to measure the decline in force expiratory volume in one second (FEV_1 that alludes to saline-induced bronchoconstriction. Oxygen saturation should be monitorec if there is any suspicion of resting hypoxemia. Oxygen supplement should be in reach for COPD patients with hypoxemia exacerbation (Paggiaro et al. 2002).

Sterile saline solution should be freshly prepared. Rescuing medications, bronchodilator (inhaled or nebulized salbutamol or other β_2-agonists) (Paggiaro et al. 2002) and other resuscitation medicines must be nearby. Sputum induction should be performed under aseptic environment. Hypertonic saline can induce bronchoconstriction in asthmatics (Smith & Anderson 1989) with unknown mechanism, possibly through the activation of mast cells (Gravelyn, et al. 1988) or sensory nerve endings (Makker & Holgate 1993).

The pretreatment with a short-acting β_2-agonists is recommended as the standard protocol to prevent excessive bronchoconstriction (Pin, et al. 1992, Wong & Fahy 1997, Jatakanon, et al. 1998) that could pose asthmatics at risk of an exacerbation. Excessive bronchoconstriction might bring about the premature termination of the induction resulting in an inadequate sputum collection. Salbutamol (200-400 µg or 2-4 puffs from a metered dose inhaler) can be used for pretreatment. Pretreatment with higher doses of salbutamol cannot provide additional advantage for the prevention of hypertonic saline-induced bronchoconstriction (Wong & Fahy 1997, Cianchetti, et al. 1999, de la Fuente, et al. 1998, Peleman, et al. 1999), but may induce more severe subsequent bronchoconstriction. Therefore, a single dose of salbutamol 200 µg is recommended both before and after the measurement of FEV_1 for 10 min. Salbutamol pretreatment does not interfere the inflammatory cell percentage in induced sputum (Cianchetti et al. 1999, Popov, et al. 1995). Regarding the effects on soluble mediators in sputum supernatant, salbutamol has no effect on eosinophil cationic protein (ECP) levels, but tends to reduce histamine concentrations (Cianchetti et al. 1999). There has been no study for the effect of salbutamol on the levels of other soluble mediators (i.e., cytokines, albumin, and neutrophil elastase) or the expression of cell activation markers as detected by immunocytochemistry. Also, the data on the comparison between different bronchodilators (β_2-agonists and anticholinergic drugs) are not available.

Monitoring pulmonary function during sputum induction is indispensable for safety precaution. However, no standardized monitoring protocol for pulmonary function has been recommended. Most studies measure FEV_1 every 5-10 min, with additional measurement if any symptom develops (Iredale, et al. 1994, Bacci, et al. 1998, Pin et al. 1992, Jatakanon et al. 1998, Wong & Fahy 1997, Maestrelli, et al. 1994). Since poor perception of dyspnea can exist while bronchospasm can occur early, the measurement of pulmonary function within the first minute of nebulization should be performed to identify supersensitive subjects. FEV_1 should be periodically monitored with an interval of ≤ 5 min during aerosol inhalation. A single measurement is appropriate if the change in FEV_1 is < 10% of the postbronchodilator FEV_1 value.

2.1 Selecting saline concentrations

The concentrations of saline solution for sputum induction varied from 0.9% to 7% in different studies. Some investigators raised the concentration in a stepwise manner (3%, 4% and 5%) during sputum induction. Saline concentration and nebulizer output might influence the safety, tolerability and success rate of the induction as well as the cellular and biochemical constituents. The 3% hypertonic saline achieved the same success rate as 3-5% given sequentially. Hypertonic saline solutions are more effective than isotonic saline for sputum induction. The latter should be reserved for patients at high risk of bronchoconstriction. There is a consensus to advocate sputum induction with 4.5% sodium chloride solution that is commercially available.

Hypertonic or isotonic saline did not elicit any significant alteration in cellular ratio or their quantity in the sputum. However, the concentration-effect relationship between different saline concentrations and the levels of most soluble mediators in the sputum supernatants remain unknown. There is no difference in the levels of ECP and histamine in the sputum supernatant induced by isotonic or hypertonic saline (Bacci, et al. 1999). Sputum supernatant osmolarity fluctuates between 70-360 mOsm as a result of the natively high variation in sputum concentrations of sodium, chloride and magnesium from individual subjects.

2.2 Selecting nebulizer

The selection of nebulizer based on type and output is important for the attainment of sputum induction. The ultrasonic nebulizer offers higher success rates than does jet nebulizer (Popov et al. 1995). The exact volume of inhaled saline solution that might be required to induce an adequate sputum sample is still unclear. The optimal duration of inhalation and the optimal output are also unclear. Longer duration might yield samples from more distal airways. Other factors include the size of aerosols and their deposited locations. The deposition at different locations might yield different sputum composition and success rates. The common practice employing ultrasonic nebulizer with an output of ~ 1 mL/min can achieve a satisfactory quantity.

2.3 Duration of nebulization

Duration of nebulization can influence the cellular components of the resulting sputum. At the early phase (0-4 min), neutrophils, eosinophils and mucin are major components in the sputum. At the later phase (16-20 min), lymphocytes, macrophages and surfactant are increasingly noticeable (Holz et al. 1998a, Gershman, et al. 1999). This pattern suggests that central airways are sampled at the early phase, whereas peripheral airways and alveoli are sampled at the later phase. The early-phase sputum may be discarded to avoid the saliva contamination. Subsequent samples may be more suitable for the analysis. Although the maximal duration of induction has not been properly studied, it depends on the conciliation between the success rate and the tolerability/safety. Shorter inhalation times (e.g. 15-20 min) have similar success rates and practicability to longer inhalation times (30 min). It is critical to keep the duration of inhalation constant between inductions in the same subject to obtain comparable results. Common practice employs a cumulative duration of nebulization for 15-20 min. Other influencing factors might include the respiratory frequency during nebulization and the patterns of inhalation (i.e., slow deep inhalation or tidal breathing) during the challenge.

2.4 Variations of expectoration

Some investigators recommend that subjects clean their oral cavity with gargle, dry with napkin, spit for saliva, and finally cough for sputum. Others argue that mouth rinsing and drying may increase oropharyngeal inflammation. Some authors encourage the use of nose clips. Sputum induction protocols are different with respect to the schedule of sputum collection. Subjects may be asked to stop inhalation at regular intervals to cough up sputum (e.g., every 5 min), or to stop only when they feel the urge to cough. Some protocols require subjects to spit saliva into one container before coughing sputum into the other. Spitting saliva before coughing sputum decreases the percentage of squamous cells in sputum by 30% and increases ECP in the supernatant by 80% (Gershman, et al. 1996). The production of a good sputum sample relies heavily on the characteristics of the individual subject rather than the technical factors.

2.5 Repeating sputum induction

Repeated sputum induction can heighten airway inflammation resulting in an iatrogenic change in cellular components. Repeating sputum induction at 8-24 h after the initial induction can increase neutrophil recruitment in the second sputum sample (Nightingale, et al. 1998, Holz, et al. 1998b). An interval of 48 h between two inductions gave comparable cell counts in normal subjects (Purokivi, et al. 2000). It is currently recommended that subsequent induction should be conducted at least 2 days apart.

2.6 Sputum induction protocols

2.6.1 Classical procedure

1. The detailed information should be provided to the patient prior to the procedure.
2. All equipments including the ultrasonic nebulizer should be checked for safety and calibrated for the output of ~ 1 mL/min.
3. The baseline FEV_1 is measured prior to the bronchodilator inhalation.
4. The bronchodilator (200 µg salbutamol) is inhaled before the commencement of sputum induction.
5. The FEV_1 is measured after the bronchodilator inhalation for 10 min.
6. Either a fixed concentration (3 or 4.5%) or increasing gradient (3, 4 and 5%) of sterile saline is nebulized as an inducer. Each nebulization lasts 5 min follows by a brief expectoration. The total duration will last ≤ 20 min. Alternatively, the continuous 20-

min nebulization may be interrupted at 1, 4, 5, 10, 15, and 20 min for expectoration. The subjects will be allowed to spit outside the schedule whenever they develop the urge to cough.

7. The FEV_1 at each interruption will be measured. The induction will be terminated if there is $\geq$ 20% fall in FEV_1 from the postbronchodilator value or the symptoms develop.

2.6.2 Customized procedure for high-risk subjects

1. The detailed information should be provided to the patient prior to the procedure.
2. All equipments including the ultrasonic nebulizer should be checked for safety and calibrated for the output of ~ 1 mL/min.
3. The baseline FEV_1 is measured prior to the bronchodilator inhalation.
4. The bronchodilator (200 µg salbutamol) is inhaled before the commencement of sputum induction.
5. The FEV_1 is measured after the bronchodilator inhalation for 10 min.
6. The 0.9% sterile saline solution will be nebulized with interruptions at 30 sec, 1, and 5 min for FEV_1 measurement as a safety precaution. If this fails to induce sputum, the 3% saline concentration will be nebulized with interruptions at 30 sec, 1, and 2 min for FEV_1 measurement. If this also fails to induce sputum, the 4.5% saline concentration will be nebulized with interruptions at 30 sec, 1, 2, 4 and 8 min for FEV_1 measurement. The induction will be terminated if there is $\geq$ 20% fall in FEV_1 from the postbronchodilator value or the symptoms develop.
7. If the subjects cannot produce spontaneous cough, they will be driven to cough and spit after 4 and 8 min.

2.7 General consideration

1. The protocol should be strictly enforced, especially the inhalation timing.
2. The induction should not be repeated within 48 h after the first induction.
3. The safety procedure must be ensured and readily accessible.

2.8 Validation of sputum

The outcome of downstream application relies on the quality of sputum. The procurement of secretions from the lower respiratory tract (sputum) induced by the inhalation of aerosol from hypertonic saline usually is contaminated with saliva. The assessment of sputum quality relies on the quantitation of contaminated squamous epithelial cells.

The collection method has been optimized to improve the induced sputum quality by the selection of viscous portion of the specimen that should minimize the contamination with squamous epithelial cells (Pizzichini, et al. 1996b). This selective collection could obtain at least two-thirds of the viable nonsquamous cells. After DTT treatment, the cellular contents of the selectively collected sample would be dispersed and were ready for cytospin examination. This selected portion usually held a similar proportion of neutrophils, eosinophils and lymphocytes. However, it contained much higher concentrations of the fluid-phase ECP than did the whole expectoration. The unaltered cellular proportion suggested that the selection for viscous portion did not alter the indices of airway inflammation. Normally, saliva contains mainly squamous epithelial cells (99%) but very low levels of ECP (2–90 ng/mL). The selected viscous portion was generally contaminated with very little squamous cells (1.2%), whereas the remaining clear portion contained a large portion of squamous cells (70%) (Pizzichini et al. 1996b). The lessening squamous cell content unmasked the inflammatory cells, resulting in better quality after the cytospins. When the squamous contamination was less than 20%, the accuracy of differential cell counts was better, as indicated by high inter and intraobserver reproducibility. The processing of cytospin is also quicker, especially when 1,000 or so cells were counted for more accurate identification of metachromatic cells and lymphocytes, the minority cellular contents. Also, the small proportion of squamous cells provides a homogeneous population of cells for examination by flow cytometry (Kidney, et al. 1996).

Alternatively, the quality of induced sputum specimen can be evaluated by the presence of high proportion of viable nonsquamous cells. Cell viability is an important requirement for accurate cell identification. With greater than 50% viability, the reproducibility of cell counts is better. Higher viability also provided an advantage for immunological staining to determine subpopulations and activation markers (Hansel, et al. 1991, Kidney et al. 1996 Vatrella, et al. 2010).

In summary, the physical selection of viscous portion from the mixed expectorate has several advantages over the whole specimen. It is almost free of squamous cells and is therefore essentially undiluted. Cells are in better condition and the concentrations of eosinophilic cationic protein are higher. If the expectorate is not processed within 2 h, a large error might be introduced. The serous portion, which is rich in squamous cells, could be homogeneously mixed with the otherwise desirable viscous portion. The reproducibility of cell counts is threatened if squamous cell contamination represents > 20% of all recovered cells.

3. The conventional practice for cytologic study

The sputum should be processed as soon as possible or within 2 h to ensure optimal cell counting and staining (Pizzichini et al. 1996b, Fahy, et al. 1993). The dithiothreitol (DTT) is used to split the disulfide bonds in mucin to release the cells (Cleland 1964). Cells anchoring to the mucus tend to get dark stain that hinders accurate identification. DTT provides more effective cellular liberation from the mucus than does phosphate buffered saline (PBS) and has no effect on cell counts.

The duration and temperature of mixing can vary between 10-30 min and 4-37°C respectively. This range of DTT exposure time at room temperature has no effect on the differential cell count (Popov, et al. 1994). The mixing of sputum with DTT can be performed with either a shaking water bath at 37°C followed by periodic aspirations, or a tube rocker at 22°C (Popov et al. 1994, Fahy et al. 1993, Spanevello, et al. 1998). The use of a plastic transfer pipette for aspiration and expulsion of sputum is not recommended since it decreases cell yield due to incomplete mixing (Popov et al. 1994, Hansel et al. 1991).

A sample filtration is strongly recommended for removing residual mucus and debris to improve slide quality. A single filtration through a 48-μm nylon mesh results in a slight reduction in the total cell count but the differential cell count remains unchanged (Efthimiadis, et al. 1996, Efthimiadis, et al. 2000). However, little is known about the effect of repeated filtrations on differential cell count.

Since the total cell count could be lessened after centrifugation (Parameswaran, et al. 2000, Rerecich, et al. 1999), it is therefore recommended that total cell count be performed prior to centrifugation to exactly obtain the original cell count. The currently automated machine is not reliable for determining total cell count and differential cell count, and is still not recommended.

Centrifugation is generally used to separate sputum cells from the fluid phase. The centrifugation force should be set between 300 - 1500 × g for 5-10 min to obtain adequate separation of the cells and the supernatant (Fahy et al. 1993, Louis, et al. 1999, Pizzichini, et al. 1996a). The storage temperatures for cells and supernatants are -20°C and -70°C respectively (Pizzichini et al. 1996a).

The optimum cell density for cytospins is 40-60 × 10^3 cells / slide that provide a more accurate estimate for cell distribution than does the smearing technique (Pizzichini et al. 1996a, Popov et al. 1994). The centrifugation force of 22 × g for 6 min is generally employed (Pizzichini et al. 1996a, Popov et al. 1994). Although this speed is below the limits of minimal cell distortion, there is a risk of losing lymphocytes at low speeds due to the dispersion to the supernatant (Fleury-Feith, et al. 1987, Mordelet-Dambrine, et al. 1984). This should be taken into account in the investigations of sputum lymphocytes.

The differential cell counts can be accomplished using either Wright's or Giemsa stain (Efthimiadis, et al. 2002). The buffers must be titrated to the optimal pH (7.1-7.2) (Efthimiadis et al. 2002) to allow accurate characterization of cells based on their optimal staining while maintaining original morphology. The differential cell count requires a minimum of 400 nonsquamous cells (Efthimiadis et al. 2002). The report should contain the relative numbers of eosinophils, neutrophils, macrophages, lymphocytes and bronchial epithelial cells expressed as a percentage of total nonsquamous cells. The percentage of squamous cells should always be reported separately.

4. Immunocytochemistry procedures

The sputum suspension is centrifuged at 300 × g for 10 min, resuspended in PBS or Hank's balance saline solution (HBSS). The cytospins are prepared on L-polylysine-coated slides to ensure minimal cell loss during multiple washing steps. The attached cells are air-dried for 10 min and fixed appropriately. After fixation, cytospins should be wrapped in foil and stored at -20°C pending for staining.

The method of fixation is critical and needs to be optimized for any particular antigen to obtain the best quality of immunostaining. The selected fixation method should allow the proper preservation of antigens / cellular morphology and the penetration of antibodies into the entire cells. The fixation regimens may include either 2 or 4% paraformaldehyde, formalin, acetone/methanol (60/40) or periodate-lysine-paraformaldehyde (PLP). The latter is a fixative for surface glycoprotein staining but may also be used for cytokine staining (McLean & Nakane 1974). For instance, PLP fixation provided better morphology of cryostat sections but poorer immunostaining than conventional acetone immersion. However, a brief acetone fixation followed by PLP fixation offered excellent morphology preservation and good quality of immunostaining (Hall, et al. 1987). PLP was proposed for the fixation of multiple membrane antigens in skin biopsies (Pieri, et al. 2002). The benefit of PLP fixation is in the preservation of cellular ultrastructures (i.e., immunoglobulins) for immunofluorescence or immunoperoxidase staining of paraffin-embedded specimens (Rantala, et al. 1985). For marker staining of induced sputum cells, PLP-sucrose provided the best results with the highest percentage of CD3$^+$ cells and a better staining quality than did the paraformaldehyde and acetone-methanol-fixed cells that provided the worst staining of CD68 (St-Laurent, et al. 2006). Simultaneous fixation and permeabilization using Ortho PermeaFix for flow cytometry are required for the best identification of intracellular antigens (i.e., eosinophil cationic protein, eosinophilic peroxidase, neutrophil myeloperoxidase) in cytocentrifuged cells (Metso, et al. 2002). Organic solvents as fixatives are not suitable while the crosslinking fixatives (e.g., paraformaldehyde) alone could not provide a complete penetration of antibodies into the cell interior (Metso et al. 2002).

To detect intracellular antigens, sputum cells must be permeabilized after the fixation with paraformaldehyde and glutaraldehyde. Permeabilization allows the antibody to gain an access to intracellular or intraorganellar antigens. Two common permeabilizing agents are organic solvents (i.e., methanol and acetone) and detergents.

The organic solvents work through dissolving lipids from cell membrane, thereby disrupting the membrane and allowing the influx of the antibodies. The ability of the organic solvents to coagulate proteins provides an additional advantage of cell fixation. The shortcoming of organic solvents is the removal of lipidic antigens or lipid associated antigens from cells.

One of the most commonly used detergents is saponin, a plant glycoside. Saponin permeabilizes cells through the removal of cholesterol, thereby puncturing holes over the membrane (Seeman, et al. 1973). Saponin can form micelle with antibody and cholesterol that facilitates the entry through the punctured holes. However, saponin cannot effectively permeabilize mitochondrial membranes and the nuclear envelope due to their low composition of cholesterol (Goldenthal, et al. 1985a). Therefore, saponin is suitable for immunostaining of intracellular membrane antigens localized over lysosomal membrane, plasma membrane, endocytic vesicles and endoplasmic reticulum (Goldenthal, et al. 1985b). However, Triton X-100 and NP-40 interfere the staining at these sites (Goldenthal et al. 1985b).

Other commonly used detergents are the non-ionic detergents such as Triton X-100 and Tween 20 (Maneechotesuwan, et al. 2010, Maneechotesuwan, et al. 2008). They carry uncharged, hydrophilic head groups of polyoxyethylene moieties. Antigens localized in mitochondria and the nucleus required Triton X-100 for their detection (Goldenthal et al. 1985b). Their shortcoming is their non-selective nature that could produce a false negative during immunostaining through the removal of proteins along with the lipids. A combination of different permeabilizing agents may be customized for each antigen (Goldenthal et al. 1985a).

Immunocytochemical staining can be performed with different varieties, including avidin/biotin complex, peroxidase/antiperoxidase and alkaline phosphatase / antialkaline phosphatase techniques. The use of immunoenzymatic techniques eliminates the need for expensive fluorescent microscopy. The alkaline phosphatase / antialkaline phosphatase method is preferable. The staining with the monoclonal antibodies on fixed slides should be titrated for appropriate concentrations and can be incubated overnight at 4°C. The secondary antibodies are then applied and the antibody/antigen complex is visualized using the alkaline-phosphatase-linked substrate, with either fast red or fast blue counterstains. Negative controls must always be included to exclude the potential false positive staining. The peroxidase staining methods are not recommended for sputum.

5. Modifications of immunocytochemistry for sputum specimen

Homogenization with low-concentration DTT (0.5 mM) would liberate the otherwise anchored cells from the surrounding mucus that facilitate the exposure of the cells to the staining antibodies (Tockman, et al. 1995). However, the use of DTT to disperse cells may hamper cellular functions (e.g., the release of elastase and myeloperoxidase (MPO) from neutrophils (van Overveld, et al. 2005)) or hinder antigenic epitopes for immunocytochemical staining. Some investigators recommended the use of paraformaldehyde to fix the sputum cells prior to the treatment with low concentration of DTT.

The method employing avidin-biotin complex (ABC) is suitable for the immunocytochemistry of sputum cells (Maneechotesuwan et al. 2008, Maneechotesuwan et al. 2010). The biotin / avidin system possesses several advantages to sputum immunocytochemistry. 1) The binding affinity of avidin to biotin is higher than that of any antibody directing against its epitope. 2) The binding of avidin to biotin is almost irreversible. 3) The multiple binding sites on each molecule (four binding sites for biotin on each avidin; two binding sites for avidin on each biotin) provide the formation for macromolecular complexes between avidin and biotinylated enzymes. The ABC-alkaline phosphatase (AP) lattice complex consists of several biotinylated alkaline phosphatase molecules cross-linked by avidin. The two biotin molecules can be joined via an avidin molecule that eventually forms a complex of avidin and biotinylated enzyme or biotinylated secondary antibody. The stoichiometry of the forming complex will contain an available biotin binding site on ABC for the binding of the biotinylated secondary antibody. The formation of the complex is developed by gradual mixing avidin and biotinylated alkaline phosphatase with predefined ratios prior to use. The ABC complex can be stable until 24 h after the formation. This technique provides multiple enzymes attaching to the antigenic site, thereby enhancing the detection sensitivity. The sensitivity provided by this method is generally higher than that obtained with the conventional peroxidase-anti-peroxidase (PAP) technique. However the size of the ABC complex can be inappropriately high that it sterically interferes the overall binding, resulting in decreasing the resolution.

The ABC method can also be modified to incorporate different enzymes that provide different chromogenic properties (Bratthauer 2010). These enzymes include alkaline phosphatase (AP). The high sensitivity of the ABC-AP system permits the detection of the small amount of antigen using higher dilution of a primary antibody. Therefore, the ABCAP system is recommended for staining situation in which high sensitivity is a prerequisite such as sputum immunocytochemistry. The advantages of these techniques lie in the availability of suitable secondary agents. The shortcoming of this system is the presence of endogenous alkaline phosphatase, which is more ubiquitous than endogenous peroxidase and is tougher to remove. However, the alkaline phosphatase can generate more color producing molecules per enzyme molecule than can peroxidase, resulting in higher sensitivity. Endogenous alkaline phosphatase can be partially blocked by incubation with 3 mM levamisole for 15 min with some remaining residual activity.

Labeling of 2 epitopes can employ the simultaneous peroxidase and AP methods. An ABC assay system can be applied to the detection of more than one antibody on an individual specimen. In double labeling, experiments having two completely different assay systems would minimize the cross-over reactivity. The first antigen can be detected with the standard ABC procedure, while the second antigen can be detected using the PAP system. The two techniques provide minimal cross-over reactivity, especially if alkaline phosphatase is employed along with the peroxidase enzyme (Gillitzer, et al. 1990).

Various sugar moieties interfere with the binding of streptavidin or avidin to biotin. The most effective inhibitory sugar is mannose, followed by other saccharides. The inhibitory action probably involves the interactions of the sugars with reactive residues at the binding sites (Houen & Hansen 1997).

6. Practical considerations

1. Commercial PBS or TBS are recommended to maximize the reproducibility. The buffer should be prepared according to the manufacturer's recommendations. The commercial buffer usually makes up 5 L of solution.
2. The primary antibody should be diluted with TBS or PBS to an optimal concentration. These should be empirically titrated on a known positive specimen. Working antibody concentrations usually lie between 10-20 µg/mL. Depending on the individual reagent, this concentration could vary considerably. The initial dilution can start at 1:10 with subsequent serial 1:10 dilutions, resulting in 10, 100, 1000, and 10,000-fold dilutions of the original antibody. Optimal staining can be obtained from this wide range of dilutions. A higher resolution for the optimal staining can be obtained through serial 1:2 dilutions of the formerly obtained wide dilution. The antibodies can be aliquoted and stored in concentrated form at -70 to –80°C indefinitely. The antibodies should be thawed once and used immediately. Refreezing antibodies should be avoided. However, the manufacturer recommends that the ABC reagent components should be stored at 4°C. The antibody can be thawed and diluted to a concentrated stock solution from which more diluted working solutions can be prepared. These stock solutions can be kept at 4-8°C for a week.
3. An antibody in either polyclonal or monoclonal format can be applied for sputum staining with distinct advantages / disadvantages. The polyclonal antibody generally provides strong signal with reasonably good specificity, but can generate some background noise. The signal strength of a monoclonal antibody depends on its qualities and affinity but its specificity for antigen binding is far better than that of the polyclonal format. The preference for a monoclonal antibody over a polyclonal antibody largely depends on the availability of a qualified antibody with matching application. The chosen antibody has to be specific with no cross reaction to other cell components. The antibody should contain a high affinity to the antigen be produced in high titer. The monoclonal antibody generally creates minimal background noise. Its high affinity can withstand multiple processing steps of staining and washing with minimal loss of the attaching antibody.
4. The incubation with 10% xenogeneic serum will mask all non-specific binding sites on the specimen. The 10-min incubation with the 10% xenogeneic serum can prevent the non-specific antibody binding. The proportion of the serum and the incubation time can be optimized to generate acceptable signal.

5. The choice of blocking serum depends on the originating species of the secondary antibody. A universal type of blocking serum can be used if the secondary antibody has multiple host species to avoid the cross-reactivity. The pooled or universal secondary antibody can be used regardless of the originating species of the primary antibody. Proteins from other sources (e.g., milk, or casein-based solutions) can be employed as blocking agents, but they may not produce better results.

6. The specimens should be kept hydrated throughout the staining procedure. The inadvertent drying can generate nonspecific antibody binding. A chamber rack can be used to prevent the flowing of antibodies away with gravity.

7. All reagents and slides should sit at room temperature before the staining. The antibodies should be fully dissolved or reconstituted. The staining results can even be better if the antibody solutions are prepared the day before and left at 4°C overnight. The incubation time and temperature can be raised to optimize the reactivity. However, this might increase the background noise too. The reaction may be slightly improved without cumulative background by the incubation with primary antibody overnight at 4°C (Clements & Beitz 1985).

8. Vigorous rinsing is recommended to lessen the background noise. However, direct splashing to the specimen surface with the wash stream may dislodge antibodies with low affinity. The stream should be rapidly running from one end of the slide, crossing the slide surface to the other end. The washing step is the most critical factor to lessen the background noise and could be extended if the background noise is heavily concerned. The inclusion of non-ionic detergent (i.e., 0.25% Triton X-100 or 0.1% Tween 20) may also lessen the background (Laitinen, et al. 1983, Maneechotesuwan et al. 2008). However, these detergents may interfere the charge interactions of antibody-antigen binding, resulting in decreasing reactivity. These detergents should be applied to poly-L-lysine-coated or charged slides to improve the hydrophobicity of these slides. The reaction can be strengthened through extending incubation times (60 min for primary and 45 min for the secondary or ABC incubation), raising chromogen concentration to prevail over the dampening effect of the detergents. The chromogen concentration may be raised from 2-5 times. If background noise is highly concerned, a preabsorption of the slide with blocking agents may be required. The antibody may be diluted in buffer containing 2% bovine serum albumin, or secondary species serum to avoid the reaction from contaminating non-specific antibodies against a serum-based constituent.

9. The formation of ABC complex takes at least 30 min to be stable. Only after the ABC complex is stably formed can the secondary antibody be added. The complex can be kept in the refrigerator for at least 72 h.

10. The counterstain should not be stronger than the principal reaction; otherwise the staining sites can be masked by darker counterstain.

11. Many substrates for alkaline phosphatase provide permanent straining with different color. A combination of 5-bromo-4-chloro-3-indolyl phosphate (BCIP) and nitro blue tetrazolium (NBT) yield permanent blue precipitates at the site of alkaline phosphatase, while the Fast Red TR/Naphthol AS-MX yields a permanent red precipitates.

12. The staining sensitivity can be increased through the use of multiple chromogens and multiple enzymes targeting numerous antigens. More substrate precipitation or greater color resolution improves the sensitivity.

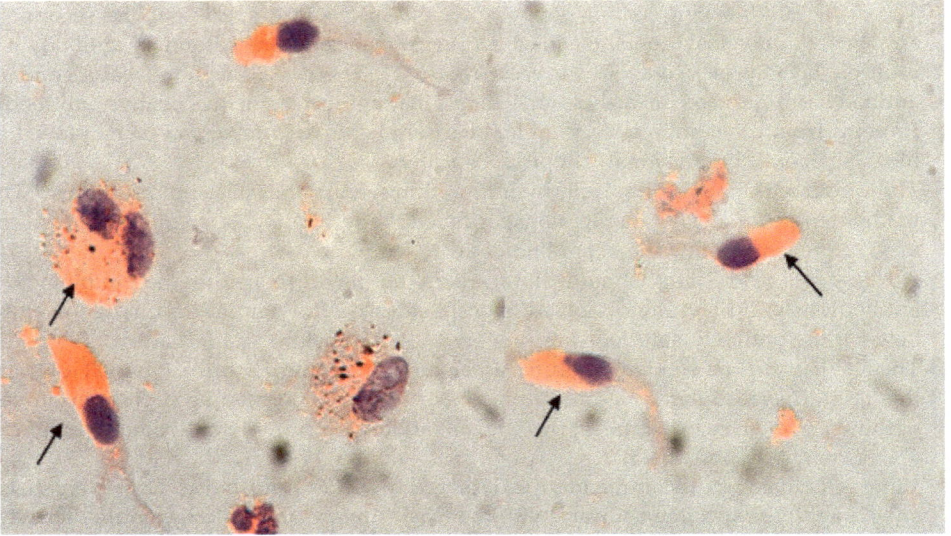

Fig. 1. The sputum cells were immunocytochemically stained for indoleamine 2, 3 dioxygenase with VECTOR® ABC-AP KIT.

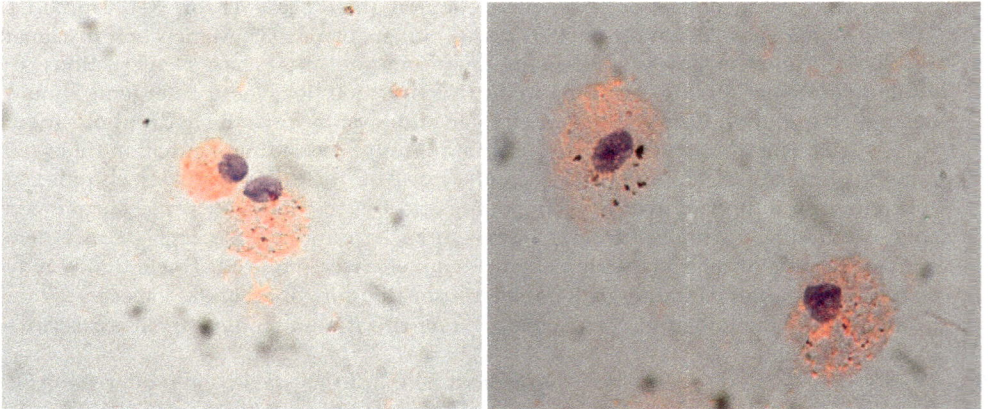

Fig. 2. The sputum cells were immunocytochemically stained for interleukin-10 with VECTOR® ABC-AP KIT.

6.1 Prototypic procedure for immunocytochemical staining of a sputum specimen (VECTASTAIN® ABC-AP KIT)

1. The cytospin slide will be removed from the freezer (-20°C) and left at room temperature for 30-60 min.
2. The cellular spot on the slides will be encircled with the wax pen.
3. The slide will be washed once with PBS for 5 min.
4. The washed slide will be placed horizontally in a humidified chamber.

5. The sputum cells will be fixed with 2% paraformaldehyde (PFA) for 5 min.
6. The slide will be completely washed once with PBS to remove residual PFA for 5 min.
7. The cell membrane will be permeabilized with 0.5% NP-40 diluted in PBS for 10 min.
8. The slide will be washed once with PBS for 5 min.
9. Nonspecific binding will be blocked with the blocking serum (normal serum) from the Vector Kit (3 drops of stock to 10 mL PBS with 0.05% tween-20) for 30 min.
10. The slide will be incubated with the primary antibody pre-diluted in PBS with 0.05% tween-20 (1:100) for 1 h.
11. The slide will be washed thrice with PBS containing 0.05% tween-20 for 5 min.
12. Incubate the slide with the biotinylated secondary antibody pre-diluted in PBS with 0.05% tween-20 (add 1 drop of stock to 10 mL PBS with 0.05% tween-20) for 45 min.
13. The VECTASTAIN® ABC-AP Reagent will be prepared by adding exactly 2 drops of Reagent A to 10 mL PBS with 0.05% tween-20, followed by adding 2 drops of Reagent B. The mixture will be immediately mixed and left at room temperature for at least 30 min before the next step.
14. The slide will be washed thrice with PBS containing 0.05% tween-20 for 5 min each.
15. The slide will be incubated with VECTASTAIN® ABC-AP Reagent for 30 min.
15. The slide will be washed once with PBS without tween-20 for 5 min.
17. The Vector® Red substrate working solution will be prepared immediately before use in test tube by adding 2 drops of Reagent 1 to 5 mL 100 mM Tris-HCl (pH 8.2-8.5) followed by a thoroughly mixing. The mixture will be subsequently reconstituted with 2 drops of Reagent 2 and 2 drops of Reagent 3 sequentially with a thoroughly mixing after each reagent addition.

7. References

Bacci, E.; Bartoli, M.; Carnevali, S. & Al., E. (1999). Eosinophil cationic protein (ECP) and histamine levels in induced sputum are not affected by hypertonic saline inhalation. *Eur Respir J*, vol. 14:Suppl. 30, p. 24.

Bacci, E.; Cianchetti, S.; Ruocco, L.; Bartoli, M.L.; Carnevali, S.; Dente, F.L.; Di Franco, A.; Giannini, D.; Macchioni, P.; Vagaggini, B.; Morelli, M.C. & Paggiaro, P.L. (1998). Comparison between eosinophilic markers in induced sputum and blood in asthmatic patients. *Clin Exp Allergy*, vol. 28, no. 10, pp. 1237-43, ISSN 0954-7894

Bratthauer, G.L. (2010). The avidin-biotin complex (ABC) method and other avidin-biotin binding methods. *Methods in molecular biology*, vol. 588, pp. 257-70, ISSN 1940-6029

Cianchetti, S.; Bacci, E.; Ruocco, L.; Bartoli, M.L.; Carnevali, S.; Dente, F.L.; Di Franco, A.;

Giannini, D.; Scuotri, L.; Vagaggini, B. & Paggiaro, P.L. (1999). Salbutamol pretreatment does not change eosinophil percentage and eosinophilic cationic protein concentration in hypertonic saline-induced sputum in asthmatic subjects. *Clin Exp Allergy*, vol. 29, no. 5, pp. 712-8, ISSN 0954-7894

Cleland,W.W. (1964). Dithiothreitol, a New Protective Reagent for Sh Groups. *Biochemistry*, vol. 3, pp. 480-2, ISSN 0006-2960

Clements, J.R. & Beitz, A.J. (1985). The effects of different pretreatment conditions and fixation regimes on serotonin immunoreactivity: a quantitative light microscopic study. *The journal of histochemistry and cytochemistry : official journal of the Histochemistry Society*, vol. 33, no. 8, pp. 778-84, ISSN 0022-1554

De La Fuente, P.T.; Romagnoli, M.; Godard, P.; Bousquet, J. & Chanez, P. (1998). Safety of inducing sputum in patients with asthma of varying severity. *Am J Respir Crit Care Med*, vol. 157, no. 4 Pt 1, pp. 1127-30, ISSN 1073-449X

Efthimiadis, A.; Pizzichini, M. & Pizzichini, E.E., Al. (1995). The influence of cell viability and squamous cell contamination on the reliability of sputum differential cell counts. *Am J Respir Crit Care Med*, vol. 151,

Efthimiadis, A.; Popov, T.; Kolendowicz, R.; Dolovich, J. & Hargreave, F.E. (1996). Increasing the yield of sputum cells for examination. *Am J Respir Crit Care Med*, vol. 149, p. A949

Efthimiadis, A.; Spanevello, A.; Hamid, Q.; Kelly, M.M.; Linden, M.; Louis, R.; Pizzichini, M.M.; Pizzichini, E.; Ronchi, C.; Van Overvel, F. & Djukanovic, R. (2002). Methods of sputum processing for cell counts, immunocytochemistry and in situ hybridisation. *Eur Respir J Suppl*, vol. 37, pp. 19s-23s, ISSN 0904-1850

Efthimiadis, A.; Weston, S.; Carruthers, S.; Hussack, P. & Hargreave, F. (2000). Induce sputum: effect of filtration on the total and differential cell counts. *Am J Respir Crit Care Med*, vol. 161, p. A853,

Fahy, J.V.; Liu, J.; Wong, H. & Boushey, H.A. (1993). Cellular and biochemical analysis of induced sputum from asthmatic and from healthy subjects. *Am Rev Respir Dis*, vol 147, no. 5, pp. 1126-31, ISSN 0003-0805

Fahy, J.V.; Wong, H.; Liu, J. & Boushey, H.A. (1995). Comparison of samples collected by sputum induction and bronchoscopy from asthmatic and healthy subjects. *Am Respir Crit Care Med*, vol. 152, no. 1, pp. 53-8, ISSN 1073-449X

Fleury-Feith, J.; Escudier, E.; Pocholle, M.J.; Carre, C. & Bernaudin, J.F. (1987). The effects of cytocentrifugation on differential cell counts in samples obtained by bronchoalveolar lavage. *Acta Cytol*, vol. 31, no. 5, pp. 606-10, ISSN 0001-5547

Gershman, N.H.; Liu, H.; Wong, H.H.; Liu, J.T. & Fahy, J.V. (1999). Fractional analysis of sequential induced sputum samples during sputum induction: evidence that different lung compartments are sampled at different time points. *J Allergy Clin Immunol*, vol. 104, no. 2 Pt 1, pp. 322-8, ISSN 0091-6749

Gershman, N.H.; Wong, H.H.; Liu, J.T.; Mahlmeister, M.J. & Fahy, J.V. (1996). Comparison of two methods of collecting induced sputum in asthmatic subjects. *Eur Respir J*, vol. 9, no. 12, pp. 2448-53, ISSN 0903-1936

Gillitzer, R.; Berger, R. & Moll, H. (1990). A reliable method for simultaneous demonstration of two antigens using a novel combination of immunogold-silver staining and immunoenzymatic labeling. *The journal of histochemistry and cytochemistry : official journal of the Histochemistry Society*, vol. 38, no. 3, pp. 307-13, ISSN 0022-1554

Goldenthal, K.L.; Hedman, K.; Chen, J.W.; August, J.T. & Willingham, M.C. (1985a). Postfixation detergent treatment for immunofluorescence suppresses localization of some integral membrane proteins. *The journal of histochemistry and cytochemistry : official journal of the Histochemistry Society*, vol. 33, no. 8, pp. 813-20, ISSN 0022-1554

Goldenthal, K.L.; Hedman, K.; Chen, J.W.; August, J.T. & Willingham, M.C. (1985b). Postfixation detergent treatment for immunofluorescence suppresses localization of some integral membrane proteins. *J Histochem Cytochem*, vol. 33, no. 8, pp. 813-20, ISSN 0022-1554

Gravelyn, T.R.; Pan, P.M. & Eschenbacher, W.L. (1988). Mediator release in an isolated airway segment in subjects with asthma. *Am Rev Respir Dis*, vol. 137, no. 3, pp. 6416, ISSN 0003-0805

Hall, P.A.; Stearn, P.M.; Butler, M.G. & D'ardenne, A.J. (1987). Acetone/periodate-lysine-paraformaldehyde (PLP) fixation and improved morphology of cryostat sections for immunohistochemistry. *Histopathology*, vol. 11, no. 1, pp. 93-101, ISSN 03090167

Hansel, T.T.; Braunstein, J.B.; Walker, C.; Blaser, K.; Bruijnzeel, P.L.; Virchow, J.C., Jr. & Virchow, C., Sr. (1991). Sputum eosinophils from asthmatics express ICAM-1 and HLA-DR. *Clin Exp Immunol*, vol. 86, no. 2, pp. 271-7, ISSN 0009-9104

Holz, O.; Jorres, R.A.; Koschyk, S.; Speckin, P.; Welker, L. & Magnussen, H. (1998a). Changes in sputum composition during sputum induction in healthy and asthmatic subjects. *Clin Exp Allergy*, vol. 28, no. 3, pp. 284-92, ISSN 0954-7894

Holz, O.; Richter, K.; Jorres, R.A.; Speckin, P.; Mucke, M. & Magnussen, H. (1998b). Changes in sputum composition between two inductions performed on consecutive days. *Thorax*, vol. 53, no. 2, pp. 83-6, ISSN 0040-6376

Houen, G. & Hansen, K. (1997). Interference of sugars with the binding of biotin to streptavidin and avidin. *Journal of immunological methods*, vol. 210, no. 2, pp. 115-23, ISSN 0022-1759

Iredale, M.J.; Wanklyn, S.A.; Phillips, I.P.; Krausz, T. & Ind, P.W. (1994). Non-invasive assessment of bronchial inflammation in asthma: no correlation between eosinophilia of induced sputum and bronchial responsiveness to inhaled hypertonic saline. *Clin Exp Allergy*, vol. 24, no. 10, pp. 940-5, ISSN 0954-7894

Jatakanon, A.; Lim, S.; Chung, K.F. & Barnes, P.J. (1998). An inhaled steroid improves markers of airway inflammation in patients with mild asthma. *Eur Respir J*, vol. 12, no. 5, pp. 1084-8, ISSN 0903-1936

Kidney, J.C.; Wong, A.G.; Efthimiadis, A.; Morris, M.M.; Sears, M.R.; Dolovich, J. & Hargreave, F.E. (1996). Elevated B cells in sputum of asthmatics. Close correlation with eosinophils. *Am J Respir Crit Care Med*, vol. 153, no. 2, pp. 540-4, ISSN 1073-449X

Laitinen, L.A.; Laitinen, A.; Panula, P.A.; Partanen, M.; Tervo, K. & Tervo, T. (1983). Immunohistochemical demonstration of substance P in the lower respiratory tract of the rabbit and not of man. *Thorax*, vol. 38, no. 7, pp. 531-6, ISSN 0040-6376

Louis, R.; Shute, J.; Goldring, K.; Perks, B.; Lau, L.C.; Radermecker, M. & Djukanovic, R. (1999). The effect of processing on inflammatory markers in induced sputum. *Eur Respir J*, vol. 13, no. 3, pp. 660-7, ISSN 0903-1936

Maestrelli, P.; Calcagni, P.G.; Saetta, M.; Di Stefano, A.; Hosselet, J.J.; Santonastaso, A.; Fabbri, L.M. & Mapp, C.E. (1994). Sputum eosinophilia after asthmatic responses induced by isocyanates in sensitized subjects. *Clin Exp Allergy*, vol. 24, no. 1, pp. 29-34, ISSN 0954-7894

Maestrelli, P.; Saetta, M.; Di Stefano, A.; Calcagni, P.G.; Turato, G.; Ruggieri, M.P.; Roggeri, A.; Mapp, C.E. & Fabbri, L.M. (1995). Comparison of leukocyte counts in sputum, bronchial biopsies, and bronchoalveolar lavage. *Am J Respir Crit Care Med*, vol. 152, no. 6 Pt 1, pp. 1926-31, ISSN 1073-449X

Makker, H.K. & Holgate, S.T. (1993). The contribution of neurogenic reflexes to hypertonic saline-induced bronchoconstriction in asthma. *J Allergy Clin Immunol*, vol. 92, no. 1 Pt 1, pp. 82-8, ISSN 0091-6749

Maneechotesuwan, K.; Ekjiratrakul, W.; Kasetsinsombat, K.; Wongkajornsilp, A. & Barnes, P.J. (2010). Statins enhance the anti-inflammatory effects of inhaled corticosteroids in asthmatic patients through increased induction of indoleamine 2, 3-dioxygenase. *J Allergy Clin Immunol*, vol. 126, no. 4, pp. 754-762 e1, ISSN 1097-6825

Maneechotesuwan, K.; Supawita, S.; Kasetsinsombat, K.; Wongkajornsilp, A. & Barnes, P.J. (2008). Sputum indoleamine-2, 3-dioxygenase activity is increased in asthmatic airways by using inhaled corticosteroids. *J Allergy Clin Immunol*, vol. 121, no. 1, pp. 43-50, ISSN 1097-6825

Mclean, I.W. & Nakane, P.K. (1974). Periodate-lysine-paraformaldehyde fixative. A new fixation for immunoelectron microscopy. *J Histochem Cytochem*, vol. 22, no. 12, pp. 1077-83, ISSN 0022-1554

Metso, T.; Haahtela, T. & Seveus, L. (2002). Identification of intracellular markers in induced sputum and bronchoalveolar lavage samples in patients with respiratory disorders and healthy persons. *Respir Med*, vol. 96, no. 11, pp. 918-26, ISSN 0954-6111

Mordelet-Dambrine, M.; Arnoux, A.; Stanislas-Leguern, G.; Sandron, D.; Chretien, J. & Huchon, G. (1984). Processing of lung lavage fluid causes variability in bronchoalveolar cell count. *Am Rev Respir Dis*, vol. 130, no. 2, pp. 305-6, ISSN 0003-0805

Nightingale, J.A.; Rogers, D.F. & Barnes, P.J. (1998). Effect of repeated sputum induction on cell counts in normal volunteers. *Thorax*, vol. 53, no. 2, pp. 87-90, ISSN 0040-6376

Paggiaro, P.L.; Chanez, P.; Holz, O.; Ind, P.W.; Djukanovic, R.; Maestrelli, P. & Sterk, P.J (2002). Sputum induction. *Eur Respir J Suppl*, vol. 37, pp. 3s-8s, ISSN 0904-1850

Parameswaran, K.; Anvari, M.; Efthimiadis, A.; Kamada, D.; Hargreave, F.E. & Allen, C.J (2000). Lipid-laden macrophages in induced sputum are a marker of oropharyngea reflux and possible gastric aspiration. *Eur Respir J*, vol. 16, no. 6, pp. 1119-22, ISSN 0903-1936

Peleman, R.A.; Rytila, P.H.; Kips, J.C.; Joos, G.F. & Pauwels, R.A. (1999). The cellular composition of induced sputum in chronic obstructive pulmonary disease. *Eu Respir J*, vol. 13, no. 4, pp. 839-43, ISSN 0903-1936

Pieri, L.; Sassoli, C.; Romagnoli, P. & Domenici, L. (2002). Use of periodate-lysine paraformaldehyde for the fixation of multiple antigens in human skin biopsies. *Eu J Histochem*, vol. 46, no. 4, pp. 365-75, ISSN 1121-760X

Pin, I.; Gibson, P.G.; Kolendowicz, R.; Girgis-Gabardo, A.; Denburg, J.A.; Hargreave, F.E. & Dolovich, J. (1992). Use of induced sputum cell counts to investigate airway inflammation in asthma. *Thorax*, vol. 47, no. 1, pp. 25-9, ISSN 0040-6376

Pizzichini, E.; Pizzichini, M.M.; Efthimiadis, A.; Evans, S.; Morris, M.M.; Squillace, D; Gleich, G.J.; Dolovich, J. & Hargreave, F.E. (1996a). Indices of airway inflammation in induced sputum: reproducibility and validity of cell and fluid-phase measurements. *Am J Respir Crit Care Med*, vol. 154, no. 2 Pt 1, pp. 308-17, ISSN 107ξ-449X

Pizzichini, E.; Pizzichini, M.M.; Efthimiadis, A.; Hargreave, F.E. & Dolovich, J. (1996b). Measurement of inflammatory indices in induced sputum: effects of selection of sputum to minimize salivary contamination. *Eur Respir J*, vol. 9, no. 6, pp. 1174-80, ISSN 0903-1936

Pizzichini, E.; Pizzichini, M.M.; Kidney, J.C.; Efthimiadis, A.; Hussack, P.; Popov, T.; Cox, C.; Dolovich, J.; O'byrne, P. & Hargreave, F.E. (1998). Induced sputum, bronchoalveolar lavage and blood from mild asthmatics: inflammatory cells, lymphocyte subsets and soluble markers compared. *Eur Respir J*, vol. 11, no. 4, pp. 828-34, ISSN 0903-1936

Popov, T.; Gottschalk, R.; Kolendowicz, R.; Dolovich, J.; Powers, P. & Hargreave, F.E. (1994). The evaluation of a cell dispersion method of sputum examination. *Clin Exp Allergy*, vol. 24, no. 8, pp. 778-83, ISSN 0954-7894

Popov, T.A.; Pizzichini, M.M.; Pizzichini, E.; Kolendowicz, R.; Punthakee, Z.; Dolovich, J. & Hargreave, F.E. (1995). Some technical factors influencing the induction of sputum for cell analysis. *Eur Respir J*, vol. 8, no. 4, pp. 559-65, ISSN 0903-1936

Purokivi, M.; Randell, J.; Hirvonen, M.R. & Tukiainen, H. (2000). Reproducibility of measurements of exhaled NO, and cell count and cytokine concentrations in induced sputum. *Eur Respir J*, vol. 16, no. 2, pp. 242-6, ISSN 0903-1936

Rankin, J.A.; Marcy, T.; Rochester, C.L.; Sussman, J.; Smith, S.; Buckley, P. & Lee, D. (1992). Human airway macrophages. A technique for their retrieval and a descriptive comparison with alveolar macrophages. *Am Rev Respir Dis*, vol. 145, no. 4 Pt 1, pp. 928-33, ISSN 0003-0805

Rantala, I.; Maki, M.; Laasonen, A. & Visakorpi, J.K. (1985). Periodate-lysine-paraformaldehyde as fixative for the study of duodenal mucosa. Morphologic and immunohistochemical results at light and electron microscopic levels. *Acta Pathol Microbiol Immunol Scand A*, vol. 93, no. 4, pp. 165-73, ISSN 0108-0164

Rerecich, T.J.; Gauvreau, G.M.; Kelly, M.M.; Hargreave, F.E. & O' Byrne, P.M. (1999). Optimization of sputum fluid phase measurements. *Am J Respir Crit Care Med*, vol. 159, p. A849,

Richter, K.; Holz, O.; Jorres, R.A.; Mucke, M. & Magnussen, H. (1999). Sequentially induced sputum in patients with asthma or chronic obstructive pulmonary disease. *Eur Respir J*, vol. 14, no. 3, pp. 697-701, ISSN 0903-1936

Seeman, P.; Cheng, D. & Iles, G.H. (1973). Structure of membrane holes in osmotic and saponin hemolysis. *The Journal of cell biology*, vol. 56, no. 2, pp. 519-27, ISSN 0021-9525

Smith, C.M. & Anderson, S.D. (1989). Inhalation provocation tests using nonisotonic aerosols. *J Allergy Clin Immunol*, vol. 84, no. 5 Pt 1, pp. 781-90, ISSN 0091-6749

Spanevello, A.; Beghe, B.; Bianchi, A.; Migliori, G.B.; Ambrosetti, M.; Neri, M. & Ind, P.W. (1998). Comparison of two methods of processing induced sputum: selected versus entire sputum. *Am J Respir Crit Care Med*, vol. 157, no. 2, pp. 665-8, ISSN 1073-449X

St-Laurent, J.; Boulay, M.E.; Prince, P.; Bissonnette, E. & Boulet, L.P. (2006). Comparison of cell fixation methods of induced sputum specimens: an immunocytochemical analysis. *J Immunol Methods*, vol. 308, no. 1-2, pp. 36-42, ISSN 0022-1759

Tockman, M.S.; Qiao, Y.; Li, L.; Zhao, G.Z.; Sharma, R.; Cavenaugh, L.L. & Erozan, Y.S. (1995). Safe separation of sputum cells from mucoid glycoprotein. *Acta cytologica*, vol. 39, no. 6, pp. 1128-36, ISSN 0001-5547

Van Overveld, F.J.; Demkow, U.; Gorecka, D.; Skopinska-Rozewska, E.; De Backer, W.A. & Zielinski, J. (2005). Effects of homogenization of induced sputum by dithiothreitol on polymorphonuclear cells. *Journal of physiology and pharmacology : an official journal of the Polish Physiological Society*, vol. 56 Suppl 4, pp. 143-54, ISSN 1899-1505

Vatrella, A.; Perna, F.; Pelaia, G.; Parrella, R.; Maselli, R.; Marsico, S.A. & Calabrese, C. (2010). T cell activation state in the induced sputum of asthmatics treated with budesonide. *Int J Immunopathol Pharmacol*, vol. 23, no. 3, pp. 745-53, ISSN 0394-6320

Voynow, J.A. & Rubin, B.K. (2009). Mucins, mucus, and sputum. *Chest*, vol. 135, no. 2, pp. 505-12, ISSN 1931-3543

Wong, H.H. & Fahy, J.V. (1997). Safety of one method of sputum induction in asthmatic subjects. *Am J Respir Crit Care Med*, vol. 156, no. 1, pp. 299-303, ISSN 1073-449X

10

The Plasticity of Pancreatic Stellate Cells Could Be Involved in the Control of the Mechanisms that Govern the Neogenesis Process in the Pancreas Gland

Eugenia Mato[1], Maria Lucas[2], Silvia Barceló[3] and Anna Novials[2]
[1]Networking Research Center on Bioengineering, Biomaterials and Nanomedicine
(CIBER-BBN), EDUAB-HSP Hospital Santa Creu i Sant Pau , Barcelona;
[2]Diabetes and Obesity Laboratory, CIBER de Diabetes y Enfermedades Metabólicas
Asociadas (CIBERDEM), Institut d'Investigacions Biomèdiques August Pi i Sunyer
(IDIBAPS) - Hospital Clínic, Universitat de Barcelona;
[3]Proteomics Unit, IIS Aragón Instituto Aragonés de Ciencias de la Salud (ICS), Unidad
Mixta de Investigación, C/Domingo Miral s/n, Zaragoza,
Spain

1. Introduction

Mammalian pancreas is a gland that plays an important role in the regulation of energy balance and nutrition. Through the synthesis and release of protein digestive enzymes and hormones, which are involved in the absorption, it uses and stores the digested nutrients. This gland divided into two compartments with exocrine and endocrine functions, together with the stroma surrounding the pancreatic parenchyma, plays important roles in the homeostasis of the body. Moreover, they are involved in the maintenance of the function of the organ, including the regenerative process observed after injury of the pancreatic tissue. However, to understand this relationship, it is necessary to understand the embryological mechanisms that control the development of the pancreatic tissue. This embryological pathway begins from the precursor cells located in the endoderm, which is able to promote the pancreatic morphogenesis after responding to specific external and internal signals. Therefore, knowledge of the different networks created by neighbouring embryonic tissues will be essential for understanding the complexity of this morphogenetic process.

The organogenesis process of the pancreas gland is originated from stem cells located in the endoderm, which have the capacity to promote the development of the exocrine and endocrine compartments, identified in the adult gland from mammals. This phenomenon follows a specific gene network activity which is regulated by specific transcription factors (Jensen J, 2004). This complex process can be summarized into three steps identified by different investigators. The first step is accomplished through the action of specific signals that are originated from the mesoderm (Sander M and German MS, 1997). In the second step, the primitive endocrine cells, which are scattered throughout the undifferentiated epithelium, proliferate and promote the primitive islets cells located in the surrounding mesenchyma. Moreover, the mesenchymals signals are important to promote the development of islet cells and increase the number of beta cells at the end of the process. All these signals also promote vascularization (Kim SK and Hebrok M, 2001; Scharfmann R 2000; Reusens B and Remacle C; 2006). In the last step, the gland is remodeled into two functional compartments (Habener JF et al. 2005). In the adult pancreas, these two compartments exhibit different physiological roles. In addition, they have an important relationship and cellular interaction.

The pancreas like other tissues is considered like a dynamic organ, able to adapt to different physiological situations, such as diabetes, obesity or in gestation. This dynamic adaptation is based on the regulation of the beta cell mass in order to maintain glucose homeostasis. There are different mechanisms that control this process, which include: apoptosis, necrosis, hypertrophy, hyperplasia and neogenesis. However, little is known about some of these processes, and in particular, the cells which are involved. In the case of the neogenesis process, many studies supported the idea that it occurs via cells which are located in, or which are associated with, the ductal epithelium of the exocrine compartment of the pancreas. One of the approaches used for investigating this hypothesis is the application of the immunocytochemical and immunohistochemical techniques. These techniques are important because they help to identify the cell population involved in the process without losing the architecture of the tissue. Moreover, they are important tools for the phenotyping of the cell population when isolated from the tissue and checked while maintained *in vitro*.

2. Historical perspective of stellate cells

In 1876 Karl von Kupffer described for the first time a new population of cells in the liver called "sternzellen " or stellate cells, due to their stellate appearance. These cells located in the space of Disse had cytoplasmatic inclusion bodies indicating to have a phagocytic function. Initially, Kupffer classified them into the "Waldeyer's perivasculare Bindgewbszellen" or reticulo-endothelial system. However, this author changed opinion and the cells were considered phagocytes and were referred to as "special endothelial cells of the sinusoids" (Kupffer C 1876). However, it was not until the beginning of the 20th century when Zimmerman described them as dendritic perisinusoidal cells surrounded by reticular fibers and named them hepatic pericytes. Later, the Japanese Anatomist Dr. Ito described a new cell population in the liver, which were located in the perisinusoidal space and contained abundant amounts of fat droplets in their cytoplasm. These cells, known as "Ito-cells" are able to store and deliver vitamin A and other liposoluble vitamins. Moreover, they are involved in the regulation of sinusoidal tone, local blood supply, and tissue repair and fibrosis. The cell presents several thick cytoplasmatic processes which are protuded directly from the perikaryon (primary process) and extended onto the outer surface of the sinusoidal entohelial cells (Ito T et al. 1951). In summary, these cells have received other names, such as: fat storing cells, pericytes, parasinusoidal, and lipocytes. Several studies demonstrated that all these cell populations shared most of their cellular and physiological characteristics and seemed to correspond to the same population. For that reason, and in order to avoid confusion, in 1996 the international community of investigators unified the nomenclature and defined

The Plasticity of Pancreatic Stellate Cells Could Be Involved in the Control of the Mechanisms that Govern the
Neogenesis Process in the Pancreas Gland
209

these cells as a "Stellate cells" (no authors listed, 1996). Soon after, Kent and Popper demonstrated that the stellate cells were linked to the pathogenesis of hepatic fibrosis (Hirosawa K and Yamada E, 1973). This important finding promoted the identification of this cell type in extrahepatic organs (pancreas, spleen, adrenal, ductus efferent and uterus) in rodent and humans (Geerts et al., 2001).

In addition, the presence of these cells in a wide variety of species, ranging from lampreys (primitive fish) to humans and in all major tissues, indicated their importance in the development of the different organs (Wake K 1987).

2.1 Stellate cells in pancreatic tissue: historical perspective

Vitamin A storing cells were first described in the pancreas by Watari, *et al.*, in 1982, using fluorescence and electron microscopy. In 1990, Ikejiri, et al., confirmed the previous results and also showed the presence of vitamin A as a autofluorescence stained in normal pancreatic sections from rats and humans. In 1997, Saotome, et al., described the presence of the myofibroblast-like cells in human pancreas, and their involvement in the extracellular matrix remodeling during the fibrosis process. However, these independent observations had not been realized to be related until 1998, when Bachem, et al., and Apte, et al., defined these two populations of cells as pancreatic stellate cells, in two different stages of activation (Quiescent and Active).

2.1.1 The embryological origin

The embryological origin of the stellate cells is unclear. Importantly, there are few studies conducted to resolve this dilemma. Most of them have been described in the liver. For that reason, different observations of these cells in liver have been extrapolated to other organs including the pancreas. However, numerous theories on the linage of these cells have been presented. The hepatic stellate cells (HSC) are proposed to be derived from mesenchymal cells that separate the pericardial and peritoneal cavities of the embryo (Morita M et al. 1998; Naito N and Wisse E 1977). However, the specific microfilaments identified in their cytoplasm and morphology, resembling the astrocyte cells from astroglia in the Central Nervous System, could also be indicating a neural-ectodermal origin (Niki T et al. 1999; Friedman SL 2000). This last observation was difficult to reconcile with the mesenchymal origin described before. Recently, the identification in bone marrow of fibroblast / myofibroblast cells, which share some HSC characteristics, suggests that stellate cells could be derived from hematopoietic stem cells (Susking DL and Muench MO, 2004; Baba S et al. 2004; Ogawa M et al 2006). In conclusion, new experimental designs are required in order to understand the embryological origin of these cells. Moreover, the possibility to use the lineage-specific promoters to drive the transgene expression could contribute to the clarification of this problem and enable the understanding of the biology of these cells.

2.1.2 Biology of pancreatic stellate cells

Pancreatic stellate cells (PSC) are located in different spaces: periacinar, perivascular and periductal of the exocrine compartment of the pancreas. They represent approximately 4% of the total cells of the gland. The cells are closely in contact with acinar, endothelial and ductal cells and establish a strict cellular communication between them through long processes containing numerous filaments and microtubules. These cells play an important role in the pancreatic pathology of the exocrine compartment of the pancreas, such as chronic pancreatitis and pancreatic cancer. In all these injury processes, PSC and HSC have shown an important phenotype transformation to a so-called activated form. In this state, the cells are able to produce large amounts of extracellular matrix proteins (EMC), fibronectin and laminin resulting in the extensive fibrosis. In this stage, the cells showed: a typical characteristic spindle –shaped, absence of the retinol in the cytoplas, the increment of the myofilaments, as in the GFAP and vimentin, as well as the presence of the new myofilament (α-SMA). Moreover, the production of multiple factors with a paracrine, autocrine and chemoattractant actions can be detected (Jasper, R 2004, Morini S et al. 2005; Omary MB et al. 2007; Kordes C et al 2009) (Fig.1 A,B). In contrast, when the cell are in the quiescent form, they present: abundant droplets of vitamin A in the cytoplasm, are less positive for desmin, vimentin, nestin and GFAP intermediate filaments, and the cytoplasmatic processes are not observed. In addition, a non-proliferative state is observed in the cells (Pinzani, M. 1995; Apte MV et al. 2003). The transitional stage of the cells was observed and the cells share some of the ultrastructural and functional characteristic for these two differentiated stages described previously.

The mechanism implicated in this transformation process is not determined yet. *In vivo* different signal transduction pathways have been described and all, including infiltrating leucocytes and damaged acinar cells, are able to initiate and maintain the activated phenotype. However, most of the information about the activation mechanism has derived from *in vitro* studies of rodent PSC maintained in culture. These cultures, initially express the molecular markers of the quiescent cells and it is easy to observe the presence of the cytoplasmic lipid droplets by oil red stain (Apte MVet al., 1988, Mato E et al 2009). However, in a short amount of time, most of the cells in the culture showed a proliferative phenotype with α-SMA and ECM protein expression. These molecules are associated with the activated phenotype (Haber PS et al. 1999). Several authors have associated this phenomenon to *in vitro* changes of Rho-ROCK pathways regulated by the actin cytoskeleton (Masamune A et al. 2003). PI 3-kinase activity is required for PDGF-stimulated PSC migration, but not cell proliferation (McCarroll JA et al. 2004). Moreover, the role of the enzymes involved in the mitogen-activated protein kinase (MAPK) family have been described : Jun N-terminal kinase JNK and p38, which are involved in the transcriptional control and PSC activation, and are mediators of signals induced by pro-inflammatory cytokines and cellular stressors (Masamune A et al. 2003). On the other hand, ligands of the nuclear receptor PPARγ (peroxisome proliferator-activated receptor γ) such as 15-deoxy-Δ12,14-prostaglandin J$_2$ and troglitazone (an antidiabetic drug of the thiazolidinedione group) stimulate maintenance of a quiescent PSC phenotype *in vitro* have been described (Masamune A et al. 2002). In summary, despite that several intracellular mediators involved in the control of the PSC activation and desactivation have been identified, most of them are unknown.

Furthermore, some authors have documented a significant increment of the PSC in the regenerative areas of the pancreas after suffering an acute pancreatitis, induced in rodent. These observations, plus the identification of the PSC positive for nestin marker, support the idea that this population could be involved in the pancreatic regeneration process (Zimmermann A et al. 2002, Ishiwata T et al 2006).

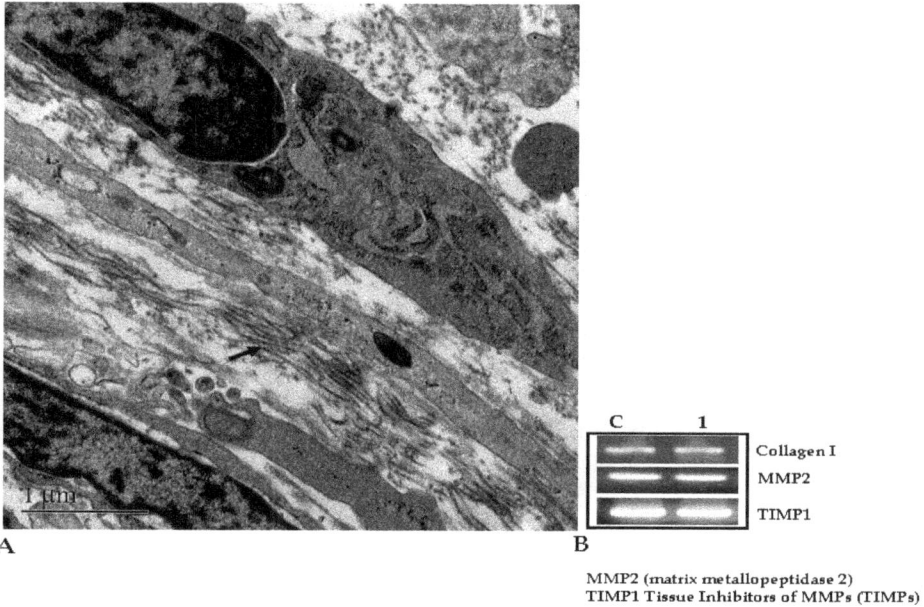

Fig. 1. A. Transmission electron micrographs of activated pancreatic stellate cells *in culture*.
The arrow show abundant collagenous fibers compatible with collagenous type I. B. RT-PCR
expression involved in the EMC remodeling (Mato E. et al., unpublished data)

3. Pancreatic progenitor cell: historical perspective

One of the important reasons to find progenitor cells in the pancreas is to cure Diabetes
Mellitus. This metabolic disorder is a common and serious disease in our society and is the
most rapidly growing chronic disease of our time. It has become an epidemic that affects
millions of people around the world. For that reason, there has been an increasing in
interest scientific community to identify the cell populations with stem or progenitor
properties in the pancreatic tissue. This finding could represent a significant therapeutic
advance in this disease.

The first description of stem and progenitors cells in adult tissue was in bone marrow and
the nervous system (Weissman IL 2000; Fuchs E and Segre JA 2000). Although it is accepted
that similar cells can exist in the other adult tissues and organs, they are not always easy to
find. One of the reasons for limited number of studies on these cells relates to the fact that
they do not have specific biological markers. Thus, finding of progenitor cells in the
pancreas is a challenge. There is some evidence in the pancreas that progenitor cells exist in
the neogenesis process, which can be induced by cellophane wrapping of the pancreas
(Rosenberg L et al. 1998), partial pancreatectomy (Bonner-Weir S et al. 1993), streptozotocin-
induced diabetes (Fernandes A et al. 1997), and also during pregnancy (Bonner-Weir S
2000). Some authors, Rosenberg in 1998 and Rafaeloff in 1997, have only associated this
phenomenon with gene (*Reg*) and proteins (islet neogenesis, INGAP) which are expressed
during the process, but not with progenitors cells. However, cell participation is possible.

Research has been launched to investigate the process of neogenesis and the cells that may be involved in this mechanism. Understanding this process will be the key since it will allow us to restore the function of the gland lost during the illness.

3.1 Progenitor cells in the pancreas tissue

3.1.1 Ductal cells

Most of the studies favor the pancreatic duct as a potential source of progenitor cells in adult pancreas (Rosenberg L 1998; Bonner-Weir S 2000). These studies are based on the information about the important role the primitive ductal epithelium has during the pancreas embryogenesis of the pancreas as a source for the islet development (Madsen OD et al. 1996, Sander M and German MS, 1997). Moreover, Gu and coworkers described the presence of endocrine cells within the adult ductal system (Gu D and Sarvetnick N 1993) and also identified beta cells associated with the human ductals (Bouwens and Pipeleers. 1998). Finally, the ability of ductal cells to expand *in vitro* and to form insulin-producing islet-like structures has also been demonstrated (Bonner-Weir S et al. 2000; Ramiya VK et al 2000).

3.1.2 Pancreatic islet as a cellular source

Another interesting hypothesis was to propose the pancreatic islet as a progenitor cell source, based on the analysis of islet regeneration in mouse pancreas models after the administration of streptozotocin. The results showed the presence of the insulin-producing cells following the injury into the adult islets. This study suggested the existence of the two types of progenitor cells, one of them expressed Glut-2 and the other coexpressed insulin and somatostatin (Guz Y et al. 2001).

Nestin-positive cells, neurogenin-3 positive cells and hormone-negative immature cells, with proliferative capacity *in vitro* has been found in rats and human islets. This supports the idea of the existence of the multipotential cells in the islet (Kodama S et al 2005. However, their participation in islet regeneration and neogenesis *in vivo* has not yet been demonstrated (Zulewski H et al. 2001). Despite the explosion in the number of *in vitro* studies that describe different types of cells with progenitor capacity within the island, there is also some critical work demonstrating that the reactivation of genes required for endocrine cell development, such as neurogenin 3, are not implicated directly in the regeneration of pancreatic tissue after pancreatectomy (Lee CS et al. 2006).

Cells with the capacity to be differentiated not only in the lineage of endocrine cells, but also in other cellular lineages, such as exocrine and glials cells, have been identified (Seaberg RM et al 2004). These progenitors could be of different origins (ductal cells or cell located inside the islets). These cells showed different molecular markers, such as "the hepatocyte growth factor receptor", c-Met. This receptor tyrosine kinase plays an important role in tumour growth by activating mitogenic signaling pathways (Seaberg RM et al 2004; Suzuki A et a ., 2004).

Other authors identified cells presenting a differentiated morphology and named then "small cells". Although these cells are positive for several pancreatic markers (PDX-1, sinaptoficin, insulin, glucagon, somatostatin, pancreatic polypeptide), they also expressed markers of undifferentiated cells, such as: alfa-fetoprotein and Bcl-2. Surprisingly, these cells were negative for nestin and cytokeratin 19, indicators of pluripotency and ductal origin. Functional analysis showed that they have the capacity to present a glucose response, but they did not respond to secretatgogues, such as IBMX (Petropavlovkaia M and RosenbergL , 2002).

3.1.3 Hematoipoietic stem cells as a progenitor cells in pancreas

Hematoipoietic stem cells have been proposed, as a new progenitor source in pancreas. In 2002, this hypothesis was formulated by Lerner, et al., who identified a population defined as Side Population, or SP, from a bone marrow origin. This SP cell population, described for the first time by Goodel MA, et al., corresponded to a small subpopulation of cells with an enriched stem cell activity and showed a "low" Hoechst 33342 dye staining pattern. Subsequent studies attributed this SP phenotype to the expression of stem cell markers sucha as MDR1 and Nestin, and also co-expressed ABCG2, an ATP-binding cassette (ABC) transporter (Zhou S, 2001). ABCG2 gene is expressed in several rodent tissues, such as in the intestine, kidney and testes (Tanaka, Y 2005). The precise physiological function of these transporters in progenitor and differentiated cells is unknown and it has been postulated that they confer protection against a number of xenobiotics, thus maintaining the regenerative capacity of the tissue (Leslie, E.M, 2005). The identification and isolation of ABCG2 positive cells in pancreatic tissue may be a new potential source of adult multipotential stem/progenitor cells, useful for the production of islet tissue for transplantation into diabetic subjects (Fetsch, PA, 2006). The presence of these cells in pancreas tissue is controversial.

3.1.4 Epithelia Mesenchyma Transition (EMT)

Finally the concept of Epithelia-Mesenchymal transition or EMT has been described during the regeneration endocrine pancreas and in the cancer development. The EMT could permit that adult cells can be differentiated into the fibroblastic-like cells as a step of transition to other cellular lineage. Recently this process has been linked with the maintenance of stem cell phenotype. However, the molecular mechanism to control the EMT process remained to be demonstrated (Gershengorn MC et al. 2004; Bonner-Weir S et al. 2004).

An explosion of publications in the last decade tried to discover what type and where the progenitors cells are localised in the pancreatic tissue. We can conclude that the number of progenitor cell types in the pancreas may not be too limited to the cells already described. It is possible that the pancreas may contain an unidentified cell population at rest, as described in oval cells in the liver, capable of initiating their proliferation during the process of neogenesis. This opens the opportunity to explore new cell populations that form the pancreatic parenchyma.

4. Immunocytochemical investigation of the role of pancreatic stellate cell as progenitor cell

The plasticity of the stellate cells phenotype during tissue injury is a proven fact and may indicate that these cells can be presented in progenitor cell features. These findings suggested a novel aspect of the stellate cell biology must be necessary investigated.

The first marker identified in HSC was nestin. Nestin, a marker for neural stem cells, was identified in HSC during the transition from the quiescent to the activated phenotype in cells maintained in culture, but no association with a progenitor role was suggested by the authors (Niki T et al. 1999). Later, other markers were identified in the HSC: CD133 (prominin-1), a glycoprotein also known in humans and rodents as a Prominin 1 (PROM1), and expressed in the adult and embryonic stem cell and Oct4 (octamer-binding transcription factor 4), also known as POU5F1 (POU domain, class 5, transcription factor 1), protein involved in the self-renewal of undifferentiated embryonic stem cells (Mizrak D et al. 2008; Niwa H et al. 2000). These two markers were able to maintain an undifferentiated phenotype without losing the ability participates in liver regeneration (Kordes C et al 2009). Finally, the HSC were able to be differentiated into endothelial or hepatocyte-like cells (Kordes eC t al. 2007; Kubota H et al 2007). Following these findings an increasing number of papers about this topic were published.

The existence and lineage of progenitor cells in the pancreas, as well as their origin and location, is a topic of debate and, although several hypotheses had been proposed, it is not yet proven. Moreover, the possibility that the PSC can act as a progenitor cell is not clear.

Nevertheless, it is also important to remark that PSCand hepatic stellate cells are identical have a common origin and both share transcriptional level, exhibiting organ-specific variations of the common transcriptional phenotype and (Bucholz M et al 2005; Omary MB et al 2007). This scenario suggests that the progenitor role for PSC could be a reality. In 2002, nestin-positive cells were identified in normal adult rat pancreas and during its regeneration. Interestingly, most of these cells presented the morphology characteristic of stellate cells. Nestin, in pancreas as in liver, was confirmed as a main marker of stellate cell activation. Other roles, including the marker of progenitor cells, were not confirmed (Lardon J 2002).

The question that needs to be addressed is whether PSC, after overexpressing some specific pancreatic transcription factors, such as Pdx1 or NeuroD1, have the ability to present the transdifferentiation process, which permits conversion into insulin-producing beta cells.

One approach to conduct these studies and broaden the possibility of unraveling the mechanisms that control self-renewal, is to explore the cell roles after their isolation and establishment of the cell culture. The first description of the stellate isolation from tissue was in 1977 by Galamos JT. The study was characterized by growth mesenchymal cells derived from liver tissue, which have probably been derived from stellate cell (Galamos JT et al. 1977). Later, density gradient centrifugation was used after *in situ* digestion of the tissue, based on their buoyancy attributable to intracellular vitamin A. The density gradient separation method remains the most widely used approach for stellate cell isolation, but criticism of this method favours the isolation of quiescent cells, which are rich in vitamin A (Friedman SK, 2008). Later, transgenic and knockout mouse models have been developed for the isolation following the standard method of murine stellate cells or for performing in situ analysis with specific stellate markers. However, one limitation of the technique is the large number of animals needed to obtain an adequate cell yield (Henderson NC et al. 2005; Kalinichenko Vet al. 2003). To solve this problem, stable cell lines obtained from human and mouse model would be an important advantage for many investigators in order to study stellate cell biology. Several methods have been described to establish from HSC cultures and pancreas cell lines, such as: long-term culture, transfection with simian virus 40 (SV40) T antigen, or ectopic expression of telomerase (Vogel S et al. 200; Murakami K et al 1995;

Apte MV et al. 1998; Kruse ML 2001; Sparmann G 2004; Masamune A et al. 2003; Satoh M et al 2002; Jesnowski R et al 1999; Löhr M et al 2001). The disadvantage of the cell lines is that they differ somewhat in their state of activation or in transcription expression and the results obtained must be validated in the *in vivo* model. Finally, the description of the cryoperservation technique for freezing primary stellate cell lines is an important advance for sharing the cells between different laboratories (Neyzen S et al. 2006).

In this context our group initiated a new research field, focusing in the identification of progenitor cell in pancreas tissue through ABCG2 transporter as a progenitor cell marker. This marker was identified as a molecular determinant of the Side-Population (SP) phenotype. However, there is no information about its expression on the pancreatic cells. Recently, overexpression of the breast cancer-resistance half-transporter protein (BCRP1) was found to be responsible for the occurrence of mitoxantrone resistance in a number of cell lines (Doyle LA et al, 1998; Miyake K et al, 1999; Litman T et al, 2000). Based on in these findings, we isolated a mitoxantrone-resistant cells population from pancreata of lactating rats by mitoxantrone selection through the ABCG2 transpoter (Fig. 2 A, B, C).

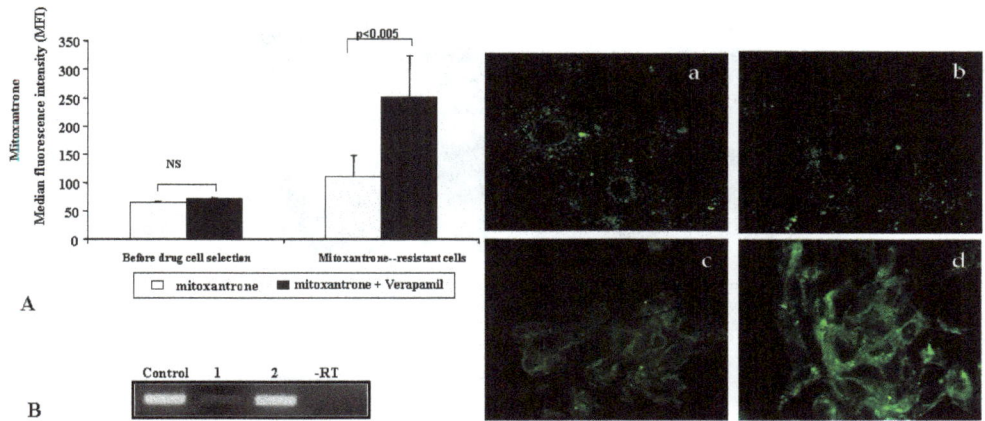

Fig. 2. ABCG2 expression, and drug uptake and retention assays in primary cell cultures (mitoxantrone-resistant cells and unselected cells). (A) One-hour drug accumulation assay with and without verapamil. The cells were preincubated with 5μM verapamil for 15 min. Subsequently, cells were treated with 8 μM mitoxantrone and assayed for drug accumulation. Each condition is the mean of three experiments ± SD. Verapamil increased the intracellular concentration of mitoxantrone in the mitoxantrone-selected drug-resistant cells. The experiment was performed in triplicate, and a representative histogram was shown. (B) The ABCG2 expression in the cells from cultures: unselected cells (line 1) and mitoxantrone-resistant cells at Stage 2 (line 2) was determined by RT-PCR. The ARIP cell line was used as a positive control of the reaction (Control), – RT corresponds to amplification in which reverse transcriptase was excluded from the reaction (negative control). (C) cells treated with mitoxantrone for 2 ' (a) and 10' (b) or treated with mitoxantrone plus verapamil (ABCG2 inhibitor) for 2′ (c) and 10′ (d) (Reproduced with permission, from Mato E. et al. Identification of a pancreatic stellate cell population with properties of progenitor cells: new role for stellate cells in the pancreas . Biochem. J. 421; 181–191© the Biochemical Society)

Next, cells were expanded, checking that the cells present in culture a fibroblast features (Fig 3 A)

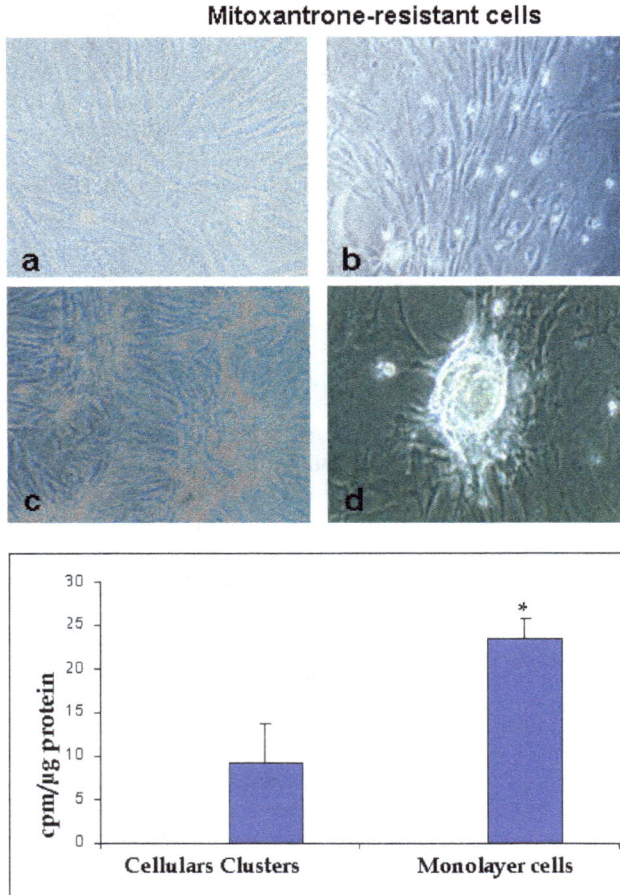

Fig. 3. Phenotype of Cell Line from mitoxantrone-resistant cell population. A The mitoxantrone-resistant cells became overgrown by cells with a fibroblastoid morphology (a,b). Spontaneously, some cells began to form three-dimensional cell clusters (c,d,e). B. Representative Histogram of the tritiated thymidine incorporation in cellular cluster and monolayer cells * p<0.05. (Reproduced with permission, from Mato E. et al. Identification of a pancreatic stellate cell population with properties of progenitor cells: new role for stellate cells in the pancreas . Biochem. J. 421 ;181–191© the Biochemical Society)

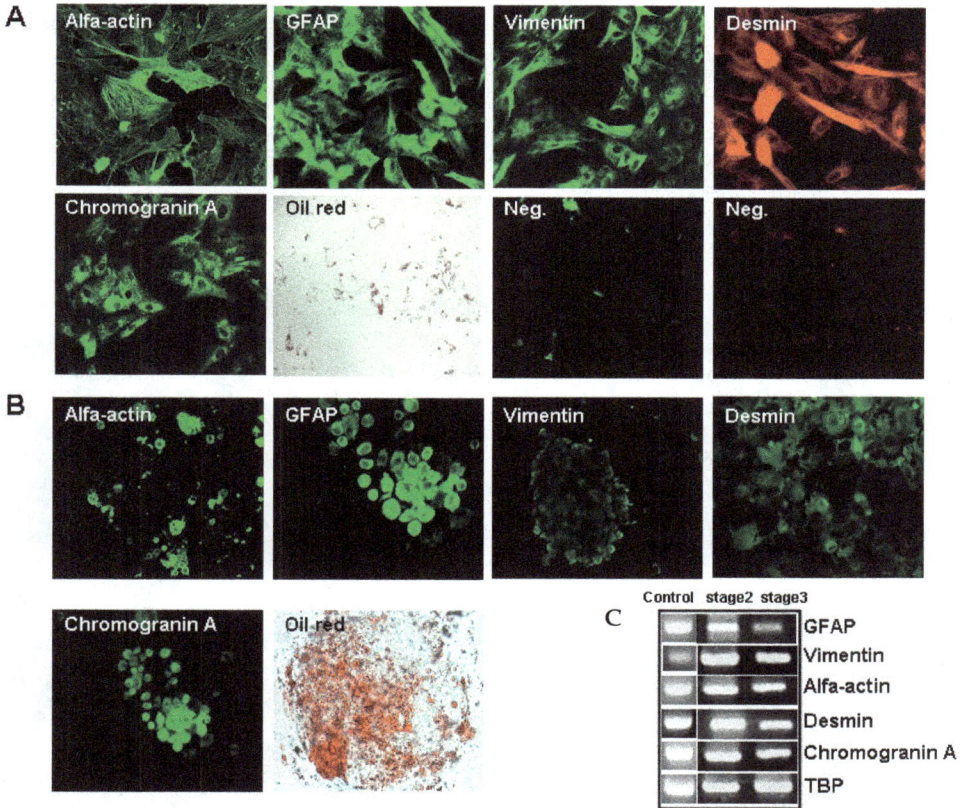

Fig. 4. Mitoxantrone-resistant cells were phenotyped by immunofluorescence and RT-PCR using pancreatic stellate markers. (A) Mitoxantrone-resistant cells at Stage 2 express the markers: alfa-Actin, GFAP, vimentin, desmin, and chromogranin A. To confirm the presence of the vitamin A stored in the fat droplets, oil red staining was performed. (B) Disaggregated from mitoxantrone-resistant cells at stage 3 were immunophenotyped for the same markers, including the oil red staining. Negative controls (Neg) were used. (X20 original magnification). (C) These results were confirmed by RT-PCR using one μg of total RNA of the mitoxantrone-resistant cells in both stages (stage 2 (monolayer cultere) and stage 3 (cellular cluster)). Control cell lines were used as a control reaction. (Reproduced with permission, from Mato E. et al. Identification of a pancreatic stellate cell population with properties of progenitor cells: new role for stellate cells in the pancreas . Biochem. J. 421 ;181–191© the Biochemical Society)

The existence of a fine balance between proliferation and differentiation process is accepted by the research community. This balance promotes the differentiation from adult stem cell to postmitotic cells through decreasing or increasing the ratio of proliferation, permitting the maintenance of the stem cell population in adult tissues (Soria B, 2001). The observation of the behavior of mitoxantrone resistant cells in culture was interesting. The results indicated that, while the cells with fibrobastoide appearance have showed a rapid and constant growth after clustering formation, they modified their behavior showing a significant reduction in their growth, without stopping completely (Fig.3, B). The results suggested the ability of the cell to be reprogrammed.

Finally the immunocharacterization of these cell cultures in monolayer and cellular cluster showed a stellate phenotype, characterised by vitamin A uptake (oil red staining) and steallate markers presence (Fig. 4 A, B).

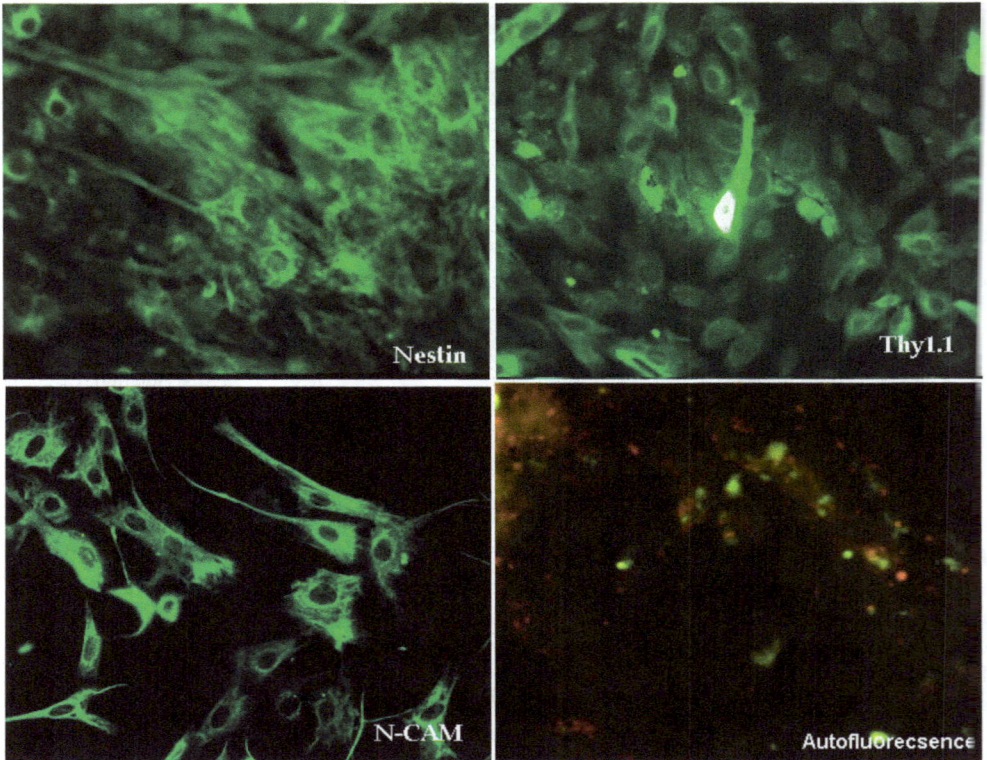

Fig. 5. Characterization of progenitor markers in mitoxantrone-resistant cell population. Nestin, Thy1.1 and N-CAM protein expression was detected by immunostaining in culture from mitoxantrone-selected drug-resistant cells (Modified with permission, from Mato E. et al. Identification of a pancreatic stellate cell population with properties of progenitor cells: new role for stellate cells in the pancreas . Biochem. J. 421 ;181–191© the Biochemical Society)

Moreover, they share markers of the adult stem cells, such as: ABCG2, Nestin, Thy1.1, and N-CAM. The latter marker participates in signal transduction and in cell type segregation as a mediator of cellular junctions during organogenesis (Esni F et al. 1999) (Fig. 5).

Little it is known about the role of Fibroblast growth factor and their receptor in stellate cells. FGF belongs to a large family of molecules that retain a high homology at the genetic level. These growth factors induce pleiotropic responses, causing effects in both embryonic development and in adult tissue (Steiling H and Werner S, 2003). Their actions are mediated by four receptors of the tyrosine kinase membrane and present different isoforms (b and c) by splicing (Itoh N and Ornitz DM, 2004). Fibroblast growth factors receptors (FGFR) have been detected over time during the development of the pancreas. In addition, their ligands, such as Fibroblast growth factor: 1, 7, 9, 10, 11, 18 (Dichmann DS et al., 2003), and the subtype of the FGFR 2, called FGFR2b, seem to have a key role in the exocrine development (Miralles F, et al. 1999). Recently, FGF7 and FGF10 have been involved in maintaining the cells in an undifferentiated stage and controlling the self-renewal of the pancreatic precursors (Elghari L et al. 2002; Norgaard GA et al, 2003). The positive gene expression for FGFR2IIIb, FGFRIII2c, FGFR1, and their specific ligands (FGF 1,7,and 10), were showed for the first time in our cell cultures (Fig. 6, Mato et al. unpublished data).

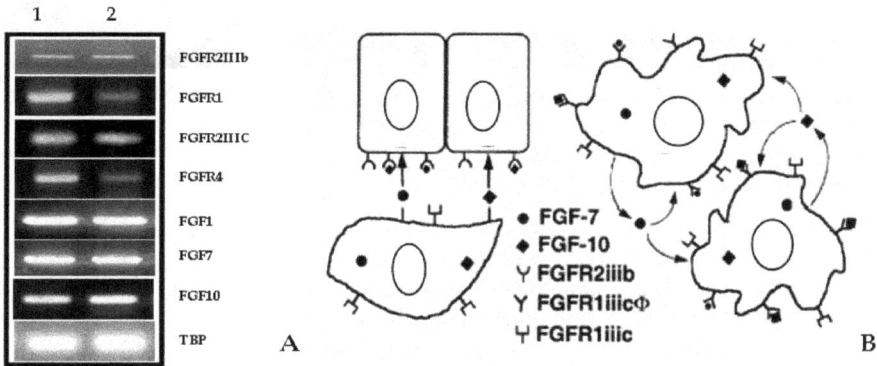

Fig. 6. Expression of the Fibroblast growth receptor and Fibroblast growth factors in the cells from cultures. Expression of FGFRIII2b, FGFR1, FGFR4, FGFR2IIIC, FGF1, FGF7, FGF10 in the cells from cultures: monolayer cells (line 1) and at clusters cells (line 2) was determined by RT-PCR. B Proposed autocrine (A) and paracrine (B) model through FGFR and their ligands of the PSC in: ductal cell, exocrine cells or themselves. (Mato E. , unpublished data)

This finding may suggest that FGFR and their ligand are involved in epithelial-mesenchymal communication of PSC and, in addition, the autocrine effect allows the maintenance of its cell population in the pancreatic tissue. On the other hand, pancreatic stellate cell do not express endocrine genes. However, during cell expansion, a spontaneous cell differentiation occurs and these cells showed a weak expression of PDX-1 in to the nucleus and the cytoplasm of the cells (Fig. 7 A, B). This gene, also known as (insulin promoter factor-1, islet/duodenum homeobox-1, somatostatin transactivating factor-1, or insulin upstream factor-1 and glucose-sensitive factor), plays a key transcription factor in the endocrine differentiation pathway and is also essential for differentiation of endocrine cells in the gastric antrum. The results suggest a transdifferentiation process. However, the molecular mechanisms of this process are unknown. In additon, few studies are investigating the effect of culture medium and additional protein components on the viability and maturation of the cells (Royer PJ et al 2006). Our results underscore the importance of defining culture medium composition in experimental procedures, in order to identify new soluble factors involved in the processes of cellular transdifferentiation.

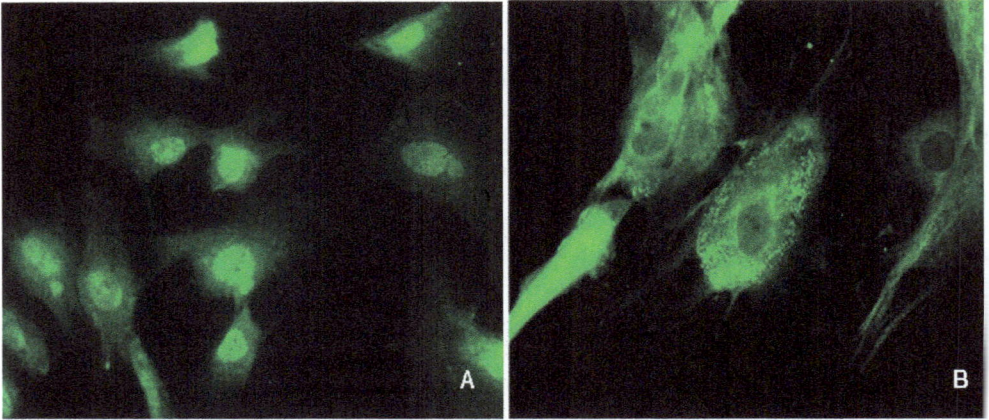

Fig. 7. Expression of Pdx-1 transcription factor in the cells from cultures (Monolayer stages). Pdx-1 protein expression was detected by immunostaining fluorescent in culture from mitoxantrone-selected drug-resistant cells. A.- Nuclear staining (X40 original magnification) B.- Cytoplasmatioc staining (X60 original magnification) (Mato E. unpublished data).

Identifying instructive signals that induce differentiation during organogenesis will be important to determine how such signalling networks are established and how they elicit multiple signalling responses in endodermal cells to activate appropriate genetic programs (Ratineau C et al 2003). Several signalling molecules have been implicated in induction of specific endodermal cell types. However, few of these factors have been examined in adult pancreatic tissue (Sttaford D et al 2006). One of these factors is GLP-1, secreted from the L-cells of the distal ileum and colon. This substance has been suggested to play an important role in increasing beta cell mass by inducing the neogenesis or transdifferentiation through the expression of Pdx-1 in ductal or islets cells (Yue F et al. 2006; Abraham EJ et al 2002; Hui H et al 2001).

Also, matrigel secreted by Engelbreth-Holm-Swarm (EHS) mouse sarcoma cells, is a gelatinous protein mixture that provides a semisolid medium that resembles the complex extracellular environment found in many tissues and is used as a substrate for three-dimensional cell culture. The addition of exendin-4 (analog to GLP-1) and matrigel to our cellular model was needed to proceed to the differentiated stages and permit detection of insulin, IAPP, glucagon, GLUT2 and the convertases PC1/3 and PC2 expression (Fig. 8 A, B). In contrast, expression of the transcription factor p48 and other exocrine genes, such as amylase, were not detected. Interestingly enough was the observation of the cytokeratin 19 (CK19) expression. These intermediary filaments present in cells of the epithelial origin, such as ductal cells, indicate that the cell could be involved in the mechanism to control the mesenchymal-epithelial transition (MET). This phenomenon consists of a promising source of cells for replacement therapies, but can also be involved in the carcinogenesis process (Mato E et al. 2009).

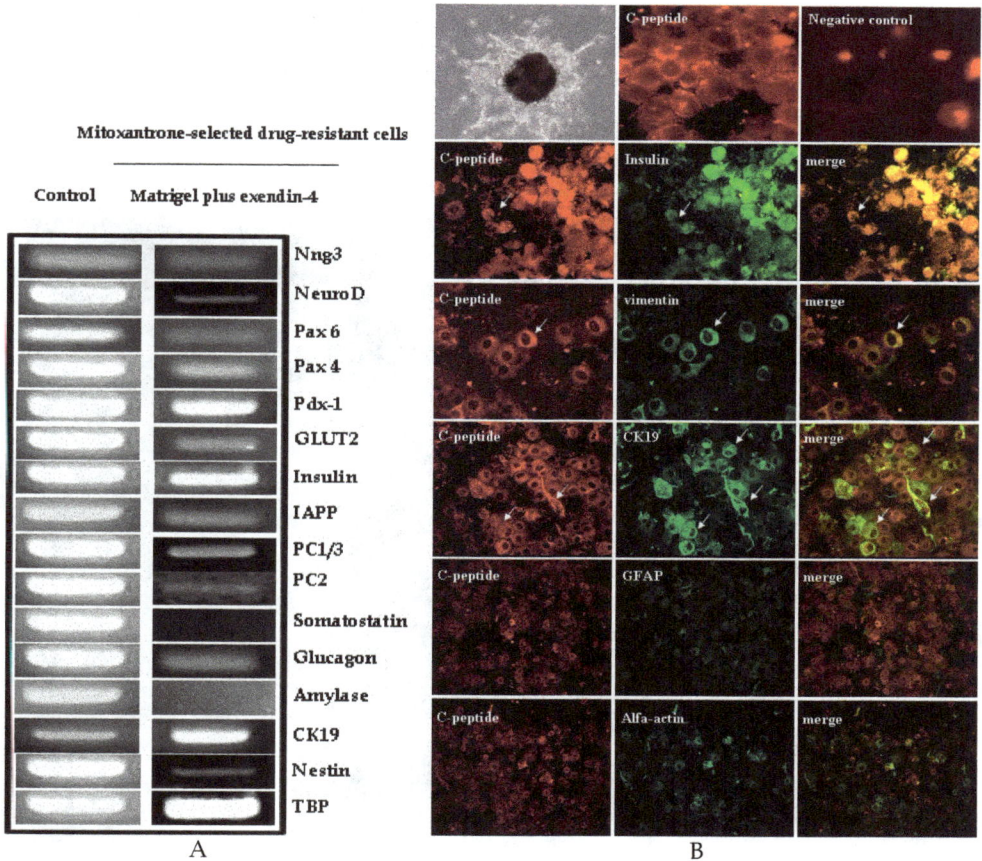

Fig. 8. Pancreatic gene expression profiles and co-immunolocalization of different markers
by cytospin-prepared cells obtained from disagregated cellular clusters after exedin-4
treatment. A.- Gene expressions profile after matrigel plus exendin-4 treatment in
mitoxantrone-resistant cell cultures. B.- Representative cellular cluster after treatment with
matrigel plus exendin-4. The markers were visualized in red: c-peptide, green: insulin,
vimentin, CK19, GFAP, alfa-actin, and yellow as the merges. The MIN-6 cells were used for
the immunohistochemistry control. (Reproduced with permission, from Mato E. et al.
Identification of a pancreatic stellate cell population with properties of progenitor cells: new
role for stellate cells in the pancreas . Biochem. J. 421 ;181–191© the Biochemical Society).

The molecular mechanisms and the receptors involved in EMT process are not indentified
yet. Most of the evidence suggests that integrin could play an important role. On the other
hand, the basement Membrane Matrix is an effective culture medium for the attachment and
differentiation of both normal and transformed anchorage dependent on epithelioid and
other cell types. The use of these three-dimensional culture systems may be particularly
relevant to such efforts by recapitulating a more physiological microenvironment (Han YP
et al. 2004; Phillips PA et al. 2003; George PC 2005). During the matrigel growth, substantial

ultrastructural changes in the cells were observed. The cells presented a smaller and more homogenous cell size with round nuclei and electron-dense homogenous chromatin, a significant increase in the number of mitochondria, lipid droplets in the cytoplasm and abundant electron-dense granules were also observed. In contrast to the cellular cluster growth in a normal condition medium, the quiescent stellate cells had a high presence of fibers compatible with collagen fibers (Fig. 9 A).

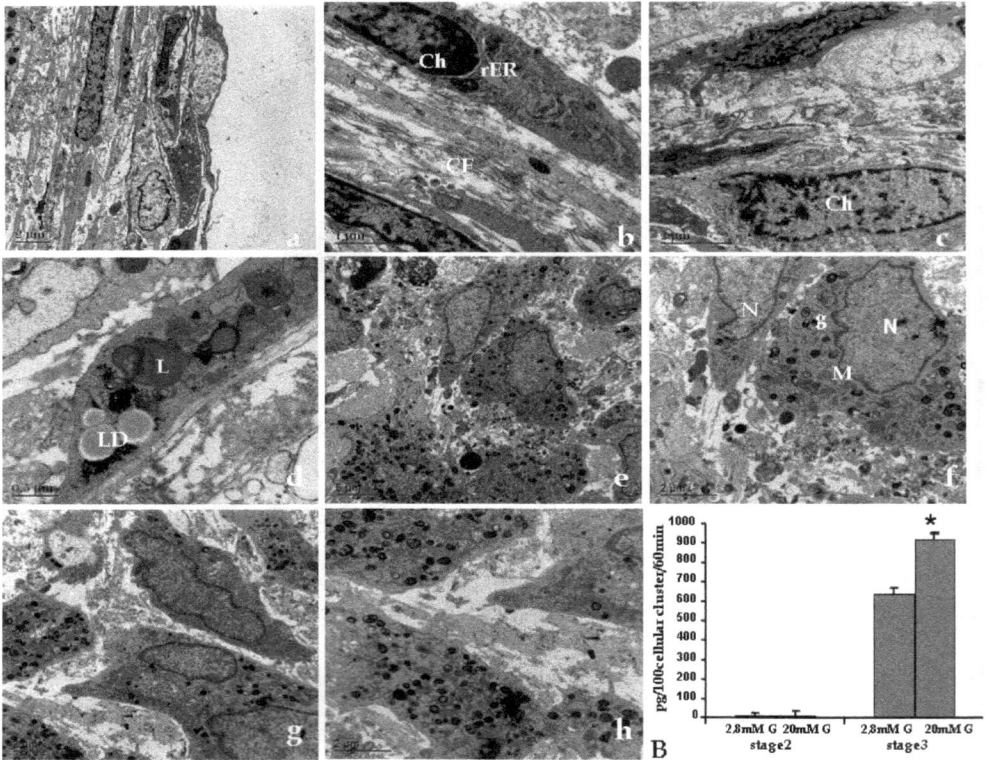

Fig. 9. Ultrastructural changes and insulin release in the Mitoxantrone-resistant cells at stage 3 after differentiation treatment with medium 3. (A) Transmission electron micrographs of undifferentiated cells (a-d) show high hypertrophy in the rough endoplasmic reticulum (rER), lipid droplets (LD), lysosomes (L) and collagenous fibers (CF). Two types of electron-dense chromatin structure were observed (Ch). However, the differentiated cells (e-h) presented a homogenous size with a round nucleus (N), at times indented, abundant mitochondria (M), and electron-dense granules in the cytoplasm were observed (g). (B) Insulin secretion after 1 hour of glucose stimulation at 20 mM vs. 2.8 mM. The results were normalized to 100 cell clusters (n=3) * p< 0.05 (employing Student's t-test) (Reproduced with permission, from Mato E. et al. Identification of a pancreatic stellate cell population with properties of progenitor cells: new role for stellate cells in the pancreas . Biochem. J. 421 ;181–191© the Biochemical Society).

Gene expression and ultrastrucutral changes detected in the cell culture growth support the idea of the ability of cells to release insulin into the medium. In this scenario, insulin secretions of several sets of cell clusters were measured by static incubation at low (2.8mM) and high (20mM) levels of glucose. Eventhough, insulin levels detected in the cell clusters were lower compared to mouse islets, an increase of 44% was detected after stimulating cellular clusters with high level of glucose. (Fig. 9 B). However, future experiments will have to demonstrate that the secretion of insulin is not only constitutive (Kuliawat R et al. 1994). Furthermore, the expression of specific markers of stellate cells remained after maintaining the cell in matrigel condition. These results may indicate the differentiation process has not been fully completed and the cells still maintained characteristics of stellate cells (Fig. 8).

An interesting strategy in order to investigate the biology of these cells is the use of proteomic approaches, since it is a useful tool for displaying protein expression patterns in the cell. For that reason, this approach has been used in active as well as quiescent stellate cells. (Kawada N et al. 2001; Pauki JA et al. 2011 (a); Paulo JA et al. 2001 (b); Wehr AYet al. 2011). In this context, the proteomic study of our cellular culture secretome was preformed. The results showed that some of these proteins have potentially great influence on the physiology of the stellate cells themselves and/or on neighbouring cells, indicating a paracrina and /or autocrine action. Moreover, we have identified some novel factors that were clustered in the differentiation/development-related proteins, such as AHNAK, Gap43, and DIXDC1 (unpublished data from Mato E et al). However, further experiments are required to investigate the interaction within these different genes.

In summary: The pancreatic stellate cells is a fascinating nonendocrine cellular model that could represent a new source of cells involved in regenerative medicine of the pancreas in the future. However, more studies are needed to understand the molecular mechanisms that control their cellular plasticity. Certainly, the use of imunocytochemical and immunohistochemical techniques, complemented with cell -tracking methods, will be important tools to unravel the role of these cells during the tissular regeneration process both in the pancreas and in the liver.

5. Acknowledgment

The authors thank Scientific and Technical Services of the University of Barcelona (SCT-UB, Campus Casanova) for technical support with electron microscopy, and Julie Shouer-Leventhal for editorial assistance. This work was partially supported by the Spanish FIS grant from the Ministry of Health - FIS PI020881, by Sardà Farriol Research Program and CIBER-BBN and CIBERDEM are ISCIII (Instituto de Salud Carlos III) projects.

6. References

Abraham E.J., Leech, C.A., Lin, J.C., et al. (2002). Insulinotropic hormone glucagons-like peptide-1 differentiation of human pancreatic islet-derived progenitor cells into insulin-producing cells. Endocrinology 143,3152-3161

Apte M.V. et al. (1998). Periacinar stellate shaped cells in rat pancreas: identification, isolation, and culture. Cancer Res.43:128–133

Apte M.V., Haber PS, Applegate TL, et al. (1998). Periacinar stellate shaped cells in rat pancreas: identification, isolation, and culture. Gut 43:128–133

Apte M.V., Haber PS, Darby SJ, et al. (1999). Pancreatic stellate cells are activated by proinflammatory cytokines: implications for pancreatic fibrogenesis. Gut 44:534–541

Bachem M.G. et al.(1998). Identification, culture, and characterization of pancreatic stellate cells in rats and humans. *Gastroenterology* 115:421–432

Baba S., Fuji H, Hirose T, et al. (2004). Commitment of bone marrow cells to hepatic stellate cells in mouse. J. Hepatol 40, 255-260

Bonner-Weir S. (2000). Perspective: postnatal pancreatic cell growth. Endocrinology 141:1926-192

Bonner-Weir S., Baxter L.A., Schuppin G.T., et al. (1993). A second pathway for regeneration of adult exocrine and endocrine pancreas. A possible recapitulation of embryonic development. Diabetes 42:1715-1720

Bonner-Weir S., Inada A., Yatoh S., et al. (2008). Transdifferentiation of pancreatic ductal cells to endocrine β-cells. Biochem Soc Trans 36: 353–356

Bouwens L. and Pipeleers D.G. (1998). Extra-insular beta cells associated with ductules are frequent in adult hunman pancreas. Diabetologia 41, 629-633

Bunting K.D. (2002). ABC Trransporters as phenotypic markers and functional regulators of stem cells. Stem Cells 20, 11-20

Buchholz M., Kestler H.A., Holzmann K., et al. (2005). Transcriptome analysis of human hepatic and pancreatic stellate cells: organ-specific variations of a common transcriptional phenotype. J Mol Med 83,795-805

Deichmann D.S., Miller C.P. Jensen J. et al.(2003). Expresión and misexpression of members of the FGF and TGFbeta familias of growth factors in the developing mouse pancreas. Dev Dyn 226,663-674

Doyle L.A., Yang W, Abruzzo LV, et al. (1998). A multidrug resistance transporter from human MCF-7 breast cancer cells. Proc Natl Acad Sci USA 95: 15665–15670

Esni, F., Taljedal, I.B., Perkl, A.K., et al. (1999). Neural cell adhesion molecule (N-CAM) is required for cell type segregation and normal ultrastructure in pancreatic islets. J. Cell. Biol. 144,325-337

Elghari L., Cras-Meneur C., Czernichow, P et al. (2002). Role for FGFR2IIIb-medianted signals in controlling panceatic endocrine progenitor cell proliferation. Proc Nat Sci USA 99, 3884-3889.

Fernandes A., King L.C., Guz Y., et al. (1997). Differentiation of new insulin-producing cells is induced by injury in adult pancreatic islets. Endocrinology 138:1750-1762

Fetsch, P.A., Abati, A., Litman, T., et al. (2006). Localization of the ABCG2 mitoxantrone resistance-associated protein in normal tissues. Cancer Lett. 235, 84-92

Friedman S.L. (2000). Molecular regulation of hepatic fibrosis, an integrated cellular response to tissue injury. J.Biol.Chem 275:2247-2250

Friedman S.L. (2008). Hepatic Stellate Cells: Protean, Multifunctional, and Enigmatic Cells of the Liver. Physiol Rev January 88; 1 125-172

Fuchs E., Segre J.A .(2000). Stem cells: a new lease on life. Cell 100:143-155

Galambos J.T., Hollingsworth MA, et al. (1977). The rate of synthesis of glycosaminoglycans and collagen by fibroblasts cultured from adult human liver biopsies. J Clin Invest 60: 107–114

Geerts, A. (2001). History, heterogeneity, development biology, and functions of quiescent hepatic stellagte cells. Semin Liver Dis 21, 311-255

Gershengorn M.C., Hardikar A. A. , Wei Ch ., et al. (2004). Epithelial-to-Mesenchymal Transition Generates Proliferative Human Islet Precursor Cells. Science 24: Vol. 306 no. 5705 pp. 2261-2264

Georges P.C. and Janmey P.A. (2005). Cell type-specific response to growth on soft materials. Journal of Applied Physiology; 98: 41547-1553

Gu D., Sarvetnick N. (1993). Epithelial cell proliferation and islet neogenesis in IFN-transgenic mice. Development 118:33-46

Guz Y., Nasir I., Teitelman G. (2001). Regeneration of pancreatic cells from intra-islet precursor cells in an experimental model of diabetes. Endocrinology 142:4956-4968

Goodell M.A., Brose K., Paradis G., et al. (1996). Isolation and functional properties of murine hematopoietic stem cells that are replicating in vivo. J Exp Med 1996; 183: 1797–806

Habener J.F., Kemp D.M. and Thomas M.K. (2005). Minireview: transcriptional regulation in pancreatic development. Endocrinology 146, 1025-1034

Haber PS, Keogh G.W., Apte M.V., et al. (1999). Activation of pancreatic stellate cells in human and experimental pancreatic fibrosis. Am J Pathol 155:1087–1095

Han Y.P., Zhou L., Wang J., et al. (2004). Essential role of matrix metalloproteinases in interleukin-1-induced myofibroblastic activation of hepatic stellate cell in collagen. J Biol Chem 279: 4820–4828

Henderson N.C., Mackinnon A.C., Farnworth S.L., et al. (2006). Galectin-3 regulates myofibroblast activation and hepatic fibrosis. Proc Natl Acad Sci USA 103: 5060–5065

Hirosawa K. and Yamada E., (1973). The localization of vitamin A in the mouse liver as revealed by electrón microscopy radioautography. J Electron Microsc (Tokyo) 22, 337-346

Hui, H., Wright, C., Perfetti, R. (2001). Glucagon-like peptide 1 induces differentiation of islets duodenal homeobox-1-positive pancreatic ductal cells into insulin-secreting cells. Diabetes 50,785 –796

Ito T. (1951). Cytological studies on stellate cells of kupffer and fat storing cells in the capillary wall of human liver (abstract). Acta Abat Jpn; 26:42

Ito T. (1973). Recent advances in the study on the fine structure of the hepatic sinusoidal wall: A review. Gunma Rep Med Sci, 6:119-163

Itoh N. and Ornitz D.M. (2004). Evolution of the Fgf and Fgfr gene families. Trends Genet 20,563-569

Ishiwata T., Kudo M, Onda M et al. (2006). Defined localization of nestin-expressing cells in L-arginine-induced acute pancreatitis. Pancreas 32,360-368

Ikejiri N. (1990). The vitamin-A storing cells in the human and rat pancreas. Kurume Med J 37:67–81.

Jasper, R. (2004). Molecular regulation of pancreatic stellate cell functions. Ml Cancer 3, 26

Jensen J. (2004). Gene regulatory factors in pancreatic development. Dev Dyn 229, 176-200

Jesnowski R., Müller P., Schareck W., et al. (1999). Immortalized pancreatic duct cells *in vitro* and *in vivo*. Ann NY Acad Sci; 880:50–65.

Kalinichenko V.V., Bhattacharyya D, Zhou Y, et al. (2003). Foxf1 +/– mice exhibit defective stellate cell activation and abnormal liver regeneration following CCl4 injury. Hepatology 37: 107–117

Kawada N. , Kristensen D.B. , Asahina K., et al. (2001). Characterization of a Stellate Cell Activation-associated Protein (STAP) with Peroxidase Activity Found in Rat Hepatic Stellate Cells· The Journal of Biological Chemistry, 276, 25318-25323

Kim S.K. and Hebrok M. (2001). Intercellular signals regulating pancreas development and function. Gene Dev 15, 111-127

Kordes C., Sawitza I., Haussinger D. (2009). Hepatic and pancreatic stellate cells in focus. Biol Chem 390:1003–1012.

Kodama S., Toyonaga T. Kondo T. et al. (2005). Enhaced expresión of PDX-1 and Nng3 by exendin-4 durin beta cell regeneration in STZ-rtrated mice. Biochem Biophys Res Commun 327, 1170-1178

Kordes C., Sawitza I., Häussinger D. (2009). Hepatic and pancreatic stellate cells in focus. Biol Chem. Oct;390(10):1003-12.

KordesC., Sawitza I., Muller-Marbach A. et al. (2007). CD133+ hepatic stellate cells are progenitor cells. Biochem Biophys Res Commun 352,410-417

Kubota H., Yao H.L. and Reid L.M .(2007). Identifiaction and characterization of vitamin A-storing cells in fetal liver: implications dor functional importance of hepatic stellate cells in liver development and hematopoiesis. Stem Cells 25,2339-2349

Kruse M.L., Hildebrand P.B., Timke C., et al. (2001). Isolation, long-term culture, anc characterization of rat pancreatic fibroblastoid/stellate cells. Pancreas;23:49–54

Kuliawat R. and Arvan P. (1994). Distinct molecular mechanism formprotein sorting within immature secretory granules of pancreatic beta-cells. J Cell Bio 126,77-86.

Lardon J., Rooman I., Bouwens L. (2002). Nestin expression in pancreatic stellate cells and angiogenic endothelial cells. Histochem Cell Biol. 2002 Jun;117(6):535-40. Epub 2002 May 14

Lee C.S., De Leon D.D., Kaestner K.H. et al (2006). Regeneration of pancreatic islets after partial pancreatectomy in mice does not involve the reactivaion of neurogenin-Ƹ. Diabetes 55,269-272

Lechner A., Leech C.A., Abraham E., et al. (2002). Nestin-positive progenitor cells derived from adult human pancreatic islets of Langerhans contain side population (SP) cells defined by expression of the ABCG2 (BCRP1) ATP-binding cassette transporter. Biochem. Biophys. Res. Commun. 293,670-674

Leslie E. M., Deeley, R.G., Cole, S. P. (2005). Multidrug resistance proteins: role of P-glycoprotein, MRP1, MRP2, and BCRP (ABCG2) in tissue defense. Toxicol. Appl. Pharmacol. 204, 216-237

Litman T., Brangi M, Hudson E, et al. (2000) The multidrug-resistant phenotype associated with overexpression of the new ABC half-transporter MXR (ABCG2). J Cell Sci 113(Part 11): 2011–2021

Löhr M., Müller P., Zauner I., et al. (2001). Immortalized bovine pancreatic duct cells become
    tumorigenic after transfection with mutant k-ras. Virchows Arch 438: 581–590

Madsen O.D., Jensen J., Blume N., et al. (1996). Pancreatic development and maturation of
    the islet cell studies of pluripotent islet cultures. Eur J Biochem 242:435-445

McCarroll J.A., Phillips P.A, Kumar R.K., et al. (2004). Pancreatic stellate cell migration: role
    of the phosphatidylinositol 3-kinase(PI3-kinase) pathway. Biochem Pharmacol
    67:1215-25

Masamune A., Kikuta K, Satoh M, Satoh K, Shimosegawa T (2003). Rho kinase inhibitors
    block activation of pancreatic stellate cells. Br J Pharmacol 140:1292-1302

Masamune A., Satoh M., Kikuta K., et al. (2003). Establishment and characterization of a rat
    pancreatic stellate cell line by spontaneous immortalization. World J Gastroenterol
    9:2751-2758

Masamune A., Satoh M., Kikuta K., et al. (2003). Inhibition of p38 mitogen-activated protein
    kinase blocks activation of rat pancreatic stellate cells. J Pharmacol Exp Ther 304:8-
    14

Masamune A., Kikuta K., Satoh M,. et al. (2002). Ligands of peroxisome proliferator-
    activated receptor-γ block activation of pancreatic stellate cells. J Biol Chem
    277:141-147

Mato E., Lucas M., Petriz J. et al. (2009). Identification of a pancreatic stellate cell population
    with properties of progenitor cells: new role for stellate cells in the pancreas .
    Biochem. J. 421 ;181–191

Mizrak D., Brittan M., Alison M.R. (2008). "CD133: Molecule of the moment". J Pathol 214
    (1): 3–9

Miralles F., Czernichow P., Ozaki K. et al. (1999). Signaling through fobroblast growth factor
    receptor 2b plays a key role 96,6267-6272

Miyake K., Mickley L, Litman T, et al. (1999). Molecular cloning of cDNAs which are highly
    overexpressed in mitoxantrone-resistant cells: demonstration of homology to ABC
    transport genes. Cancer Res 59: 8–13

Morita M. et al. (1998). Analysis of the sinusoidal endothelial of the feta rat liver a sinusoidal
    enthelial cell specific antibodyt, SE-1. Cell Struct Funct 23:341-348

Morini S., Carotti S, Carpino G, et al.(2005). GFAP expression in the liver as an early marker
    of stellate cells activation. J Anat Embryol. 110(4):193-207

Murakami K., Abe T., Miyazawa M., et al. (1995). Establishment of a new human cell line,
    LI90, exhibiting characteristics of hepatic Ito (fat-storing) cells. Lab Invest 72: 731–
    739

Naito N. and Wisse E. (1977). Observation on the fine structure and cytochemistry of
    sinusoidal cells in fetal and neonatal rat liver, In: Wisse E., Knook D., ed. Kupffer
    Cells and Other Lever Sinusoidal Cells. Amstenrdam: Elvesier/North Holland
    Biochemicak Press; 497-505

Neyzen S., Van de Leur E., Borkham-Kamphorst E., et al. (2006). Cryopreservation of
    hepatic stellate cells. J Hepatol 44: 910–917

Niki T., Pekny M., Hellemans K., et al. (1999). Class VI intermediate filament protein nestin
    is induced during activation of rat hepatic stellate cells. Hepatology. 29(2):520-7

Niwa H., Miyazaki J., Smith A.G. (April 2000). "Quantitative expression of Oct-3/4 defines differentiation, dedifferentiation or self-renewal of ES cells". Nat. Genet. 24 (4): 372–6

No authors listed (1996). Hepatic stellate cell nomenclature. Hepatology 23(1):193

Norgaard G.A., Jensen J.N., Jensen J. (2003). FGF10 signaling maintauns the pancreatic progenitor cell state revealing a novel role of Notch in organ development. Dev Bio 264,323-338.

Ogawa M., La Rue A.C., Drake C.J. et al (2006). Hematopoietic origin of fibroblasts/myofibroblasts: Its patholophysiologic implications. Blood 108, 2893-2896

Omary M.B., Lugea A., Lowe A.W., et al. (2007). The pancreatic stellate cell: a star on the rise in pancreatic diseases. J Clin Invest 117: 50–59

Paulo J.A., Urrutia R., Banks PA, et al. (2011). Proteomic analysis of a rat pancreatic stellate cell line using liquid chromatography tandem mass spectrometry (LC-MS/MS). Proteomics. Sep 25

Paulo J.A., Urrutia R., Banks P.A., et al. (2011). Proteomic Analysis of an Immortalized Mouse Pancreatic Stellate Cell Line Identifies Differentially-Expressed Proteins in Activated vs Nonproliferating Cell States. J Proteome Res. 2011 Oct 7;10(10):4835-44

Petropavlovkaia M. and Rosenberg .L (2002). Identification and characterization of small cells in the adult pancreas: potential progenitor cells? Cell Tissue Res 310, 51-58.

Phillips PA, Wu M J , Kumar RK , et al. (2003). Cell migration: a novel aspect of pancreatic stellate cell biology Gut 52:677-68.2

Pinzani, M. (1995). Novel insights into the biology and pahysiology of the Ito cell. Pharmacol Ther 66, 387- 412

Ramiya V.K., Maraist M., Arfors K.E., et al. (2000). Reversal of insulin-dependent diabetes using islets generated in vitro from pancreatic stem cells. Nature Med 6:278-282

Rafaeloff R., Pittenger G.L., Barlow S.W., et al. (1997). Cloning and sequencing of the pancreatic islet neogenesis associated protein (INGAP) gene and its expression in islet neogenesis in hamsters. J Clin Invest 99:2100-2109

Ratineau, C., Duluc, I., Pourreyron, C., et al. (2003). Endoderm- and mesenchyme-dependent commitment of the differentiated epithelial cell types in the developing intestine of rat. Differentiation 71,163-169

Reusens B. and Remacle C. (2006). Programming of the endocrine pancreas by the early nutricional environment. Int J Biochem Cell Bio 38, 913-922

Royer, P.J., Tanguy-Royer, S., Ebstein, F., et al. (2006). Culture medium and protein supplementation in the generation and maturation of dendritic cells. Scandinavian Journal of Immunology 63,401-409

Rosenberg L. (1998) Induction of islet cell neogenesis in the adult pancreas: the partial duct obstruction model. Microsc Res Tech 43:337-346

Sander M. and German MS (1997). The beta cell transcription factors and development of the pancreas J Mol Med 75,327-340

Satoh M., Masamune A, Sakai Y, et al. (2002). Establishment and characterization of a simian virus 40-immortalized rat pancreatic stellate cell line. Tohoku J Exp Med 198:55-69

Sparmann G, Hohenadl C, Tornoe J, et al. (2004). Generation and characterization of immortalized rat pancreatic stellate cells. Am J Physiol Gastrointest Liver Physiol;287:G211–G219

Scharfmann R. (2000). Control of early development of the pancreas in rodent and humans: implications of signals from the mesenchyme. Diabetologia 43, 1083-1092

Seaberg R.M., Smukler SR, Kieffer TJ et al (2004). Clonal identification of multipotent precursors from adult mouse pancreas that generate neural and pancreaticv lineages.Nat Biotechnol 22 1115-1124

Soria B., Skoudy A., Martin F. (2001). From stem cells to beta cells: new strategies in cell therapy of diabetes mellitus. Diabetologia 44:407-415

Steiling H. and Werner S. (2003) Fibroblast growth factors: key players in epithelial morphogenesis, repair and cytoprotection. Curr Opin Biotechnol 14, 533-537

Susking D.L. and Muench M.O., (2004). Searching for common sgtem cells of the hepatic and hematopoietic systems in the human fetal liber: CD34+ cytokeratin 7/8+ cells express markers for stallte cells. J Hepatol 40, 261-268.

Suzuki A., Nakauchi H., Taniguchi H. (2004). Prospective isolation of multipotent pancreatic progenitors using flow-cytometric cell sorting. Diabetes.53(8):2143-52

Sttaford, D., White, R.J., Kinkel, M.D., et al. (2006). Retinoids signal directly to zebrafish endoderm to specify insulin-expressing beta-cells. Development 133,949-956

Tanaka Y., Slitt A.L.,Leazeer T.M., et al. (2005). Tissue distribution and hormonal regulation of the breast cancer resistence protein (Bcrp/Abcg2) in rats and mice. Biocehm.Biophys.Res.Commun. 326, 181-187

Von Kupffer C. Ueber Sternzellen der leber. In: Abdruck aus Verhandlungen der Anatomischem Gesellschaft auf der 12 Versammlung in Kiel vom 17-20 April 1898, ed von Bardeleben K., pp80-86 (Gustav Fischer, Jena)

Vogel S., Piantedosi R., Frank J., et al. (2000). An immortalized rat liver stellate cell line (HSC-T6): a new cell model for the study of retinoid metabolism in vitro. J Lipid Res 41: 882–893

Watari N., Hotta Y., Mabuchi Y. (1982). Morphological studies on a vitamin A-storing cell and its complex with macrophage observed in mouse pancreatic tissues following excess vitamin A administration. Okajimas Folia Anat. Jpn.58:837–858

Wehr A.Y., Furth E.E., Sangar V., et al. (2011). Analysis of the human pancreatic stellate cell secreted proteome. Pancreas. 2011 May;40(4):557-66

Weissman I.L. (2000) Stem cells: units of development, units of regeneration, and units in evolution. Cell 100:157-168.

Yue F., Cui, L., Johkura, K., et al. (2006). Glucagon-like peptide-1 differentiation of primate embryonic stem cells into insulin-producing cells.Tissue Engineering 12,2105-2115

Zimmermann A., Gloor B., Kappeler A. et al. (2002). Pancreatic stellate cells contribute to regeneration early after acute necrotising pancreatitis in humans. Gut 51,574-578

Zulewski H., Abraham E.J., Gerlach M.J., et al. (2001). Multipotential nestin-positive stem cells isolated from adult pancreatic islets differentiate ex vivo into pancreatic endocrine, exocrine, and hepatic phenotypes. Diabetes 50:523-533

Zhou S., Schuetz J.D., Bunting K.D., et al. (2001). The ABC transporter Bcrp1/ABCG2 is expressed in a wide variety of stem cells and is a molecular determinant of the side-population phenotype. Nat Med 7: 1028–34

11

The Schwann Cell-Axon Link in Normal Condition or Neuro-Degenerative Diseases: An Immunocytochemical Approach

Alejandra Kun[1,2] et al.[*]
Department of Proteins & Nucleic Acids,
Instituto de Investigaciones Biológicas Clemente Estable (IIBCE), Montevideo,
Uruguay

1. Introduction

Peripheral nerve axons of mammals have been demonstrated to contain ribosomes (Court et al., 2008, 2011, Kun et al. 2007, Li et al. 2005a and 2005b, Sotelo et al. 1999), as well as specific mRNAs that have been shown to concentrate in specific peripheral axonal domains (Koenig & Martin 1996, Koenig et al., 2000, Sotelo-Silveira et al., 2006, 2008), the so called Periaxoplasmic-Ribosomal-Plaques (PARPs). Two possible origins have been proposed to supply mRNAs and ribosomes to axons and PARPs: a) from neuronal body axonal transport, or b) Schwann cell to axon trans-cellular transfer (Court et al. 2008, 2011, Sotelo-Silveira et al. 2006, Sotelo et al., to be published elsewhere). We showed that Schwann cell provide newly synthesized RNA (Bromouridine -BrU- labeled RNA) to the axon by a transcellular transfer process. This newly synthesized RNA was provided by Schwann cell nucleus and transported to the axon throughout Schmidt-Lanterman Incisures, and/or Nodes of Ranvier using the actin network, using molecular motors such as Myosin-Va. This was found in normal regenerating nerves disconnected from their neuronal body of origin, meaning that the only possible origin of this axonal RNA is the Schwann cell (to be published elsewhere).

The transfer of mRNAs and ribosomes from Schwann cell to the axon in normal or regenerating nerve fibers we found, make us think about which role it may play in neurodegenerative diseases. Mice models of Charcot-Marie-Tooth (CMT, Trembler-J mouse, Patel 1992, Suter 1992), as well as human nerve samples of CMT patients, were analyzed in

[*] Gonzalo Rosso[1], Lucía Canclini[1], Mariana Bresque[1], Carlos Romeo[1], Karina Cal[1], Aldo Calliari[1,3], Alicia Hanuz[1], José Roberto Sotelo-Silveira[4,5] and José Roberto Sotelo[1]
1 *Department of Proteins & Nucleic Acids, Instituto de Investigaciones Biológicas Clemente Estable (IIBCE), Montevideo, Uruguay*
2 *Biochemical Section, School of Science, Universidad de la República (UdelaR), Montevideo, CSIC Project, UdelaR, Uruguay*
3 *Biophysics Area, School of Veterinary, (UdelaR), Montevideo, Uruguay*
4 *Cell and Molecular Biology Department, School of Sciences, (UdelaR), Montevideo, Uruguay*
5 *Department of Genetics, Instituto de Investigaciones Biológicas Clemente Estable (IIBCE), Montevideo, Uruguay*

order to study the metabolic characteristics of this Schwann cell-axon relationship may have to the pathogenesis of this important human illness.

CMT is the most frequent genetic peripheral neuropathy (1/2500 prevalence, Berciano & Combarros, 2003, Inherited Peripheral Neuropathies Mutation Database: www.molgen.ua.ac.be/cmtmutations). This chronic progressive illness has two possible origins: axonal, CMT-II (neurofilament, KIF 1B or Rab7 protein mutations, among others), or Schwann cell, CMT-I (PMP-22, Conexin 32, P0 protein mutations, among others, Mersiyanova et al, 2000, Pérez-Ollé et al, 2002, Verhoeven et al, 2003, Zhao et al, 2001). Regardless the initial alteration, all CMT end in a functional axonopathy, which emphasize the importance of Schwann cell-axon relationship in the context of the gene expression of both cells. This local supply of transcripts may be altered in CMT, probably causing the final pathologic phenotype.

The histological analysis of CMT (I or II) patients' nerves showed the conventional myelin or axonal typical alterations (onion bulbs, axonal ovoids, internode shortening, fiber diameter variations, paranodal remyelination, etc.), plus a marked increase of axonal sprouting (myelinopathies). Molecular composition of CMT1 human patients, normal rats, PMP-22 mutant mice nerves, as well as mice organotypic dorsal root ganglia culture, was characterized here. More and more, mutant animals, transgenic animals, transfected cell culture, or cell culture obtained from any of these animal types are used to unravel the pathogenesis of important human diseases. The present paper contribute to the understanding of human Charcot-Marie-Tooth syndrome, because as we will describe below we found abnormal distribution of mutant PMP-22 transcript and protein, but also an irregular accumulation of ribosomes on altered Schwann cells and axons.

2. Schwann cell organization. Characterization of normal human and rat Schmidt Lanterman Incisures (SLI)

Whole mount of normal teased human and rat peripheral nerve fibers preparation let us know the normal interrelation between glia cell and axons in PNS. Teased fibers of Human Sural nerves and rat Sciatic nerves, (immunostained by floating), permitted to characterize the molecular expression of both nerve cells, Schwann cell and neuron (axonal domain). Internodal non-compact myelin mainly represented by Schmidt Lanterman Incisures (SLI) has been clearly characterized in this type of whole mount. Confocal single stacks show the SLI immunoreactivity with antibodies against tubulin, vimentin, Myelin Associated Glycoprotein (MAG) and ribosomes, in human sural nerve teased fibers (arrows in Figure 1, A green, B, C and F, respectively). Ribosomes are present in Schwann cell cytoplasm, Nodes of Ranvier and SLI (Figure 1, F). Nucleic acids have been also found in SLI of human fibers, identified by a fluorescent specific probe (Yoyo-1) as can be seen in Figure 1 D. Central axonal area in human longitudinal fiber appear strongly stained with anti-Neurofilament-200kDa (Figure 1, E,I. Vimentin seems most evident in external Schwann cell and SLI cytoplasm (Figure 1, C). Filamentous actin, recognized by Phalloidin coupled to Alexa 546, shows a moderate signal in Schwann cell cytoplasm and axoplasm, and a more vigorous signal in SLI (Figure 1, A red). A similar signal pattern has been seen in rat sciatic teased fibers (Figure 1, J, K, L). A three dimensional reconstruction of confocal stacks series from a whole mount single fiber (rat), immunostained with anti-ribosomal antibody, let as to identify the spiral funnel-like path of

SLI. The whole reconstruction was 90° rotated to show the SLI image (Figure 1, M), enreached in ribosomes. The same SLI path image is outlined in Figure 1, N. The fine structure of myelinated sural human fibers show a well organized SLI (Figure 1 H, clear arrow). The compact myelin (Figure 1 H, white asterisk), surround the non compact myelin of SLI (Figure 1, H clear arrow). The external SLI domain is characterized by a well structured autotypical adherents type junction (Figure 1, H asterisk). Near this region, it can be seen the external mesaxón (Figure 1, H, black arrow). The external Schwann cell cytoplasm (Figure 1, H, eSc) appear clear, with evident cytoskeleton. Boxed area in H is enlarged in I. Close to the axolema, Schwann cell cytoplasm among the non compact myelin membranes of the SLI (Figure 1, I, asterisk), show the presence of a Multivesicular Body (Figure 1, I arrow). Near this region, the compact myelin, appear devoid of cytoplasm. Internodal regions constitute the largest domain of contact between the glial cell and axon. However, the molecular characterization of internodal transcellular interactions is barely known. Non-compact myelin is found in paranodal regions and Schmidt-Lanterman Incisures, which traverse diagonally compact myelin. It is postulated that, under normal conditions, the presence of cytoplasm in SLI would ensure protein turnover and vesicular trafficking (lysosomes, vesicles of endoplasmic reticulum) for the homeostasis of essential myelin and cell domains distant within the glial cell itself. These glial "shortcut" have been largely described in literature. The internodes SLI's number varies, depending on the species, axonal diameter and physiological conditions (Cajal, 1928, Ghabriel, et al., 1979a and b, 1980a and b, 1981, 1987; Hiscoe, 1947; MacKenzie et al., 1984; Robertson, 1958,). In the present work SLI have been characterized throughout the molecular expression of cytoskeleton components (actin, tubulin, vimentin) and adhesion molecules (MAG), showing that the SLI cytoplasm have well organized "roads" devoted to traffic. The presence of multi-vesicular bodies at the SLI have been seen in the past (Hall &Williams, 1970) as has been also described here, revealing an active vesicular metabolism in the cytoplasm of non compact internodal myelin. The expression of nucleic acid especially ribosomal RNA, indicate that SLI and Nodes of Ranvier have local traffic of translational machinery. Why ribosomes could be transported to these domains? One of the possible answer is the local synthesis of glial and myelin proteins. However, we noted other possible roles related to axon-glia homeostasis and axonal maintenance (Court et al., 2008; Kun et al., 2007, Sotelo-Silveira et al., 2006), which could be altered in pathological conditions. Some of our more recent results contribute to this hypothesis. Indeed, we found a SLI local expression of heavy neurofilament subunit mRNA and his final product, the corresponding protein, in normal and pathological conditions (manuscript en redaction). However it is important to highlight that trans-cellular traffic between Schwann cell and axon especially through SLI implies vesicular transport system inside SLI and a trans-endocytosis mechanism between both nerve cells. That would mean a different set of adhesion and signaling molecules in each part of this process, where MAG seems to be one of the candidates molecules involved. MAG expression is specific to myelin-forming cells in the early process of myelination. One of its functions is to promote initial interactions in the process of fastening the first layer of myelin around axons (inner mesaxon), and further development of myelin. But the level expression of MAG is relatively high, suggesting other possible roles. Among them, a particularly important role for MAG is the receptor binding axonal ligand (protein-ganglioside complex), which could activate intra-axonal signal transduction cascades necessary for the maintenance and survival of myelinated axons (Quarles, 2007).

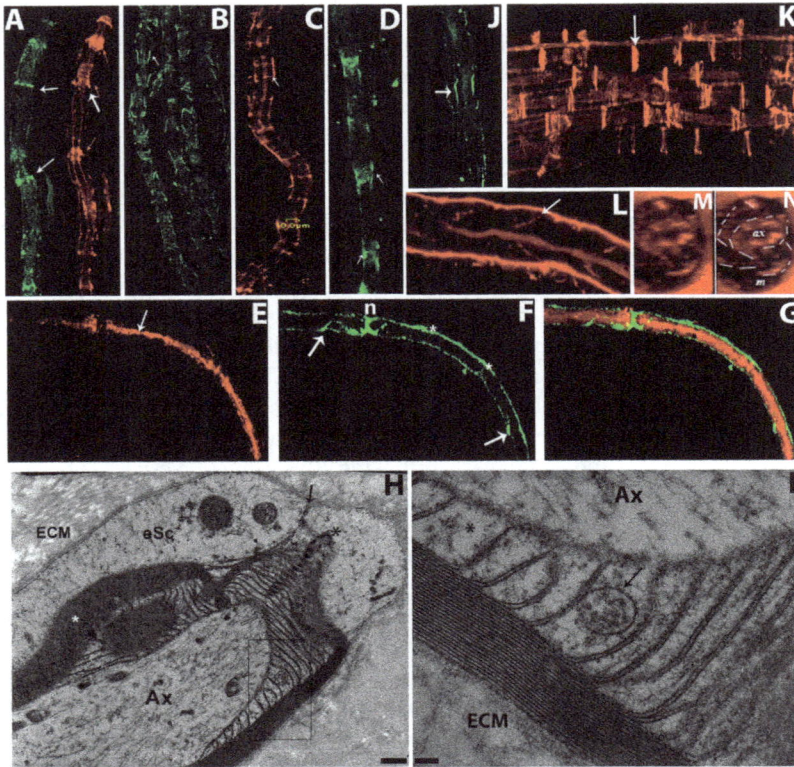

Fig. 1. Molecular characterization of normal Schmidt Lanterman Incisures (SLI) and Nodes of Ranvier (NR), in human and rat peripheral fibers.

Human SLI have been studied by immuno-confocal microscopy (A-G) and conventional electron microscopy (H and I). Human sural teased fibers were immunostained and observed by confocal microscopy. Single confocal planes show SLI (arrows) enriched in Myelin Associated Glycoprotein (A, green), Actin (A, red), Tubulin (B) Vimentin (C) and nucleic acid (D, Yoyo-1) and a few NR. Among them, the ribosomes (F, green) are present in Schwann cell cytoplasm (asterisk), in Nodes of Ranvier (F, n) and SLI (arrows). Central axonal domain appear strongly stained with anti-Neurofilament -200 kDa (E in red, arrow), merged image is shown in G showing how ribosomes are entering to the axon throughout NR (demonstraded by Z stacks analysis, not shown here) and SLI. Actin (K, in red), nucleic acids (J, YoYo-1, in green) and ribosomes (L in red) are also present in SLI of rat sciatic teased fibers. Three dimensional reconstruction of panel L, in which Z-stacks series was rotated 90° to show the SLI funnel spiral path image in M, enriched in ribosomes. The SLI path is outlined, in the same image, in N (ax, axoplasm; m, myelin). TEM, the ultrastructure of semi-longitudinal section of human sural fiber reveal a well organized SLI (H, clear arrow). In H, black asterisk shows the external part of SLI adherens junction, black arrow indicates the external mesaxon and white asterisk indicate compact myelin. The external Schwann cell cytoplasm (H, eSc) and the extra cellular matrix (H, ECM) are also indicated. Boxed area in H is enlarged in I. The presence of

Schwann cell cytoplasm among the non compact myelin membranes of the SLI (I, asterisk) might be observed, meanwhile it is completely absent in adjacent compact myelin. A Multivesicular Body (MVB) is present in the SLI cytoplasm (I, arrow). Bar in H represents 400nm, bar in I represents 100nm.

3. Schwann cell and axonal ribosomes have been identified in normal rat sciatic nerve fibers. Post-embedding immuno-gold staining

Normal rat sciatic nerve fibers have been explored to highlight the distribution of translational machinery, based on ribosomal recognition. An expected pattern of Schwann cell ribosomes (internal positive control) has been found by indirect postembedding immunogold staining (Figure 2, B), where it is possible to recognize immunocomplex in Schwann cell cytoplasm (Figure 2, B, box 8) and in myelinic region (Figure 2, B, boxes 8 and 9). Immunocomplex, recognized by gold particles are also present in the axoplasmic domains (Figure 2, A and B, box 10). Axoplasmic immunogold complexes (Figure 2, A, box 1) have a diverse distribution involving mitochondria (Figure 2, A, M) and their proximity (Figure 2, A, box 6). The immunocomplex seems to be associated to the cytoskeleton (Figure 2, A, boxes 2, 4 and 6), or in a multivesicular like-body (Figure 2, A, box 5). Axoplasmic polysomes not associated to cytoskeleton, has been found linked to immunogold particles (Figure 2, A, box 3). The samples were not osmificated previously to LRW inclusion, but exposed to osmium vapor, after immunostaining. Because that, the membranes are poorly countersained in LRW. The ultrathin sections were incubated with uranyl acetate and lead citrate as usual in Transmission Electron Microscocopy (TEM).The uranyl solution recognize the amino and phosphate groups of proteins and nucleic acid, while lead citrate is osmium-enhancing and bind to hydroxyl groups (also including phosphate groups). The absence of osmium staining before the inclusion, also result in a weak lead salt counterstaining. The Nano-gold particles have 15 nm in diameter.

TEM experimental evidences have demonstrated the presence of ribosomes in normal peripheral axons in different species including invertebrates (Kun et al., 1998 and 2007; Koenig and Martin, 1996; Zelena, 1972a, 1972b and 1970; Martin et al., 1989; Sotelo et al, 1999;). Axonal translational machinery could play an important role in structural and physiological maintenance, especially considering renewal and turn-over of proteins. Considering the Schwann cell as a positive internal control (as well as mitochondria, since the polyclonal antibody recognized also prokaryotic ribosomes), comparing their ribosome immunoreactions, axonal ribosomes appear associated to cytoskeleton in free polysomes-like complex. We have proposed that the axoplasmic ribosomes could have two possibles origins: the neuronal cell body and the neighbor Schwann cell (Kun et al. 2007; Sotelo-Silveira et al., 2006,). Peripheral normal axons run unusual long distances in cellular scale. However, its unusual geometry does not affect the homogeneity of the axoplasm. Cajal (1928), proposed that Wallerian degeneration was somehow in equilibrium with axon regeneration, and now we know that the axon is a dynamic steady state. Meanwhile, in hereditary peripheral neurodegenerative diseases (CMT, the most frequent peripheral human neuropathy), there is a chronic condition where the "normal" equilibrium is never reached. In addition, the chronic condition makes the displacement of the equilibrium always worst, because cellular repair mechanisms are insufficient to reverse the illness progress. In those conditions, the repair mechanisms are permanently activated. The resultant neuro-pathological phenotype, as occur with the normal phenotype, emerges from the integration of both Schwann cell and the axons (Aguayo et al, 1977; Salzer et al., 2008; Suter and Scherer, 2003,).

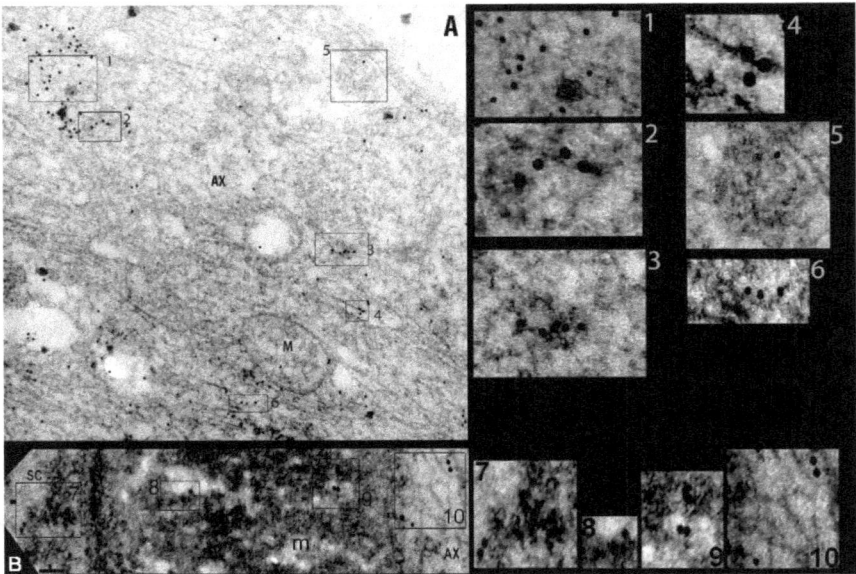

Fig. 2. TEM, normal rat sciatic nerve, postembedding ribosome immuno-gold staining.

Ultrathin transversal sections of rat sciatic nerve included in LRW hydrophilic resin was immunostained using a specific polyclonal antibody against ribosomes (raised in rabbit Kun et al. 2007), recognized in time by a goat-antirabbit gold conjugated antibody. It can be observed in A the presence of single or grouped axoplasmic immunocomplex, recognized by gold particles. The immunocomplex are framed in different boxes (boxes 1, 2, 3, 4, 5 and 6), some close to mitochondria (box 6) and within it (M, mitochondria), associated to the cytoskeleton (boxes 2, 4, 5 and 6), or in a Multivesicular like-Body (box 5). The polysomes domain are clearly decorated by immunogold particles (box 3).The boxed areas in A are enlarged in numbered boxes on the right hand (1 to 6). Different transversal fiber domains show immunogold staining (B, box 7, 8, 9 and 10). The material was included without conventional posfixation or staining in block (osmium tetroxide, uranil acetate). After imunostaining, the ultrathin sections were exposed to osmium vapors (to specially emphasize the membrane structure) and counterstained with uranyl acetate and lead citrate as usually used in TEM. The bar in A and B represent 100nm, each gold particle has 15nm in diameter. AX, axoplasm; M, mitochondria; m, myelin; SCc, Schwann cell cytoplasm.

4. Schwann cells' ribosomes are involved in axonal sprouting of human CMT-1 sural nerve fibers. Axonal sprouting is promoted by myelin compaction decrease

The presence of ribosomes has been used to recognize the Schwann cell citoplasmic domains in CMT-1 human patient whole mount of sural nerve teased fibers, immunostained by floating. A longitudinal confocal tridimensional reconstruction of one of that fibers is showed in Figure 3, A and B (original image is showed in A, but specific cellular domains have been outlined in B). Delaminated myelin (dM) let the Schwann cell cytoplasm (Figure 3, B, SCc) expand among their de-compacted layers, that appears strongly stained with ribosomes (Figure 3 A and B, red signal). Some of the ribosomal signals are also present at the axoplamic region (Figure 3B, merged yellow, asterisks) in the main axon (Figure 3 B, mAx) and in the origin of the new-born axon-sprout 1 (Figure 3B, oS1). Next to the main axon, a bulk of ribosomes also surround axonal sprouting path (Figure 3B, R). The axonal domains are identified by phosphorylated neurofilament protein (NF-P) signal (Figure 3, green). An inhomogeneous arrangement of NF-P signal distribution (characteristic of this pathology) is showed in the main axon (Figure 3B, mAx) and in the lateral sprout axons 1 and parallel axon sprout 2 and 3 (Figure 3 B, S1, S2 and S3). A similar pattern of expression is showed in other single human CMT-1 sural teased fiber showed in Figure 3 C, D and E. A most homogeneous distribution of NF-P signal is observed in that fiber. The main axon and their sprouting, NF-P signaled (Figure 3 C, green), also expand among the delaminated myelin, highly decorated by ribosomes (Figure 3 D, red signal). It is a longitudinal image of a conventional "onion bulb" typical diagnosis image, with specific molecule expression.

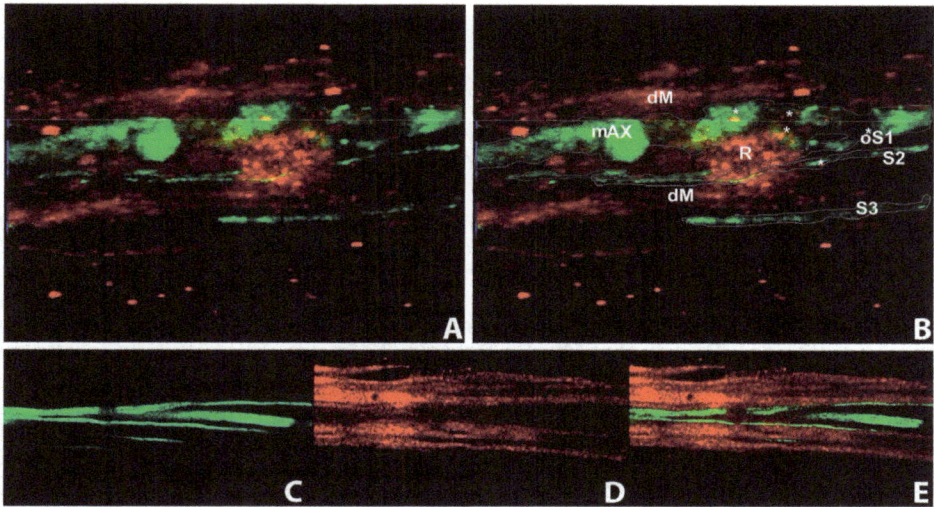

Fig. 3. Ribosomes and Phosphorylated Neurofilaments in Human CMT-1 sural teassed fiber. A. The original image is shown in A, while the specific cellular domains and components have been outlined and indicated in B. The image show a longitudinal single teased fiber from a human CMT1 sural nerve (confocal tridimensional reconstruction). The main axon (mAx) shows an inhomogeneous distribution of phosphorylated neurofilaments (green) and a lateral axoplasmic sprouting (S1) originated from the main axon (oS1). The delaminated myelin (dM) allows the Schwann cell cytoplasm (SCc) to expand among their membranes strongly inmunostained by the ribosomal antibody (red). The presence of ribosomes are also evident at the axoplasmic level (asterisks, merged color) in the main axoplasm and in the new-born axoplasm (sprouting, S1). A big conglomerate of ribosomes (R) appears close to the main axoplasm and it is crossed by the (S1). Sproutings running in parallel to the main axon can be also seen in the same image (S2, S3). A human teased sural fiber from other

CMT1 patient is showed in C, D and E. The axoplasm is also identified by the phosphorylated neurofilament (green, C) and the Schwann cell cytoplasm by the ribosomes (red, D). The image shows a longitudinal "onion bulb"-like organization (E, merged image). Also are evident the thin sprouts originated from the main axon, passing throughout the layers of delaminated myelin.

Alteration in PMP-22 represent 70% of myelinopathy (Young et al., 2003), in demyelinating human peripheral fibers, we observed axonal sprouting consequently with the lack of inhibitory effect of myelin and normal Schwann cell (Shen et al., 1998; De Bellard et al., 1996). This axonal growing is in close relation with structural and functional changes, as observed among myelin disassembly, axonal sprouting. Trembler-J mice (*Tr-J*) develop a neurodegenerative phenotype that is validated as an animal model of CMT1A (Devaux and Scherer, 2005; Sereda and Nave, 2006; Sidman et al, 1979). Gene mapping indicates that the primary defect is a mutation resulting in a leucine (16) to proline substitution in PMP-22 (Suter et al, 1992). The same amino acid substitution was found in a human family who suffered CMT1A (Valentijn et al, 1992). This substitution prevents normal protein folding, insertion into the membrane and normal myelination. Heterozygous mice (Trj/+) show a spastic paralysis and generalized tremor. While homozygous mice (TrJ/TrJ), show more severe peripheral myelin deficiency, causing death before weaning (Henry et al, 1983; Henry and Sidman, 1983; Suter et al, 1992). The main changes affecting the SNP, begins to be evident from post-natal day 20 (P20), showing a characteristic body tremor and the impossibility of abduction and extension of the hind legs (from P11), as we recently described (Rosso et al, 2010). Typically the mutation is confirmed by individual genotyping of mice (mainly by PCR-RFLP; Notterpek et al, 1997; Fortun et al, 2005; Khajavi et al, 2007, Rosso et al, 2010).

5. Schwann cell express a mutate pmp22 gen in CMT-1A animal model (Trembler mouse). Peripheral Myelin Protein-22 in vitro distribution is altered in Trembler J mice

The PMP-22 is a myelin protein whose mutation is characteristic of CMT1A human peripheral illness. Trembler J mice is an animal model to study CMT1A caused by *pmp22* mutation. The Schwann cell expression of *pmp22* (green) has been observed in normal (Figure 4, WT, A and B) and heterozygous (Figure 4, TrJ, D and E) organotypic culture of embryonic dorsal root ganglion (DRG). The wild type (WT, +/+) Schwann cell *pmp22* RNA is observed at cytoplasm and in perinuclear domains, excluding the nucleoplasm, that appear mostly empty of ISH signal (Figure 4, asterisk in A and B). The opposite is observed in Trembler J heterozygous (HZ, TrJ/+) DRG, where the *pmp22* expression is mostly concentrated into nuclear and perinuclear domains (Figure 4, asterisk in D and E). When the PMP-22 expression was analyzed in adult sciatic nerves, the immunocytochemical signal also appears concentrated in the nucleoplasm of HZ teased fibers (Figure 4F, green, asterisk). The adult WT teased sciatic fibers showed a cytoplasmic and perinuclear distribution, excluding the nucleoplasm (Figure 4C). Axoplasmic and Schwann cell domains are also counter stained with anti NF-68 antibody (blue signal) and Phalloidin-Alexa 545 (red signal) in Figure 4, C and F. In TrJ/+ mice the whole expression of *pmp22* is concentrated at Schwann cell nuclei region (Figure 4D, 4E and 4F).

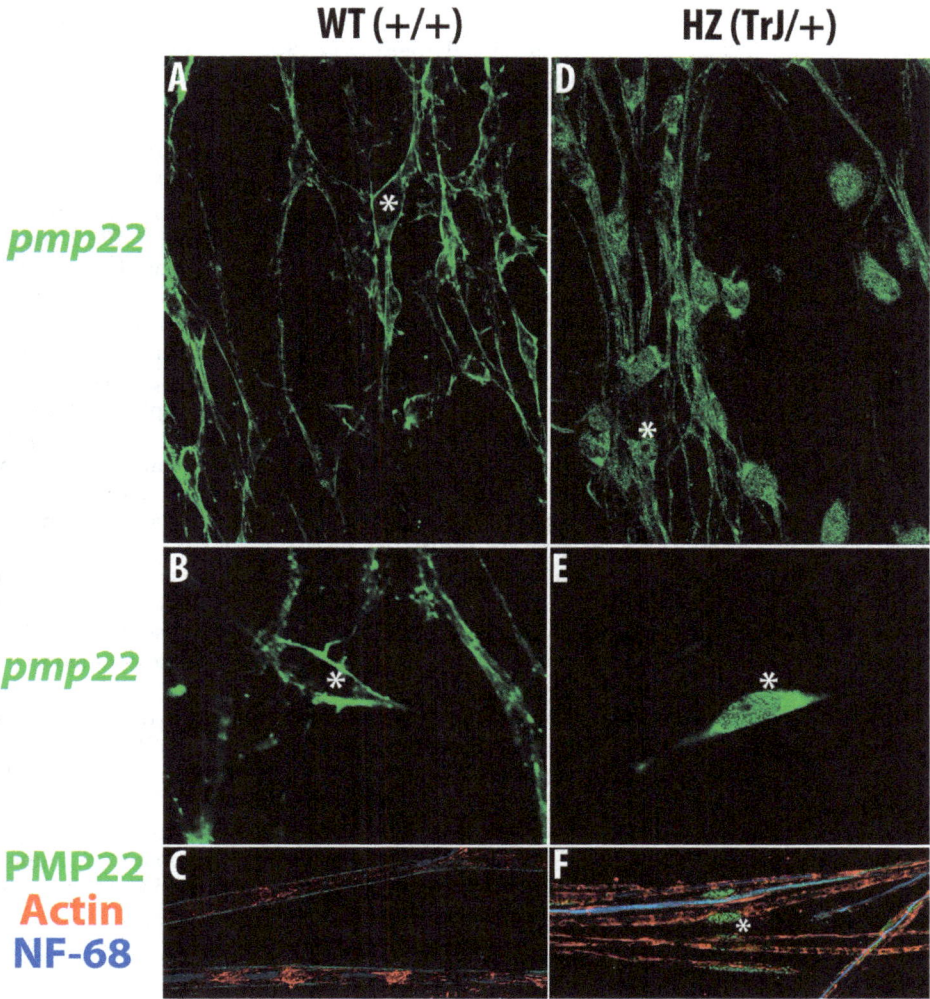

Fig. 4. Peripheral *pmp22* expression. The expression of peripheral myelin protein 22 gene might be observed in normal (WT, +/+, A and B) and heterozygous (HZ TrJ/+, D and E) cultured E13 DRGs. In normal genotype (+/+), the Schwann cell transcript distribution show a peri-nuclear cytoplasmic arrangement, remaining the nucleoplasm almost empty of In Situ Hybridization (ISH) signal (asterisk in A and B); meanwhile in the Schwann cell from mutants (TrJ/+), the whole nuclear domain appear filled of *pmp22* ISH signal (asterisk in D and E). B and E are different regions at higher magnification. C and F show immunocytochemical recognition of proteins: PMP-22 (green), Actin (red) and Neurofilament 68kDa subunit (blue) in WT +/+ (C) and HZ TrJ/+ (F) in teased sciatic nerve fibers from adult mice. As we showed for *pmp22* transcript the mutant HZ TrJ/+ Schwann cell nuclei not only contain the RNA, but the product of its translation PMP-22 protein (F, green). The opposite occurs in WT (A, B and C).

In myelinating Schwann cells from adult Trembler-J mice, PMP-22 protein accumulates in cytoplasmic aggregates, ER-Golgi compartments and is associated with other proteins in endosomes and lysosomes suggesting high levels of protein degradation (Suter and Snipes, 1995a, 1995b). In normal Schwann cells, ~80% of the newly-synthesized PMP-22 is degraded within 30 min by the proteasome, likely due to inefficient folding (Notterpek et al 1999a; Ryan et al., 2002). The proteasome is a multi-catalytic complex involved in a variety of cellular processes, including the degradation of short-lived proteins (Goldberg, 2003). PMP-22 is a short-live molecule, that form aggregates when the proteasome is inhibited or the protein is mutated (Fortun et al., 2007). It is conceivable that the amount of PMP-22 targeted for degradation is increased in the gene duplication and point mutations disease, which could overwhelm the proteasome and lead to the accumulation of miss-folded proteins along the secretory pathway. Removal of pre-existing PMP-22 aggregates is assisted by autophagy, and chaperones/autopaghy induction can suppress accumulation of PMP-22 aggregates. The expression of pmp22, observed in WT (+/+) DRG cultures, occurs from early stages of culture, showing a discrete distribution in any glia cytoplasm, including perinuclear region, correlating strongly with PMP-22 expression (data not shown). Adjacent axons show the presence of phosphorylated neurofilament, typically with discontinuous distribution (data not shown). However, in HZ (TrJ/+) DRGs, the pmp22 distribution is concentrated almost exclusively in glial nuclear and perinuclear regions, being absent from the glial myelin domains along the axon. In adults TrJ/+ fibers the nuclear and perinuclear PMP-22 expression confirm a pattern of mutated pmp22 expression conserved from early myelination to adult stage. PMP-22 signals are granular and form a bulk of molecules, similars to those described as aggresomes. It has been described that aggresome formation is accompanied by redistribution of cytoskeleton components. Intermediate filaments, play key role forming a condense cage surrounding a pericentriolar aggregated and ubiquitinated proteins. A growing number of disease-associated proteins have been found to accumulate in aggresomes, including peripheral myelin protein 22 (PMP-22) (Notterpek et al., 1999; Ryan et al., 2002), huntingtin (Waelter et al., 2001), parkin, and alpha-synuclein (Junn et al., 2002). However, the presence of pmp22 transcript and PMP-22 protein signals in nuclear domains during myelinogenesis and in TrJ/+ adult fibers, respectively (Fig. 4), suggest other type of alterations. Would they be, a) an altered post-transcriptional regulation, and/or b) an altered motor protein link, potentially involved in pathogenesis and illness consolidation.

6. Experimental procedures

6.1 Animal care and maintenance

Rattus norvegicus (Sprague Dawley) were obtained from the IIBCE colony. TremblerJ (B6. D2-Pmp22 <Tr-j>/J, Jackson Laboratory, USA) mice colony was started in 2008 (CSIC Grant-Universidad de la República 2005-2009). These mice carries a spontaneous mutation in Peripheral Myelin Protein-22 (PMP-22). All mice are recorded, numbered and genotyped following the method previously described (Rosso et al, 2010). Harems are formed after determining the stage of the female estral cycle. Pregnancy starts is determined by the examination of vaginal exudates and controlled by weekly weight of females. The colony, have now 100 animals, living in isolated cages, stored in isolated rooms.

All animals are maintained under controlled temperature and light cycle. Water and food are supplied ad libitum. The animal housing conditions are in agreement with the National

Committee for Animal Care and Maintenance (CHEA-Universidad de la República-Uruguay, *www.chea.udelar.edu.uy*)

6.2 Trembler J mice genotyping

Genomic DNA was extracted using a phenol–chloroform based method. DNA concentration was determined by spectrophotometry. PCR was performed on 200ng of DNA using specific primers (forward: 5'-GTTCCAAAGGCAAAAGATGTTC-3'; reverse: 5'-AACAATAAT CCCAAACCACACTTC-3') that flanked the mutation site. PCR products were digested with BfaI (Fermentas) for 2h at 37 °C and separated by 6% polyacrylamide gel electrophoresis (PAGE). The digestion products were stained with AgNO3. The BfaI digestion of the amplified fragment from the wild-type allele produced two fragments of 221 and 500 basepairs (bp).The TrJ mutation results in a loss of the BfaI site. Consequently, the amplified fragment obtained using the TrJ allele as template was visualized as a single band of 721bp.

6.3 DRGs organotypic culture

Thirteen day mice embryos (E-13) were euthanized and Dorsal Root Ganglia were collected for organotypic culture in appropriated media (Neurobasal (Invitrogen) complemented by 0,20 ml/ml B27, 0,01µg/ml Nerve Growth Factor (NGF) and 2mM glutamax). The culture was done at 37°C under 5% CO_2 controlled atmosphere. The cultured medium was changed every 48 hours. At 16 days of dorsal root ganglia culture, the culture media was complemented with 50µg/ml Ascorbic Acid, to promote *in vitro* myelination.

6.4 Fixation

Rats and mice were euthanized under pentobarbital anesthesia following the Uruguayan Committee for Ethical Animal Experimentation (CHEA in Spanish). Rats were euthanized by intracardiac perfusion of fixative (4% paraformaldehyde (PFA) for confocal microscopy and 0.25% glutaraldehyde was added for Electron Microscopy (EM) in PHEM (25mM HEPES, 10 mM EGTA, 60 mM PIPES, 2mM MgCl2 , adjusted to pH 7,2- 7,6, with KOH) after heparinization. Sciatic nerves were excised immediately after perfusion (cut in 2-mm pieces), pre-immersed in the same fixative solution for 2 hr, then washed in PHEM for 2 hr (gentle stirring), changing solution every 10 min. Mice sciatic nerves were excized and fixed by immersion in 3% PFA in PHEM, 30 minutes, at 4°C. The samples were then washed 6X5 minutes with gentle stirring. Cultured ganglia were fixed by 10 min immersion in 2% PFA in PHEM. Throughly washed in PHEM 1 hour (6 X 10 min). The samples followed different pre-treatement before in situ molecular studies.

6.5 Human samples

Clinically diagnosed patients with a family history of polyneuropathy, presenting a chronic sensory-motor polyneuropathy were included in the present study. Electrophysiological studies were performed to confirm the nature of the demyelinating neuropathy prior to their inclusion. Alcoholics, diabetics, exposed to toxic or neurotoxic drugs and/or bearers of systemic diseases patients were specifically excluded. Electrophysiological studies were: nerve conduction velocity analysis, electromyogram and quantification of motor units. Patients meeting the inclusion criteria have been informed of the study and samples were taken only after they signed an Informed Consent to participate in the CSIC I+D (2005-2009) Project (Grant from Universidad de la República, Uruguay). Human control was obtained from Human whole donors from the "Instituto Nacional de Donantes y Trasplantes de células, Tejidos y Organos (INDT)". Briefly, a portion of each sural nerve biopsy from human CMT patient, used for histopathological diagnosis was fixed in 3% PFA diluted in PHEM, 30 minutes at 4°C with stirring. Fixative was washed with PHEM (6X10 minutes). When it was possible, the epineuria was removed to facilitate the ulterior treatments.

6.6 Electron Microscopy

Nerves pieces were processed for postembedding immunostaining. (Bozzola & Russel, 1998; Vazquez Nin, 2001). Briefly, nerve pieces were dehydrated increasing concentration of alcohol, until pure alcohol. Thereafter were embedded in hydrophilic resine LR-white (LRW) by increasing its concentration. When they were in pure LRW (overnight at room temperature), the blocks were polymerized, under anhydrous conditions, in two steps: 24 hs at 45°C and 24 hours at 60°C. Ultrathin sections (60-100 nm) were obtained by ultramicrotomy and collected in nickel grids, without film. Indirect immunostaining was followed. The grids were placed exposing the ultrathin section to different series of drops on Parafilm extended layer. The general processes are described under "*Immunostaining*". After that, sections were exposed to Osmium Tetroxide vapours and counterstained with Uranyl Acetate and Lead Citrate as usual in Transmission Electron Microscopy (TEM).

7. Immunocytochemistry

7.1 Pre treatment

7.1.1 Collagen digestion

Epineuria from intact nerve mice were dissected and the extracellular matrix was unstructured by collagenase digestion with Collagenase XII (Sigma), 0.3mg/ml dissolved in PHEM without EGTA (25mM HEPES, 60 mM PIPES, MgCl2, pH 7,2- 7,6) with 5mM CaCl2 final concentration, 1 hour, at 37°C. The enzymatic activity was stopped by cold washing in buffer PHEM (3X10 minutes).

7.1.2 Teasing

After collagenase digestion, rat and mouse sciatic nerves were placed on a cold slide and were mechanically teased under stereoscopic microscope using blunt needles to prevent fibers tearing. Further treatments were performed over floating fibers.

7.1.3 Resin relaxing

Ultrathin LRW sections were incubated 10 minutes in sodium periodate 0.56M in water at room temperature (RT) to relaxing the resin and washed thereafter with PHEM (6X5 minutes), to remove resins.

7.1.4 Permeabilization

The permeabilization must be a single event that will balance the benefit of increased accessibility of intracellular epitopes for immunocytochemistry or ISH with the hindrance of the consequent deterioration of the structure to be recognized. Cell membranes were permeabilized once with Triton X-100 0.1% in PHEM, at variable times (10 minutes for cell cultures to 30 minutes for whole fibers and LRW ultrathin sections), stirring at RT. Excess detergent was removed by successive washes in buffer PHEM (3x5 minutes). In the procedures that follows no detergent was used. No detergent was used in human samples.

7.1.5 Aldehyde blocking

To reduce background from aldehydes and ketones free groups (generated by fixative or belonging to the cellular structure) they were blocked by incubation with sodium borohydride 0.1% in water for 10 minutes (in cell cultures) or 20 minutes (in intact nerve), at RT. The remains of borohydride are eliminated by repeated washings with buffer PHEM at RT and gentle stirring.

7.1.6 Unspecific antigen blocking

In indirect immunostaining, nonspecific antigen reactions were blocked using 5% normal serum from the animal species in which the anti antibody was raised, dissolved in incubation buffer (IB, 0.150 mM glycine, 0.1% Bovine Serum Albumin in PHEM), 30 minutes at 37°C. After that, the tissue or cells were immediately incubated with the specific antibody.

7.1.7 Immunostaining

After pre-treatment, the procedure was done basically as described by Kun et al (2007) and Sotelo et al. (1999), with some modifications. The sections were incubated 1 h at 37°C with the specific antibodies in IB (see above) in a wet chamber. Washed 3X5 min, at 37°C with IB and incubated over 45 minutres at 37°C with the anti-antibodies. Fluorophores photobleaching was avoided maintaining the slices in dark. The non binding antibodies were eliminated by washing with IB (3X5 min) at 37°C and PHEM (3X5 minutes) at RT. Slides for confocal microscopy were mounted with Prolong Gold Antifade (Invitrogen). The ultrathin sections were counter-stained as described in "*Electron Microscopy Section*".

7.1.8 Specific antibodies and fluorescent probes

The specific antibodies used in the present work, were: polyclonal anti-Ribosomes antibody (Kun et al, 2007) work dilution 1:500, monoclonal anti phosphorylated Neurofilament (Stemberger) work dilution 1:2500; monoclonal anti-Neurofilament-200 (phosphorylated and non phosphorylated) from Sigma, working dilution 1:800. Human anti-P ribosomal protein (Immunovision) working dilution 1:100, polyclonal anti-Myelin Associated Glycoprotein S/L (MAG, Chemicon) working dilution 1:150; monoclonal anti-Vimentin (Sigma) working dilution 1:400; Alexa-546 Phaloidin (Invitrogen) working dilution 1:40. YoYo-1 (Invitrogen) 1:1000; DAPI (Sigma) working dilution 1:1000.

7.1.9 Secondary-antibodies

The anti-antibodies used in the present work were: Goat anti-Mouse Alexa 488 (A11029, Invitrogen), working dilution 1:2000, Goat anti Rabbit Alexa 546 (A11030, Invitrogen), working dilution 1:2000, Goat anti-Mouse CY5 (Chemicon) working dilution 1:800; Goat anti-Rabbit 15nm gold conjugated (Aurion) working dilution 1:80.

7.2 *In situ* hybridization

The cDNA encoding mouse *pmp-22* mRNA region (425-538) was cloned in the PST-19 plasmid. The plasmid was linearized with HindIII or EcoRI and used as template by *in vitro* transcription reactions. Single-stranded RNA probes were transcribed using SP6 or T7 RNA polymerase according to manufacturer's instructions and labeled with digoxigenin-UTP (Roche). The samples were first permeabilized as indicated in "Permeabilization Section". The endogenous peroxidase was blocked with 0.03% H_2O_2 diluted in PHEM during 1 hour at RT, changing the solution every 15 minutes to refresh the offer of hydrogen peroxide. The samples were then washed with PHEM (3X5 min). The pre-hybridization condition was performed to avoid unspecific probe binding. The prehybridization and hybridization was done in the same condition: incubation with hybridization solution (10% dextran sulfate, 0.1mg/ml tRNA, 0.5mg/ml salmon sperm DNA, 50% formamide, 4XSSC) during two hours at 50°C without probes (prehybridization) or with sense/antisense probes. Immediately before hybridization, digoxigenin labeled transcripts were denatured 3 min at 95°C and fast returning to 4°C thereafter; 5 min to denature the RNA, avoiding the RNA folding. The non hybridized RNA probes were eliminated by repeated washing with decreasing saline concentration (stringency increase), until 0,25X SSC. After that, the hybrid was fixed with fresh prepared 2% PFA in PHEM and gently washed with PHEM (3X5minutes). The hybrid was recognized by sheep anti-digoxigenin antibody conjugated to peroxidase. The immunocomplex were developed by the Tyramide labelling kit (Roche) giving a fluorescent complex (in 520 nm). General methods were adapted and applied to our conditions (Morel, et al. 2001 Sambrook & Russell, 2001). After that, immunostaining was applied as described in "Immunostaining".

8. Conclusions

1. Schmidt-Lanterman Incisures and Nodes of Ranvier are pathways for ribosomes and RNAs to be related to axonal function.

2. CMT human patients' nerves showed inhomogeneous neurofilament arrangements in axons. Myelin is delaminated. Big groups of ribosomes are present near the irregular newborn axonal sprouting. Ribosomes have been found also in axoplasm.

3. The presence of *pmp22* transcript and PMP-22 protein signals in nuclear domains during myelinogenesis and in TrJ/+ adult fibers described here (Fig. 4), suggest other type of alterations. They would be, a) altered post-transcriptional regulation? and/or b) altered motor protein link, both potentially involved in pathogenesis and illness consolidation.

9. Acknowledgements

CSIC-Universidad de la República, Montevideo, Uruguay, PEDECIBA, MEyC, ANII, FIRCA-NIH. The authors especially thank Dr. Timothy De Voogd (Professor at the University of Cornell, NY, USA) for his careful reading and correction of the present manuscript.

10. References

Aguayo, AJ, Atiwell, M, Trecarten, J, Perkins, S, Bray, M. 1977. Abnormal myelination in transplanted Trembler mouse Schwann cells. Nature 265: 73-75.

Berciano, J. and Combarros, O. 2003. Hereditary Neuropathies. Current Opinion in Neurology, 16: 613-622.

Bozzola, JJ & Russell, LD, 1998. Electron Microscopy. Principles and techniques for Biologists, ISBN 0763701920. 2nd edition, Jones & Bartlett Publischers, Sudbury, Massachusetts.

Depillar ME, Tang S, Mukhopadhyay G, Shen YJ, Filbin MT. 1996.

Myelin-associated glycoprotein inhibits axonal regeneration from a variety of neurons via interaction with a sialoglycoprotein. Mol Cell Neurosci.;7(2):89-101.

Cajal, SRY. 1928. Degeneration and regeneration of the nervous system, volumenes I y II Oxford University Press London.

Cornbrooks CJ, Mithen F, Cochran JM, Bunge RP.1982.Factors affecting Schwann cell basal lamina formation in cultures of dorsal root ganglia from mice with muscular dystrophy. Brain Res. ;282(1):57-67.

Court FA, Hendriks WT, MacGillavry HD, Alvarez J, van Minnen J. 2008. Schwann cell to axon transfer of ribosomes: toward a novel understanding of the role of glia in the nervous system. J Neurosci.;28(43):11024-9.

Court FA, Midha R, Cisterna BA, Grochmal J, Shakhbazau A, Hendriks WT, Van Minnen J. 2011.Morphological evidence for a transport of ribosomes from Schwann cells to regenerating axons. Glia. 2011 Oct;59(10):1529-39. doi: 10.1002/glia.21196. Epub 2011 Jun 8.

Devaux J.J. & Scherer SS , 2005 .Altered Ion Channels in an Animal Model of Charcot-Marie-Tooth Disease Type IA, The Journal of Neuroscience, February 9, • 25(6):1470 – 1480.Fortun J, Li J, Go J, Fenstermaker A, Fletcher BS, Notterpek L. 2005- Impaired proteasome activity and accumulation of ubiquitinated substrates in a hereditary neuropathy model. J Neurochem. ;92:1531–1541.

Fortun J, Verrier JD, Go JC, Madorsky I, Dunn WA, Notterpek L. 2007.The formation of peripheral myelin protein 22 aggregates is hindered by the enhancement of autophagy and expression of cytoplasmic chaperones.Neurobiol Dis. ;25(2):252-65.

Ghabriel, MN and Allt G. 1987. Incisures of Schmidt-Lanterman, Progress in Neurobiology vol 17,25-58.

Ghabriel MN, Allt G. 1981. Incisures of Schmidt-Lanterman. Prog Neurobiol1,7(1-2):25-58.

Ghabriel MN, Allt G. 1980a. Schmidt-Lanterman incisures. II. A light and electron microscope study of remyelinating peripheral nerve fibres.Acta Neuropathol. 52(2):97-104.

Ghabriel, MN and Allt G. 1980b Schmidt-Lanterman incisures I. A light an electron microscopy study of remyelinating peripheral nerve fibres. Acta Neuropathol.52, 85-95.

Ghabriel MN, Allt G. 1979a. The role of Schmidt-Lanterman incisures in Wallerian degeneration. II. An electron microscopic study.Acta Neuropathol. 48(2):95-103.

Ghabriel, MN and Allt G. 1979b. The role of Schmidt-Lanterman ncisures in Wallerian degeneration. I. A quantitative teased fibre study. Acta Neuropathol. 48, 83-93.

Goldberg AL. 2003. Protein degradation and protection against misfolded or damaged proteins. Nature.;426(6968):895-9. Review.

Hall, SM & Williams, PL. 1970. Studies on the incisures of Schmidt and Lanterman. J. Cell Sci. 6, 767-791.

Inherited Peripheral Neuropathies Mutation Database, http://www.molgen.ua.ac.be/cmtmutations/

Henry, E.; Cowen, J.; Sidman, R. (1983) Comparison of Trembler and Trembler- J Phenotypes: varying severity of peripheral hypomyelination. Journal of Neuropathology and Experimental Neurology, 42, 688-706.

Henry EW, Sidman RL. 1983.The murine mutation trembler-J: proof of semidominant expression by use of the linked vestigial tail marker.J Neurogenet.;1(1):39-52.

Hiscoe, HB, 1947. Distribution of nodes and incisures in normal and regenerated nerve fibers. Anat. Rec 99, 447-475.

Junn E, Lee SS, Suhr UT, Mouradian MM. 2002. Parkin accumulation in aggresomes due to proteasome impairment.J Biol Chem 6;277(49):47870-7.

Khajavi M, Shiga K, Wiszniewski W, He F, Shaw CA, Yan J, Wensel TG, Snipes GJ, Lupski JR. 2007- Oral curcumin mitigates the clinical and neuropathologic phenotype of the Trembler-J mouse: a potential therapy for inherited neuropathy. Am J Hum Genet.;81(3):438-53.

Koenig E & Martin R. 1996. Cortical plaque-like structures identify ribosome-containing domains in the Mauthner cell axon. J Neurosci.;16(4):1400-11.

Koenig E, Martin R, Titmus M, & Sotelo-Silveira J. 2000. Cryptic Peripheral Ribosomal Domains Distributed Intermittently along Mammalian Myelinated Axons. The Journal of Neuroscience, November 15, 2000, 20(22):8390–8400.

Kun A, J. C. Benech, A. Giuditta & J.R. Sotelo. 1998. Polysomes are present in the squid giant axon: an Immuno Electron Microscopy. ICEM-14, Electron Microscopy, 1998, Vol. I:825-826.

Kun, A, Otero, L., Sotelo-Silveira, J. & Sotelo, J. 2007. Ribosomal distribution in axón of mammalian myelinated fibers. J.Neuroscience Research 85:2087-2098

Li YC, Li YN, Cheng CX, Sakamoto H, Kawate T, Shimada O & Atsumi S .2005. Subsurface cisterna-lined axonal invaginations and double-walled vesicles at the axonal-myelin sheath interface. Neurosci Res.;53(3):298-303.

Li YC, Cheng CX, Li YN, Shimada O, & Atsumi S. 2005.Beyond the initial axon segment of the spinal motor axon: fasciculated microtubules and polyribosomal clusters. Anat; 206(6):535-42.

MacKenzie ML, Ghabriel MN & Allt G. 1984. Nodes of Ranvier and Schmidt-Lanterman incisures: an in vivo lanthanum tracer study.J Neurocytol. 13(6):1043-55.–

Martin R, Fritz W & Giuditta A. 1989. Visualization of polyribosomes in the postsynaptic area of the squid giant synapse by electron spectroscopic imaging. J Neurocytol.; 18(1):11-8.

Mersiyanova IV, Perepelov AV, Polyakov AV, Sitnikov VF, Dadali EL, Oparin RB, Petrin AN & Evgrafov OV, 2000. A new variant of Charcot-Marie-Tooth disease type 2 is probably the result of a mutation in the neurofilament-light gene. Am J Hum Genet 67: 37-46.

Morel G, Cavalier A & Williams L. 2001. In situ hybridization in electron microscopy CRC Press Boca New York Washington, D.C. ISBN 0849300444, 9780849300448. Editors Morel G, Cavalier A, Williams L.

Notterpek L, Ryan MC, Tobler AR & Shooter EM. 1999. PMP-22 accumulation in aggresomes: implications for CMT1A pathology. Neurobiol Dis. ;6(5):450-60.

Notterpek L, Shooter EM & Snipes GJ. 1997. Upregulation of the endosomal-lysosomal pathway in the trembler-J neuropathy. J Neurosci.;17(11):4190-200.

Patel PI, Roa BB, Welcher AA, Schoener-Scott R, Trask BJ, Pentao L, Snipes GJ, Garcia CA, Francke U, Shooter EM, Lupski JR & Suter U. 1992.The gene for the peripheral myelin protein PMP-22 is a candidate for Charcot-Marie-Tooth disease type 1A. Nat Genet.;1(3):159-65.

Pérez-Ollé, R, Leung, CL, Liem, RKH. 2002. Effects of Charcot-Marie-Tooth linked mutation of the neurofilament light subunit on intermediate filament formation. Journal of Cell Science 115:4937-4946.

Quarles RH. 2007.Myelin-associated glycoprotein (MAG): past, present and beyond.J Neurochem.100(6):1431- 48.

Ryan TE, Patterson SD. 2002.Proteomics: drug target discovery on an industrial scale.Trends Biotechnol.;20(12 Suppl):S45-51. Review.

Robertson JD. 1958. The ultrastruture of Schmidt Lanterman clefts and related shearing defects of the myelin sheath. J. Biophis. Biochem. Cytol., 4, 39-46. 48.

Rosso G, Cal K, Canclini L, Damián JP, Ruiz P, Rodríguez H, Sotelo JR, Vazquez C, Kun A. 2010. Early phenotypical diagnoses in Trembler-J mice model. J Neurosci Methods;190(1):14-9.

Salzer JL, Brophy PJ, Peles E. 2008. Molecular Domains of Myelinated Axons in the Peripheral Nervous System. Glia 56:1532–1540

Sambrook Y, Russell DW. 2006. Molecular Cloning A laboratory manual. Third edition. Cold Spring Harbor Laboratory Press. ISBN 0879697717

Sereda M, Griffiths I, Pühlhofer A, Stewart H, Rossner MJ, Zimmerman F, Magyar JP, Schneider A, Hund E, Meinck HM, Suter U, Nave KA. 1996. A transgenic rat model of Charcot-Marie-Tooth disease. Neuron.;16(5):1049-60.

Shen YJ, DeBellard ME, Salzer JL, Roder J, Filbin MT. 1998.Myelin-associated glycoprotein in myelin and expressed by Schwann cells inhibits axonal regeneration and branching.Mol Cell Neurosci. 1998 Sep;12(1-2):79-91.

Sidman, R.; Cowen, J.; Eicher, E. 1979. Inheredit muscle and nerve diseases in mice: A tabulation with commentary. Ann NY Academy Science, 317, 497-505.

Small JR, Ghabriel MN, Allt G 1987. The development of Schmidt-Lanterman incisures: an electron microscope study.J Anat.150:277-86.

Snipes GJ, Suter U.1995. Molecular basis of common hereditary motor and sensory neuropathies in humans and in mouse models.Brain Pathol. ;5(3):233-47.

Snipes GJ, Suter U. 1995.Molecular anatomy and genetics of myelin proteins in the peripheral nervous system. J Anat.;186 (Pt 3):483-94. Review.

Sotelo JR, Kun A, Benech JC, Giuditta A, Morillas J, Benech CR.1999. Ribosomes and polyribosomes are present in the squid giant axon: an immunocytochemical study.Neuroscience.; 90(2):705-15.

Sotelo-Silveira JR, Calliari A, Kun A, Koenig E, Sotelo JR. 2006.RNA trafficking in axons.Traffic.;7(5):508-15. Review.

Suter,U. and Scherer, S. 2003, Disease mechanisms in inherited neuropathies. Nature Rev Neurosc 4:714-726.

Suter U, Moskow JJ, Welcher AA, Snipes GJ, Kosaras B, Sidman RL, Buchberg AM, Shooter EM.1992. A leucine-to-proline mutation in the putative first transmembrane domain of the 22-kDa peripheral myelin protein in the trembler-J mouse.Proc Natl Acad Sci U S A.; 15;89(10):4382-6.

Valentijn LJ, Baas F, Wolterman RA, Hoogendijk JE, van den Bosch NH, Zorn I, Gabreëls-Festen AW, de Visser M, Bolhuis PA. 1992.Identical point mutations of PMP-22 in Trembler-J mouse and Charcot-Marie-Tooth disease type 1A. Nat Genet. ;2(4):288-91.

Vallat JM, Funalot B. 2010 Charcot-Marie-Tooth (CMT) disease: an update.Med Sci (Paris). 2010 Oct;26(10):842-7.

Verhoeven K, De Jonghe P, Coen K, Verpoorten N, Auer-Grumbach M, Kwon JM, FitzPatrick D, Schmedding E, De Vriendt E, Jacobs A, Van Gerwen V, Wagner K, Hartung HP, Timmerman V. 2003. Mutations in the Small GTP-ase Late Endosomal Protein RAB7 Cause Charcot-Marie-Tooth Type 2B Neuropathy. Am J Hum Genet 72: 722-727

Waelter S, Boeddrich A, Lurz R, Scherzinger E, Lueder G, Lehrach H, Wanker EE. 2001. Accumulation of mutant huntingtin fragments in aggresome-like inclusion bodies as a result of insufficient protein degradation. Mol Biol Cell.;12(5):1393-407.

Young P, Suter U. 2003.The causes of Charcot-Marie-Tooth disease. Cell Mol Life Sci. ;60(12):2547-60.

Zelená J. 1970.Ribosome-like particles in myelinated axons of the rat. Brain Res.; 1;24(2):359-63.

Zelená J. 1972. Ribosomes in the axoplasm of myelinated nerve fibres.Folia Morphol (Praha).;20(1):91-

Zelená J. 1972. Ribosomes in myelinated axons of dorsal root ganglia. Z Zellforsch Mikrosk Anat. 1972;124(2):217-29.

Zhao C, Takita J, Tanaka Y, Setou M, Nakagawa T, Takeda S, Yang HW, Terada S, Nakata T. Takei Y, Saito M, Tsuji S, Hayashi Y, Hirokawa N. 2001. Charcot-Marie-Tooth disease type 2A caused by mutation in a microtubule motor KIF1Bbeta. Cell 1;105(5):587-97.

12

Immunocytochemistry in Early Mammalian Embryos

Hesam Dehghani

Embryonic and Stem Cell Biology and Biotechnology Research Group, Research Institute of Biotechnology, and Department of Basic Science, Faculty of Veterinary Medicine, Ferdowsi University of Mashhad, Mashhad, Iran

1. Introduction

The preimplantation period of mammalian development hosts very important cellular and molecular events. This period starts with the fertilization of oocyte by sperm, a process that reprograms the highly differentiated nuclei of these germ cells, and leads to the generation of a totipotent one-cell embryo. Then, the embryo performs cleavage divisions with short cell cycles to quickly increase its cell number. During this period, the genome of the preimplantation embryo manifests profound changes in nuclear and chromatin organization, histone modifications, and transcriptional activity. These genome alterations are also coupled to cell signaling pathways and their regulatory effects. The final product of the preimplantation development is a multi-cellular blastocyst containing three types of cells, epiblasts, hypoblasts, and trophoblast cells [1].

To study and understand the biology of preimplantation embryos, different techniques have been used. The paucity of cells and the difficulties associated with the preparation and production of preimplantation embryos have been the main limiting factors for the application of a wide range of experimental techniques. Thus, what is known about early embryos today is mainly the results of the use of a few experimental techniques and their adapted modifications. These include DNA and RNA amplification techniques, transcript labeling, *in situ* hybridization of DNA and RNA, gene manipulation studies, and light, electron, and immunofluorescence microscopy techniques.

The application of each technique has revealed a specific aspect of preimplantation developmental biology. Table 1 summarizes and compares the contributions of different experimental techniques applied on preimplantation mammalian embryos. In the rest of this chapter, I will focus on the immunocytochemical staining of embryos and its different applications in preimplantation development.

2. Contribution of immunocytochemistry to understanding the biology of preimplantation mammalian embryos

Application of immunocytochemistry (ICC) on preimplantation embryos has provided invaluable information on different aspects of preimplantation development. I will briefly

Technique	Knowledge contribution	Example references
Conventional and quantitative RT-PCR	Evaluation of the transcription of individual genes	[2-8]
Gene expression profiling (microarray)	Large-scale evaluation of the expression of genes	[9-14]
Electron microscopy techniques	Studying the ultrastructural organization of the embryonic cells	[15-19]
Labeling of nascent transcripts	Quantification of transcriptional activity	[20-24]
In situ hybridization of DNA and RNA	Intracellular localization of chromosomes and transcripts	[25-28]
Gene knockout and knock down techniques	Studying the function of individual genes	[29-38]
Immunocytochemistry	Intracellular localization of proteins Quantitative evaluation of the expression of proteins Identification of protein modifications Evaluation of the activity of certain signaling pathways	[20, 31, 39-45]

Table 1. The major Experimental techniques applied to study the preimplantation embryos.

review the applications of ICC for localization of proteins, for studying the modifications of chromatin and alteration of chromatin organization, and for analyzing cell signaling pathways in preimplantation embryos.

2.1 Cellular and intra-cellular localization of proteins

During preimplantation development, it is very important to identify whether a given protein is expressed, where in the cell it is localized, in which blastomeres it is expressed, and when its expression is eliminated. All of this information relate to the function of protein during preimplantation development. Immunocytochemistry has been an indispensable technique to reveal this information. Application of an alternative Western blotting will not provide any information on the intracellular localization of the protein or the types of expressing cells.

Looking at more than two decades of research on Oct4 clearly shows that what we know on the role of this transcription factor in pluripotency, has all started from this immunocytochemical observation that this protein is differentially expressed in the mouse preimplantation embryonic cells [46]. While it had been previously revealed that it has a strong transcriptional activator effect in the inner cell mass of the preimplantation embryo [47] and it is transcribed in these cells [48], it was its protein localization (using specific antibodies and ICC procedure) that convincingly illustrated its relationship to stemness and pluripotency. A number of later functional studies also used ICC to reveal the function of Oct4 during preimplantation development and pluripotency [49, 50]. The same route of discovery has been traveled for other stemness genes [51].

Using immunocytochemistry and confocal microscopy we have been able to reveal the subcellular distribution and to analyze the relative amount of ten isozymes of PKC (alpha, betaI, betaII, gamma, delta, epsilon, eta, theta, zeta, iota/lambda) and a PKC-anchoring protein, receptor for activated C-kinase 1 (RACK1), between the two-cell and blastocyst stages of mouse preimplantation development [39]. In a functional study, we used the same principle to analyze the relative amount of each PKC isozyme within each blastomere and relate this to the transcriptional activity of the 4-cell mouse embryo [20]. Thus for a given protein in the preimplantation embryo, ICC technique can be applied to study its differential expression between embryonic blastomeres, to identify its intracellular localization within individual blastomeres, and also to semi-quantitate its expression. Recently, using fluorescently-labeled specific antigen binding fragments (Fabs), it has been shown that it is possible to monitor the distribution and global level of endogenous histone modifications in living blastomeres without disturbing cell growth and embryo development [52].

2.2 Identification of histone modifications and the study of nuclear organization

The last two decades has witnessed a considerable number of research efforts using ICC to identify a variety of post-translational modifications on histones and to analyze the expression of chromatin-remodeling factors in preimplantation embryos (Table 2). Immunocytochemical detection and localization of nuclear subdomains (Figure 1), histone modifications, enzymes responsible for these modifications, different histone variants, distinct chromatin remodeling factors, and the status of transcription in preimplantation stages of development (Figure 2), has provided ample evidence and knowledge on the biology of chromatin during preimplantation development (Table 2).

In a very close subject, ICC procedure has also been applied to investigate the organization of chromatin, the architecture of nucleus, and the formation of sub-nuclear compartments by ultra-structural studies in preimplantation embryos. In fact, the correlative fluorescence and electron microscopy technique has allowed the ultra-structural identification of nuclear entities which are identified and tagged by immunocytochemistry [15, 53, 54] (Figure 3). As it has been shown in the figure, immunocytochemical detection of a chromocenter domain immuno-stained with CREST antibody is indispensable for finding and imaging it under the electron microscope. The same principle has been used to identify a sub-nuclear compartment immunocytochemically, and to study its ultra-structure, e.g. localizing fibrillarin by ICC to identify nucleolus in the nucleus of preimplantation embryos for ultra-structural analysis [19, 55-57].

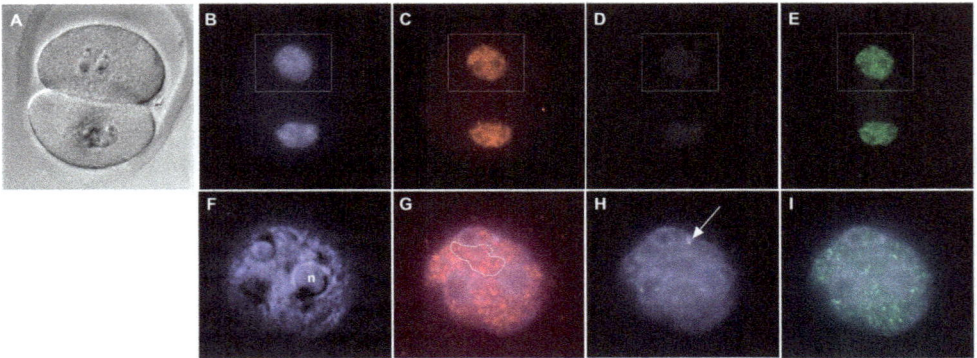

Fig. 1. Immunocytochemistry and confocal imaging of a two-cell mouse embryo to evaluate the function of nucleus. Top row contains confocal images from an optical slice of a two-cell mouse embryo which has been immunolabeled and stained with different antibodies and imaged in different channels. The bottom row contains the merged images of the top nucleus in different channels. A two-cell embryo (A; DIC image) contains two nuclei that are not very chromatin-condensed by DAPI staining (B). Nucleoli (n) in the magnified nucleus in F (the merged image of A and B) show very thin rim of fairly condensed chromatin. Immunolabeling of RNA polymerase II (phosphorylated at serine 5 of its CTD) shows a hyperactive transcription (C). A highly transcribed region of nucleus has been marked in G (the merged image of B and C). Immunolabeling with CREST antibody reveals centromeres (D), which are mainly located at the edge of nucleoli in H (the merged image of B and D). White arrow in H, shows a CREST-labeled spot. Immunolabeling with the antibody against acetylated lysine of H3 histone reveals regions of "open" chromatin (E) which are distributed throughout the nucleus (I, the merged image of B and E).

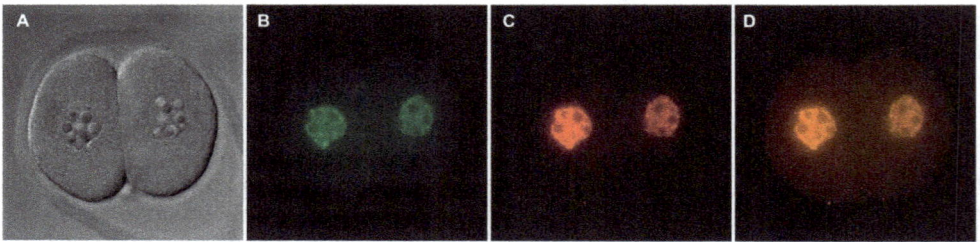

Fig. 2. Immunocytochemical localization of hyperactive transcription domains in a two-cell stage mouse embryo. A) DIC image; B) Immunolabeling with the antibody against the acetylated lysine of histone H3; C) Immunolabeling with the antibody against RNA polymerase II (phosphorylated at serine 5 of its CTD); D) A merged image of B and C. The yellow color in D represents nuclear domains which contain acetylated H3K9 and RNA pol II, indicating that transcription is occurring in chromatin domains with a relaxed state, where a large number of acetylated histone H3K9 moieties are present.

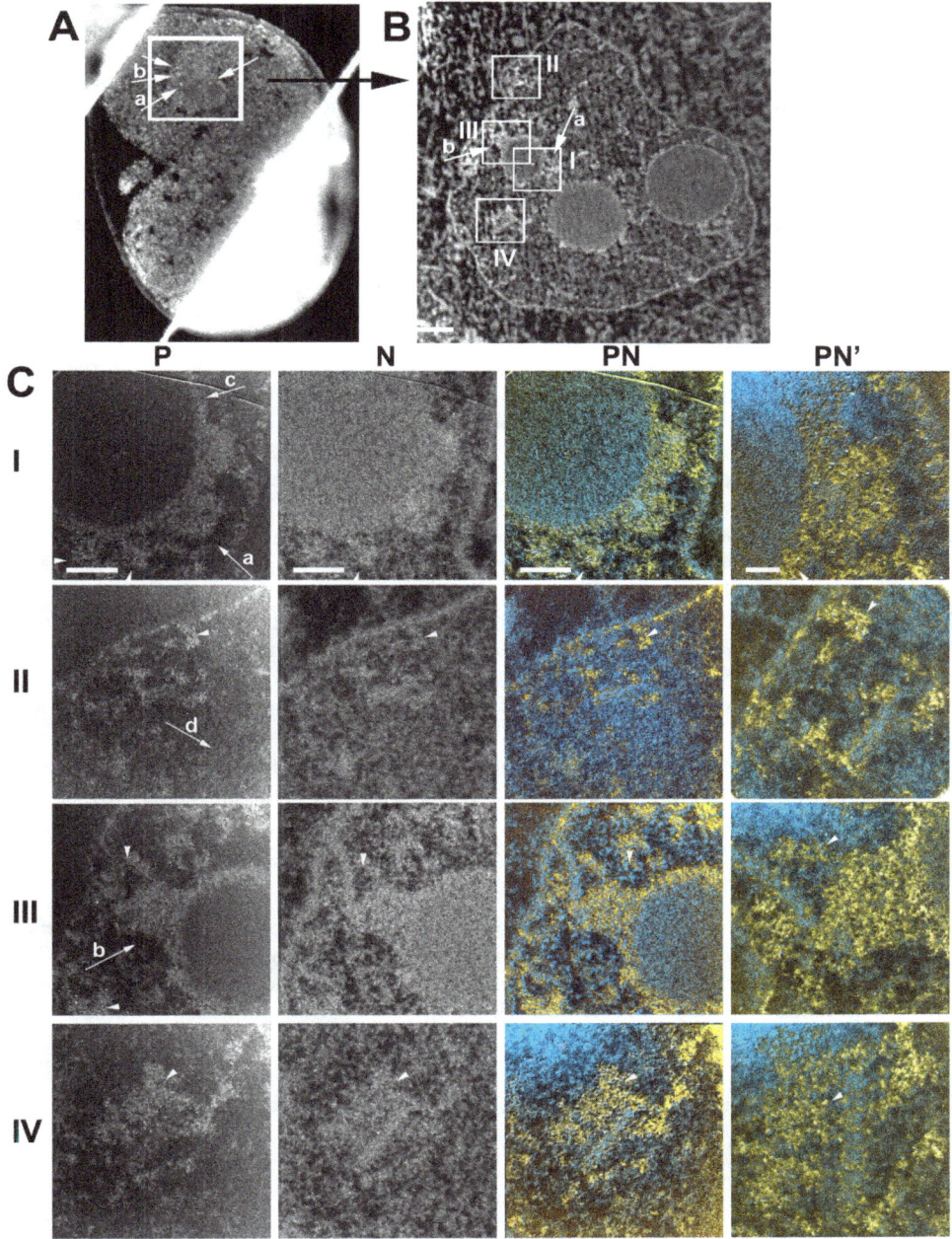

Fig. 3. Chromatin organization in the two-cell stage preimplantation mouse embryo. **A)** Fluorescence image of a physical section spanning through the nucleus (shown by the white

box) of a two-cell stage embryo. White arrows point to the centromeres which are immunostained with CREST antibody. **B)** The rectangular region in panel A has been imaged by low magnification electron spectroscopic imaging (ESI; 155 keV phosphorus-enriched)[15, 53, 54, 58, 59]. Three different-sized nucleoli with very homogenous mass are noticeable in this nucleus. Arrows 'a' and 'b' point to the centromeres designated similarly in panel A. The scale bar is 2 nm. **C)** Different regions of the nucleus in panel B have been imaged with higher magnification ESI. Columns P, N, PN, and PN' denote images of phosphorus map, nitrogen map, overlay of phosphorus and nitrogen maps, and higher magnification overlays of phosphorus and nitrogen maps, respectively. The segmentation of signals in PN and PN' permits the visualization of chromatin fibers as yellow, while non-chromosomal proteins due to their relatively low N:P ratio content are in blue color. White arrowheads point to the representative gold-tagged histone H3 (methylated at lysine 9) molecules which are accumulated at different areas of the nucleus. Scale bars for columns P, N, and PN are 500 nm, and scale bar for column PN' is 200 nm. (C-I) Nucleoli comprise a homogeneous structure with scarce amounts of ribonucleoprotein (weak signal in P map), but large amounts of protein (strong signal in N map). A more condensed patch of chromatin at the edge of nucleolus (arrow a) is highly positive for K9-methylated H3, while a very thin layer of chromatin at the edge of nucleolus (arrow c) does not show accumulation of this signal. The area shown by 'arrow a' which is designated similarly in panels A and B corresponds to a chromocenter. **(C-II)** A very thin layer of condensed chromatin (as 30nm fibers) at the nuclear envelope which in some parts is positive for K9-methylated H3 blends in with the open lattice of 10nm chromatin fibers (shown by arrow d). The open lattice is filled with large amounts of non-chromosomal proteins shown as blue in PN image. The relation of chromatin and non-chromosomal proteins is better visualised in the higher resolution/magnification image of PN'. **(C-III)** Patches of condensed chromatin at the edge of nucleolus and in the vicinity of nucleolus (white arrowheads) are positive for K9-methylated H3, but only the area at the edge of nucleolus (shown by arrow b) corresponds to the chromocenter 'b' in panels A and B. **(C-IV)** Non-centromeric condensed chromatin (as 30 nm fibers and positive for K9-methylated H3) is surrounded by dispersed network of 10nm fibers.

2.3 Evaluation of the activity of certain signaling pathways

Immunocytochemsitry has also been used to discover the presence of many components of signaling pathways including Wnt, hedgehog, receptor tyrosine kinase, and PKC in preimplantation embryos. These studies based on imaging and localization of specific proteins has clearly established a framework for future functional studies. In Table 3 some of these studies have been summarized.

3. Immunocytochemistry of oocytes and preimplantation mammalian embryos

3.1 Harvesting oocytes and preimplantation embryos

Depending to the species, oocytes can be acquired and preimplantation embryos can be produced in different ways. In mouse, it is very easy to harvest from oviduct and uterus, the oocytes and embryos grown in vivo to certain stages of preimplantation development. It is also possible to harvest embryos at early cleavage stages and grow them in culture medium

Findings	Implication	Example references
Lack of the constitutive heterochromatin markers histone H4 trimethyl Lys20 (H4K20me3) and chromobox homolog 5 (HP1α); the presence of heterochromatin markers, H3K9me3, 5-methyl cytosine (5MC), HP1β, H3K27me3, H4K20me1 and H4K20me2	Heterochromatin is in an immature state in mouse preimplantation embryos	[60]
Presence of the acetylated forms of H3K9 and H3K27	H3K27 acetylation is important for normal embryonic development	[52]
Relatively higher expression in oocytes and early cleavage stage embryos of methionine adenosyltransferase 1A protein up to the 8-cell stage compared with the morulae and blastocyst stages	nutrient-sensitive epigenetic regulation and perturbation may be performed through specific enzymes at the earliest stages of preimplantation development	[61]
Embryos at 2-, 4-, and 8-cell stages lack macroH2A except in residual polar bodies. MacroH2A protein expression reappears in embryos after the 8-cell stage and persists in morulae and blastocysts, where nuclear macroH2A is present in both the trophectodermal and inner cell mass cells.	Normal embryos execute three to four mitotic divisions in the absence of macroH2A prior to the onset of embryonic macroH2A expression. Embryos made by somatic nuclear transfer utilize the same chromatin remodeling mechanisms.	[62, 63]
HDAC1 is expressed in preimplantation embryos , where its expression inversely correlates with changes in the acetylation state of histone H4K5 during preimplantation development	HDAC1 is involved in the formation of a chromatin-mediated transcriptionally repressive state that initiates in the late two-cell embryo	[31]
ICC of late zygotes shows that constitutive heterochromatin is only maternally labeled by H3K9me3 and HP1β	In early embryos, Suv39h-mediated H3K9me3 constitutes the dominant maternal transgenerational signal for pericentric heterochromatin formation	[34]
After fertilization, level of H3K79me2 and H3K79me3 modifications rapidly decrease, and the hypomethylated state is maintained at the interphase (before the blastocyst stage), except for a transient increase in H3K79me2 at mitosis (M phase). H3K79me3 is not detected throughout preimplantation, even at M phase	Elimination of H3K79 methylation after fertilization is involved in genomic reprogramming	[64]
p150CAF-1 is expressed in preimplantation embryos and loss of p150CAF-1 function leads to early developmental arrest and alteration of heterochromatin organization	Chromatin assembly machinery is involved in controlling the spatial organization and epigenetic marking of the genome in early embryos	[42]

Table 2. Immunocytochemical identification and analysis of some histone modifications and chromatin remodeling factors in preimplantation embryos.

Findings	Pathway*	Example references
Expression of protein kinase C isoforms in each stage of preimplantation development	Activation of PKC through G-protein coupled receptors	[39, 65]
Expression of Hh receptor PTCH1 and co-receptor SMO	Signaling events mediated by the Hedgehog family	[66]
Expression of β-catenin	Wnt signaling network	[67-69]
Presence of Aurora C in cleavage-stage embryos	Signaling by Aurora kinases	[70]
Expression of proteins in MAPK pathway	p38 MAPK signaling pathway	[71]
IRS-1 is expressed in all cell lineages of the peri-implantation mouse embryo and mediates some effects of insulin and IGFs at this stage.	Insulin pathway	[72]
Expression and localization of beta 1, beta 5 and alpha 6 integrins and ZO-1 and E-cadherin proteins	E-cadherin signaling pathway & integrin family cell surface interactions	[73, 74]
Strong expression of c-MYC signal in the nucleus of growing and fully grown oocytes as well as in preimplantation embryos before the morulae stage	C-MYC pathway	[75]
The p 85 and p110 subunits of PI3K and Akt are expressed from the 1-cell through the blastocyst stage of murine preimplantation embryo development	The PI3K/Akt pathway	[76]

*Name of pathways have been adapted from NCI-Nature Pathway Interaction Database [77]

Table 3. Components of signaling pathways immunocytochemically identified in preimplantation embryos.

(in vitro culture; IVC). In addition, the early embryos could be produced by in vitro fertilization (IVF) of oocytes, and subsequently cultured in vitro. The most practical method to acquire bovine embryos is through IVF followed by IVC.

Superovulation of female mouse: In mouse, whether we need oocytes, in vivo grown embryos, or in vitro fertilized and cultured embryos, the female mice required to be superovulated. In response to a hormonal regimen, 3 weeks-old female mice produce the highest number of oocytes (metaphase II stage) and embryos. This is believed to be related to the lack of reproductive cycles and an inactive state of hypothalamic-hypophysial-gonadal axis at this age. The acquired number of harvested oocytes and embryos after superovulation is also largely affected by the strain and maintenance (nutritional and light-dark cycle) conditions.

To induce superovulation of female mice, the following steps need to be taken.

1. Mice should be kept in a 12 hour light-dark cycle in a properly ventilated room with a temperature of 22-26°C.
2. Administration of hormones is performed by intra-peritoneal injection of female mice at 3-weeks age. If the mice are bred in the same facility, then the first injection time would be two days after weaning from mother. However, if mice will be transferred to the facility from another location, then the first injection time would be after a two-day acclimatization period.
3. Human chorionic gonadotropin (hCG) and pregnant mare serum gonadotropin (PMSG), which are in the lyophilized powder form, should be dissolved in sterile saline solution (0.9% NaCl) under a laminar hood. The final concentration is 5 IU per 0.1 ml. Once all the powder in each vial has been dissolved, 0.5 ml of each solution should be drawn into individual insulin syringes and immediately placed in -80°C freezer.
4. Each 3 weeks-old female mouse is injected intra-peritoneally at 14pm on day -3 with 0.1 ml of PMSG (5 IU). The syringe containing the hormone should be removed from freezer and brought to the ambient temperature 15 minutes before injection.
5. On day -1 at 12pm (46 hours after PMSG injection), each injected mouse will be injected again with 0.1 ml of hCG (5 IU) intra-peritoneally. If harvesting of embryos is intended, each female mouse after injection should be placed in the cage of individual males (Note 1) for overnight mating. However if oocyte recovery is anticipated, the females are returned to their own cage after second injection.

Harvesting oocytes: In the morning of day 0, oocytes can be recovered from oviducts of injected females. Oviducts are flushed with M2 medium (Sigma-Aldrich, St Louis, MO, USA) as previously described [78]. It should be noted that the oocytes at this stage are surrounded by layers of granulosa cells. Thus, to perform ICC and properly localize and image specific proteins in oocytes, the granulosa cells need to be digested away. Otherwise, it will not be possible to properly image oocyte itself, especially when an epi-fluorescence microscope is used for imaging.

Harvesting embryos: To harvest embryos, the injected female is placed in male's cage for overnight mating. Presence of a copulation (vaginal) plug the next morning (on day 0), would be an indication for mating. Embryos at different stages of preimplantation development can be harvested at different time points. Table 4 represents approximate time points for the recovery of embryos at different stages of mouse development.

3.2 Immunocytochemistry

Oocytes or embryos do not attach to the slides or coverslips. Thus, the ICC procedure on harvested oocytes or embryos is somewhat different from the ICC procedure performed on cells grown on coverslips or cells attached to slides. During the procedure, oocytes and

Stage	Day	Time
One-cell stage	0 (The day after hCG injection and mating*)	10-12 am
Two-cell stage (most likely at G2 phase of cell cycle)[79]	1	9 am (45 hours after hcG injection) (33 hours post coitum*)
Four-cell stage (G1 or S phase of the cell cycle)[79]	1	4 pm (52 hours after hcG injection) (40 hours post coitum)
Eight-sixteen cell stage	2	9 am (69 hours after hcG injection) (57 hours post coitum)
Morulae stage	2	4 pm (76 hours after hcG injection) (64 hours post coitum)
Early blastocyst	3	9 am (93 hours after hcG injection) (81 hours post coitum)

* When male and females are placed in a cage for mating in an evening, the 12:00 midnight is arbitrarily chosen as the time of mating.

Table 4. Approximate time points for the recovery of embryos at different stages of mouse preimplantation development.

embryos should be manually transferred between different media containing fixative, permeabilizing agent, or antibodies. Use of depression slides as container and a stereomicroscope would facilitate the procedure. Pipettors (e.g. 20 μl) or mouth-controlled pipet devices [78] are used for the transfer, while embryos are watched under the stereomicroscope.

The following procedure is a prototype to perform ICC (using fluorescent secondary antibodies) on oocytes and embryos. For simplicity, only embryos (not oocytes) are referred to in the procedure.

1. **Washing:** Wash embryos in 200 μl of PBS twice. This will involve the quick transfer of the harvested embryos into the depression slides containing PBS. Under the stereomicroscope, the embryos could be counted and screened for fragmented or abnormal morphology.

2. **Fixation:** Transfer embryos into 200 μl of 4% paraformaldehyde in PBS and incubate at room temperature for 20 minutes. After fixation, the embryos are washed in PBS three times (of 5 minutes each) at room temperature. At this step, the embryos can be stored in PBS at 4°C overnight.

3. **Permeabilization** (Note 2): Transfer fixed embryos into 200 μl of 0.5% Triton X 100 in PBS and incubate for 5 minutes at room temperature. Wash the permeabilized embryos in PBS three times (of 5 minutes each) at room temperature.

4. **Incubation in primary antibody:** Transfer embryos into 200 μl of primary antibody (diluted in PBS). Incubate in a humid chamber for 2 hours at room temperature or overnight at 4°C. Wash the embryos in PBS three times (of 5 minutes each) at room temperature (Note 3).

5. **Incubation in secondary antibody:** Transfer embryos into 200 μl of secondary antibody (diluted in PBS). Incubate in a humid chamber for 1 hour at room temperature or overnight at 4°C. Wash the embryos in PBS three times (of 5 minutes each) at room temperature. If the antibody is conjugated to a fluorescent tag, then the incubation and washing steps should be performed at dark (Note 4).

6. **Mounting:** During the mounting procedure, the embryos should be placed in a small volume (20 μl) of mounting medium in the circle on the slide (Figure 4) (Note 5). First, place the mounting or anti-fade medium in the circle. Then, transfer the embryos into the middle of medium. Eyelash probe could be used to move embryos into the middle of circle. Let the embryos sink to the bottom of the medium. Place a coverslip very carefully on the circle on the slide, trying not to move embryos toward the edges of the circle. Seal around the edges of coverslip with nail polish. The mounted embryos can be examined right away or stored at 4°C.

Fig. 4. Making circles of nail polish on the slide for mounting of immunostained embryos.

7.   **Microscopic examination:** Depending to the type of secondary antibody used and the available equipment, the embryos can be imaged using light, epi-fluorescence, or confocal laser scanning fluorescence microscopy.

3.3 Notes

Note 1. Male mice reach sexual maturity at the age of 8 weeks. It is important that after weaning the individual male pups to be kept in separate cages. It is believed that keeping several male pups together in one cage, except in the dominant male, may suppress their hormonal maturity. It is also important to place one injected female in the male's cage. Male mouse should not be placed in female's cage. Only one female and not more should be placed in the male's cage. The day after mating males and females should be separated again.

Note 2. Permeabilization is only necessary when an intracellular antigen or protein is to be detected. For immunocytochemical detection of proteins or antigens which are localized on the cellular membrane, a permeabilization step is not performed.

Note 3. Permeabilization and incubation of embryos in primary antibody causes them to sink toward and occasionally adhere the bottom of depression slides. This makes the transfer of embryos between different containers very difficult. Eyelash probe (commercially supplied or homemade by gluing an eyelash to a needle) would be an indispensible device for these situations. With this device under a stereomicroscope, it would be very easy to detach the embryos from the bottom of depression slides and guide them toward the transfer pipette.

Note 4. Different secondary antibodies may be used. If the secondary antibody is conjugated to biotin, alkaline phosphatase, or horseradish peroxidase, different substrates are used to reveal antigen-primary antibody-secondary antibody complexes and different procedures are followed before the mounting step.

Note 5. Placing coverslip directly onto a slide with embryos in between will cause the physical rupture and burst of embryos. Thus, it is very important to produce a space between slide and coverslip. For this purpose, small circles (with a diameter of 5mm) are made on the slide by nail polish. We use an insulin syringe (attached to its needle) filled with nail polish to make the circles with defined edges. When the circle of nail polish is dried, the space in the middle will be used for mounting of embryos.

4. Conclusion

The mammalian preimplantation development contains a highly regulated series of cellular and molecular events that are necessary for normal cell growth, cell division and differentiation. Our understanding of the mechanisms involved in these events has significantly increased in recent years, while much remains to be learned about the mechanisms involved in controlling growth and proliferation, transcriptional control and cell fate decisions. Immunocytochemistry has had and remains to have a significant role for the discovery of these events. In this chapter, its contribution to our current understanding of the different aspects of preimplantation development has succinctly reviewed. In addition, the ICC procedure has been elaborated.

5. Acknowledgments

Work in the author's laboratory is supported by grants from Ferdowsi University of Mashhad. I would like to thank Professor David Bazett-Jones and members of his laboratory, especially Ren Li, Reagan Ching, and Kashif Ahmed for making it possible to begin the ultra-structural analysis of preimplantation embryos, which with no doubt will be a source of important clues on the regulation of transcription and differentiation. I apologize to those colleagues whose publications due to space limitations could not be cited.

6. References

[1] Rossant, J. Lineage development and polar asymmetries in the peri-implantation mouse blastocyst. *Semin Cell Dev Biol*, 15, 5 (Oct 2004), 573-581.

[2] Falco, G., Stanghellini, I. and Ko, M. S. Use of Chuk as an internal standard suitable for quantitative RT-PCR in mouse preimplantation embryos. *Reprod Biomed Online*, 13, 3 (Sep 2006), 394-403.

[3] Fiddler, M., Abdel-Rahman, B., Rappolee, D. A. and Pergament, E. Expression of SRY transcripts in preimplantation human embryos. *Am.J.Med.Genet.*, 55, 1 1995), 80-84.

[4] Fiorenza, M. T. and Mangia, F. Quantitative RT-PCR amplification of RNA in single mouse oocytes and preimplantation embryos. *Biotechniques*, 24, 4 (Apr 1998), 618-623.

[5] Nowak-Imialek, M., Wrenzycki, C., Herrmann, D., Lucas-Hahn, A., Lagutina, I., Lemme, E., Lazzari, G., Galli, C. and Niemann, H. Messenger RNA expression patterns of histone-associated genes in bovine preimplantation embryos derived from different origins. *Mol.Reprod.Dev.*, 75, 5 2008), 731-743.

[6] May, A., Kirchner, R., Muller, H., Hartmann, P., El Hajj, N., Tresch, A., Zechner, U., Mann, W. and Haaf, T. Multiplex RT-PCR Expression Analysis of Developmentally Important Genes in Individual Mouse Preimplantation Embryos and Blastomeres. *Biol Reprod.* 2009 Jan; 80(1): 194-202.

[7] Cui, X. S., Shen, X. H. and Kim, N. H. High mobility group box 1 (HMGB1) is implicated in preimplantation embryo development in the mouse. *Mol Reprod Dev*, 75, 8 (Aug 2008), 1290-1299.

[8] Ebrahimian, M., Mojtahedzadeh, M., Bazett-Jones, D. and Dehghani, H. Transcript isoforms of promyelocytic leukemia in mouse male and female gametes. *Cells Tissues Organs*, 192, 6 2010), 374-381.

[9] Beyhan, Z., Ross, P. J., Iager, A. E., Kocabas, A. M., Cunniff, K., Rosa, G. J. and Cibelli, J. B. Transcriptional reprogramming of somatic cell nuclei during preimplantation development of cloned bovine embryos. *Dev.Biol.*, 305, 2 2007), 637-649.

[10] Cheon, Y. P., Li, Q., Xu, X., DeMayo, F. J., Bagchi, I. C. and Bagchi, M. K. A Genomic Approach to Identify Novel Progesterone Receptor Regulated Pathways in the Uterus during Implantation. *Mol.Endocrinol.*, 16, 12 2002), 2853-2871.

[11] Dobson, A. T., Raja, R., Abeyta, M. J., Taylor, T., Shen, S., Haqq, C. and Pera, R. A. The unique transcriptome through day 3 of human preimplantation development. *Hum Mol Genet*, 13, 14 (Jul 15 2004), 1461-1470.

[12] Hamatani, T., Daikoku, T., Wang, H., Matsumoto, H., Carter, M. G., Ko, M. S. and Dey, S. K. Global gene expression analysis identifies molecular pathways distinguishing

blastocyst dormancy and activation. *Proc.Natl.Acad.Sci.U.S.A.*, 101, 28 2004), 10326-10331.

[13] Hamatani, T., Ko, M., Yamada, M., Kuji, N., Mizusawa, Y., Shoji, M., Hada, T., Asada, H., Maruyama, T. and Yoshimura, Y. Global gene expression profiling of preimplantation embryos. *Hum Cell*, 19, 3 (Aug 2006), 98-117.

[14] Zeng, F. and Schultz, R. M. RNA transcript profiling during zygotic gene activation in the preimplantation mouse embryo. *Dev.Biol.*, 283, 1 2005), 40-57.

[15] Ahmed, K., Dehghani, H., Rugg-Gunn, P., Fussner, E., Rossant, J. and Bazett-Jones, D. P. Global chromatin architecture reflects pluripotency and lineage commitment in the early mouse embryo. *PLoS One*, 5, 5 2010), e10531.

[16] Cremer, T. and Zakhartchenko, V. Nuclear architecture in developmental biology and cell specialisation. *Reprod Fertil Dev*, 23, 1 2011), 94-106.

[17] Svarcova, O., Strejcek, F., Petrovicova, I., Avery, B., Pedersen, H. G., Lucas-Hahn, A., Niemann, H., Laurincik, J. and Maddox-Hyttel, P. The role of RNA polymerase I transcription and embryonic genome activation in nucleolar development in bovine preimplantation embryos. *Mol Reprod Dev*, 75, 7 (Jul 2008), 1095-1103.

[18] Kikuchi, K., Ekwall, H., Tienthai, P., Kawai, Y., Noguchi, J., Kaneko, H. and Rodriguez-Martinez, H. Morphological features of lipid droplet transition during porcine oocyte fertilisation and early embryonic development to blastocyst in vivo and in vitro. *Zygote.*, 10, 4 2002), 355-366.

[19] Laurincik, J., Thomsen, P. D., Hay-Schmidt, A., Avery, B., Greve, T., Ochs, R. L. and Hyttel, P. Nucleolar proteins and nuclear ultrastructure in preimplantation bovine embryos produced in vitro. *Biol.Reprod.*, 62, 4 2000), 1024-1032.

[20] Dehghani, H., Reith, C. and Hahnel, A. C. Subcellular localization of protein kinase C delta and epsilon affects transcriptional and post-transcriptional processes in four-cell mouse embryos. *Reproduction*, 130, 4 (Oct 2005), 453-465.

[21] Aoki, F., Worrad, D. M. and Schultz, R. M. Regulation of transcriptional activity during the first and second cell cycles in the preimplantation mouse embryo. *Dev. Biol.*, 1811997), 296-307.

[22] Aoki, F., Hara, K. T. and Schultz, R. M. Acquisition of transcriptional competence in the 1-cell mouse embryo: requirement for recruitment of maternal mRNAs. *Mol Reprod Dev*, 64, 3 (Mar 2003), 270-274.

[23] Harrouk, W., Khatabaksh, S., Robaire, B. and Hales, B. F. Paternal exposure to cyclophosphamide dysregulates the gene activation program in rat preimplantation embryos. *Mol.Reprod.Dev.*, 57, 3 2000), 214-223.

[24] Kanzler, B., Haas-Assenbaum, A., Haas, I., Morawiec, L., Huber, E. and Boehm, T. Morpholino oligonucleotide-triggered knockdown reveals a role for maternal E-cadherin during early mouse development. *Mech Dev*, 120, 12 (Dec 2003), 1423-1432.

[25] Baart, E. B., van den Berg, I., Martini, E., Eussen, H. J., Fauser, B. C. and Van Opstal, D. FISH analysis of 15 chromosomes in human day 4 and 5 preimplantation embryos: the added value of extended aneuploidy detection. *Prenat Diagn*, 27, 1 (Jan 2007), 55-63.

[26] Rubio, C., Rodrigo, L., Perez-Cano, I., Mercader, A., Mateu, E., Buendia, P., Remohi, J, Simon, C. and Pellicer, A. FISH screening of aneuploidies in preimplantation embryos to improve IVF outcome. *Reprod.Biomed.Online.*, 11, 4 2005), 497-506.

[27] Baart, E. B., Van Opstal, D., Los, F. J., Fauser, B. C. and Martini, E. Fluorescence in situ hybridization analysis of two blastomeres from day 3 frozen-thawed embryos followed by analysis of the remaining embryo on day 5. *Hum Reprod*, 19, 3 (Mar 2004), 685-693.

[28] Verlinsky, Y., Cieslak, J., Evsikov, S., Galat, V. and Kuliev, A. Nuclear transfer for full karyotyping and preimplantation diagnosis for translocations. *Reprod.Biomed.Online.*, 5, 3 2002), 300-305.

[29] Fedoriw, A. M., Stein, P., Svoboda, P., Schultz, R. M. and Bartolomei, M. S. Transgenic RNAi reveals essential function for CTCF in H19 gene imprinting. *Science*, 303, 5655 (Jan 9 2004), 238-240.

[30] Lykke-Andersen, K., Gilchrist, M. J., Grabarek, J. B., Das, P., Miska, E. and Zernicka-Goetz, M. Maternal Argonaute 2 Is Essential for Early Mouse Development at the Maternal-Zygotic Transition. *Mol Biol Cell*(Aug 13 2008).

[31] Ma, P. and Schultz, R. M. Histone deacetylase 1 (HDAC1) regulates histone acetylation, development, and gene expression in preimplantation mouse embryos. *Dev Biol*, 319, 1 (Jul 1 2008), 110-120.

[32] Nganvongpanit, K., Muller, H., Rings, F., Gilles, M., Jennen, D., Holker, M., Tholen, E., Schellander, K. and Tesfaye, D. Targeted suppression of E-cadherin gene expression in bovine preimplantation embryo by RNA interference technology using double-stranded RNA. *Mol Reprod Dev*, 73, 2 (Feb 2006), 153-163.

[33] Dehghani, H., Narisawa, S., Millan, J. L. and Hahnel, A. C. Effects of disruption of the embryonic alkaline phosphatase gene on preimplantation development of the mouse. *Dev Dyn*, 217, 4 (Apr 2000), 440-448.

[34] Puschendorf, M., Terranova, R., Boutsma, E., Mao, X., Isono, K., Brykczynska, U., Kolb, C., Otte, A. P., Koseki, H., Orkin, S. H., van Lohuizen, M. and Peters, A. H. PRC1 and Suv39h specify parental asymmetry at constitutive heterochromatin in early mouse embryos. *Nat Genet*, 40, 4 (Apr 2008), 411-420.

[35] Zou, G. M., Thompson, M. A. and Yoder, M. C. RNAi knockdown of transcription factor Pu.1 in the differentiation of mouse embryonic stem cells. *Methods Mol Biol*, 4072007), 127-136.

[36] Yagi, R., Kohn, M. J., Karavanova, I., Kaneko, K. J., Vullhorst, D., DePamphilis, M. L. and Buonanno, A. Transcription factor TEAD4 specifies the trophectoderm lineage at the beginning of mammalian development. *Development*, 134, 21 (Nov 2007), 3827-3836.

[37] Xie, Y., Wang, Y., Sun, T., Wang, F., Trostinskaia, A., Puscheck, E. and Rappolee, D. A. Six post-implantation lethal knockouts of genes for lipophilic MAPK pathway proteins are expressed in preimplantation mouse embryos and trophoblast stem cells. *Mol Reprod Dev*, 71, 1 (May 2005), 1-11.

[38] Truchet, S., Chebrout, M., Djediat, C., Wietzerbin, J. and Debey, P. Presence of permanently activated signal transducers and activators of transcription in nuclear interchromatin granules of unstimulated mouse oocytes and preimplantation embryos. *Biol.Reprod.*, 71, 4 2004), 1330-1339.

[39] Dehghani, H. and Hahnel, A. C. Expression profile of protein kinase C isozymes in preimplantation mouse development. *Reproduction.*, 130, 4 2005), 441-451.

[40] Yamazaki, T., Kobayakawa, S., Yamagata, K., Abe, K. and Baba, T. Molecular dynamics of heterochromatin protein 1beta, HP1beta, during mouse preimplantation development. *J.Reprod.Dev.*, 53, 5 2007), 1035-1041.

[41] Ohnuma-Ishikawa, K., Morio, T., Yamada, T., Sugawara, Y., Ono, M., Nagasawa, M., Yasuda, A., Morimoto, C., Ohnuma, K., Dang, N. H., Hosoi, H., Verdin, E. and Mizutani, S. Knockdown of XAB2 enhances all-trans retinoic acid-induced cellular differentiation in all-trans retinoic acid-sensitive and -resistant cancer cells. *Cancer Res*, 67, 3 (Feb 1 2007), 1019-1029.

[42] Houlard, M., Berlivet, S., Probst, A. V., Quivy, J. P., Hery, P., Almouzni, G. and Gerard, M. CAF-1 is essential for heterochromatin organization in pluripotent embryonic cells. *PLoS.Genet.*, 2, 11 2006), e181.

[43] Dodge, J. E., Kang, Y. K., Beppu, H., Lei, H. and Li, E. Histone H3-K9 methyltransferase ESET is essential for early development. *Mol.Cell Biol.*, 24, 6 2004), 2478-2486.

[44] Burns, K. H., Viveiros, M. M., Ren, Y., Wang, P., Demayo, F. J., Frail, D. E., Eppig, J. J. and Matzuk, M. M. Roles of NPM2 in chromatin and nucleolar organization in oocytes and embryos. *Science*, 300, 5619 2003), 633-636.

[45] Sarmento, O. F., Digilio, L. C., Wang, Y., Perlin, J., Herr, J. C., Allis, C. D. and Coonrod, S. A. Dynamic alterations of specific histone modifications during early murine development. *J Cell Sci*, 117, Pt 19 (Sep 1 2004), 4449-4459.

[46] Palmieri, S. L., Peter, W., Hess, H. and Scholer, H. R. Oct-4 transcription factor is differentially expressed in the mouse embryo during establishment of the first two extraembryonic cell lineages involved in implantation. *Dev Biol*, 166, 1 (Nov 1994), 259-267.

[47] Scholer, H. R., Balling, R., Hatzopoulos, A. K., Suzuki, N. and Gruss, P. Octamer binding proteins confer transcriptional activity in early mouse embryogenesis. *Embo J*, 8, 9 (Sep 1989), 2551-2557.

[48] Scholer, H. R. Octamania: the POU factors in murine development. *Trends Genet.*, 7, 10 1991), 323-329.

[49] Nichols, J., Zevnik, B., Anastassiadis, K., Niwa, H., Klewe-Nebenius, D., Chambers, I., Scholer, H. and Smith, A. Formation of pluripotent stem cells in the mammalian embryo depends on the POU transcription factor Oct4. *Cell*, 95, 3 (Oct 30 1998), 379-391.

[50] Haraguchi, S., Saga, Y., Naito, K., Inoue, H. and Seto, A. Specific gene silencing in the pre-implantation stage mouse embryo by an siRNA expression vector system. *Mol Reprod Dev*, 68, 1 (May 2004), 17-24.

[51] Rossant, J. The impact of developmental biology on pluripotent stem cell research successes and challenges. *Dev Cell*, 21, 1 (Jul 19 2011), 20-23.

[52] Hayashi-Takanaka, Y., Yamagata, K., Wakayama, T., Stasevich, T. J., Kainuma, T., Tsurimoto, T., Tachibana, M., Shinkai, Y., Kurumizaka, H., Nozaki, N. and Kimura, H. Tracking epigenetic histone modifications in single cells using Fab-based live endogenous modification labeling. *Nucleic Acids Res*, 39, 15 (Aug 1 2011), 6475-6488.

[53] Bazett-Jones, D. P., Li, R., Fussner, E., Nisman, R. and Dehghani, H. Elucidating chromatin and nuclear domain architecture with electron spectroscopic imaging. *Chromosome Res*, 16, 3 2008), 397-412.

[54] Fussner, E., Ahmed, K., Dehghani, H., Strauss, M. and Bazett-Jones, D. P. Changes in Chromatin Fiber Density as a Marker for Pluripotency. *Cold Spring Harb Symp Quant Biol*(Dec 7 2010).

[55] Biggiogera, M., Burki, K., Kaufmann, S. H., Shaper, J. H., Gas, N., Amalric, F. and Fakan, S. Nucleolar distribution of proteins B23 and nucleolin in mouse preimplantation embryos as visualized by immunoelectron microscopy. *Development.*, 110, 4 1990), 1263-1270.

[56] Baran, V., Mercier, Y., Renard, J. P. and Flechon, J. E. Nucleolar substructures of rabbit cleaving embryos: an immunocytochemical study. *Mol.Reprod.Dev.*, 48, 1 1997), 34-44.

[57] Hyttel, P., Laurincik, J., Rosenkranz, C., Rath, D., Niemann, H., Ochs, R. L. and Schellander, K. Nucleolar proteins and ultrastructure in preimplantation porcine embryos developed in vivo. *Biol Reprod*, 63, 6 (Dec 2000), 1848-1856.

[58] Dehghani, H., Dellaire, G. and Bazett-Jones, D. P. Organization of chromatin in the interphase mammalian cell. *Micron*, 36, 2 2005), 95-108.

[59] Dellaire, G., Nisman, R. and Bazett-Jones, D. P. Correlative light and electron spectroscopic imaging of chromatin in situ. *Methods Enzymol*, 3752004), 456-478.

[60] Wongtawan, T., Taylor, J. E., Lawson, K. A., Wilmut, I. and Pennings, S. Histone H4K20me3 and HP1alpha are late heterochromatin markers in development, but present in undifferentiated embryonic stem cells. *J Cell Sci*, 124, Pt 11 (Jun 1 2011), 1878-1890.

[61] Ikeda, S., Namekawa, T., Sugimoto, M. and Kume, S. Expression of methylation pathway enzymes in bovine oocytes and preimplantation embryos. *Journal of experimental zoology. Part A, Ecological genetics and physiology*, 313, 3 (Mar 1 2010), 129-136.

[62] Chang, C. C., Ma, Y., Jacobs, S., Tian, X. C., Yang, X. and Rasmussen, T. P. A maternal store of macroH2A is removed from pronuclei prior to onset of somatic macroH2A expression in preimplantation embryos. *Dev Biol*, 278, 2 (Feb 15 2005), 367-380.

[63] Chang, C. C., Gao, S., Sung, L. Y., Corry, G. N., Ma, Y., Nagy, Z. P., Tian, X. C. and Rasmussen, T. P. Rapid elimination of the histone variant MacroH2A from somatic cell heterochromatin after nuclear transfer. *Cellular reprogramming*, 12, 1 (Feb 2010), 43-53.

[64] Ooga, M., Inoue, A., Kageyama, S., Akiyama, T., Nagata, M. and Aoki, F. Changes in H3K79 methylation during preimplantation development in mice. *Biol Reprod*, 78, 3 (Mar 2008), 413-424.

[65] Pauken, C. M. and Capco, D. G. The expression and stage-specific localization of protein kinase C isotypes during mouse preimplantation development. *Dev.Biol.*, 223, 2 2000), 411-421.

[66] Nguyen, N. T., Lo, N. W., Chuang, S. P., Jian, Y. L. and Ju, J. C. Sonic hedgehog supplementation of oocyte and embryo culture media enhances development of IVF porcine embryos. *Reproduction*, 142, 1 (Jul 2011), 87-97.

[67] Xie, H., Tranguch, S., Jia, X., Zhang, H., Das, S. K., Dey, S. K., Kuo, C. J. and Wang, H. Inactivation of nuclear Wnt-beta-catenin signaling limits blastocyst competency for implantation. *Development*, 135, 4 (Feb 2008), 717-727.

[68] Kemler, R., Hierholzer, A., Kanzler, B., Kuppig, S., Hansen, K., Taketo, M. M., de Vries, W. N., Knowles, B. B. and Solter, D. Stabilization of beta-catenin in the mouse

zygote leads to premature epithelial-mesenchymal transition in the epiblast. *Development*, 131, 23 (Dec 2004), 5817-5824.

[69] Wang, Q. T., Piotrowska, K., Ciemerych, M. A., Milenkovic, L., Scott, M. P., Davis, R. W. and Zernicka-Goetz, M. A genome-wide study of gene activity reveals developmental signaling pathways in the preimplantation mouse embryo. *Dev Cell*, 6, 1 (Jan 2004), 133-144.

[70] Avo Santos, M., van de Werken, C., de Vries, M., Jahr, H., Vromans, M. J., Laven, J. S., Fauser, B. C., Kops, G. J., Lens, S. M. and Baart, E. B. A role for Aurora C in the chromosomal passenger complex during human preimplantation embryo development. *Hum Reprod*, 26, 7 (Jul 2011), 1868-1881.

[71] Wang, Y., Wang, F., Sun, T., Trostinskaia, A., Wygle, D., Puscheck, E. and Rappolee, D. A. Entire mitogen activated protein kinase (MAPK) pathway is present in preimplantation mouse embryos. *Dev.Dyn.*, 231, 1 2004), 72-87.

[72] Puscheck, E. E., Pergament, E., Patel, Y., Dreschler, J. and Rappolee, D. A. Insulin receptor substrate-1 is expressed at high levels in all cells of the peri-implantation mouse embryo. *Mol Reprod Dev*, 49, 4 (Apr 1998), 386-393.

[73] Bloor, D. J., Metcalfe, A. D., Rutherford, A., Brison, D. R. and Kimber, S. J. Expression of cell adhesion molecules during human preimplantation embryo development. *Mol.Hum.Reprod.*, 8, 3 2002), 237-245.

[74] Dubey, A. K., Cruz, J. R., Hartog, B. and Gindoff, P. R. Expression of the alphav integrin adhesion molecule during development of preimplantation human embryos. *Fertil.Steril.*, 76, 1 2001), 153-156.

[75] Suzuki, T., Abe, K., Inoue, A. and Aoki, F. Expression of c-MYC in nuclear speckles during mouse oocyte growth and preimplantation development. *J Reprod Dev*, 55, 5 (Oct 2009), 491-495.

[76] Riley, J. K., Carayannopoulos, M. O., Wyman, A. H., Chi, M., Ratajczak, C. K. and Moley, K. H. The PI3K/Akt pathway is present and functional in the preimplantation mouse embryo. *Dev Biol*, 284, 2 (Aug 15 2005), 377-386.

[77] Schaefer, C. F., Anthony, K., Krupa, S., Buchoff, J., Day, M., Hannay, T. and Buetow, K. H. PID: the Pathway Interaction Database. *Nucleic Acids Res*, 37, Database issue (Jan 2009), D674-679.

[78] Nagy A, Gertsentein M, Vintersten K and R, B. *Manipulating the Mouse Embryo: A Laboratory Manual*. Cold Spring Harbor Laboratory Press, Cold Spring Harbor, NY 2003.

[79] Pratt, H. P. M. and Monk, M. *Isolation, culture and manipulation of preimplantation mouse embryos*. IRL Press, City, 1987.

13

Immuno-Glyco-Imaging in Plant Cells: Localization of Cell Wall Carbohydrate Epitopes and Their Biosynthesizing Enzymes

Marie-Laure Follet-Gueye et al.[*]
University of Rouen/Laboratoire « Glycobiologie et Matrice Extracellulaire Végétale »/UPRES EA 4358, Institut Fédératif de Recherche Multidisciplinaire sur les Peptides 23/Plate-forme de Recherche en Imagerie Cellulaire de Haute Normandie (PRIMACEN), Mont-Saint Aignan Cedex; France

1. Introduction

Plant cells are surrounded by a carbohydrate-rich compartment called the cell wall. This compartment plays a central role in growth, development, sexual reproduction and defence (Vicré et al., 2005; Mollet et al., 2007; Hématy et al., 2009; Seifert & Blaukopf, 2010). It is a carbohydrate-based structure composed mostly of polysaccharides (~90%) and glycoproteins (~10%). Polysaccharides include cellulose, pectin domains such as homogalacturonans (HG) with different levels of methylesterification, xylogalacturonan, rhamnogalacturonan-I (RG-I) and rhamnogalacturonan- II (RG-II) and hemicelluloses (depending on the species, xyloglucan [XyG] or arabinoxylan) (Liepman et al., 2010), whereas cell wall proteins are either enzymatic such as glucanases (Gilbert, 2010) or non-enzymatic such as arabinogalactan proteins (AGP) (Ellis et al., 2010, Driouich & Baskin, 2008).

Unlike cellulose microfibrils, complex polysaccharides (pectins and hemicelluloses) are assembled within the Golgi apparatus (GA) and are shuttled by GA-derived secretory vesicles to the cell wall (Driouich et al., 1993; Chevalier et al., 2010). The synthesis of these glycomolecules requires the action of a set of Golgi glycosyltransferases, in addition to nucleotide sugar transporters and nucleotide sugar interconversion enzymes (Sandhu et al., 2009).

[*] Mollet Jean-Claude[1], Vicré-Gibouin Maïté[1], Bernard Sophie[1], Chevalier Laurence[2], Plancot Barbara[1], Dardelle Flavien[1], Ramdani Yasmina[1], Coimbra Silvia[3] and Driouich Azeddine[1]
[1] University of Rouen/Laboratoire « Glycobiologie et Matrice Extracellulaire Végétale »/UPRES EA 4358, Institut Fédératif de Recherche Multidisciplinaire sur les Peptides 23/Plate-forme de Recherche en Imagerie Cellulaire de Haute Normandie (PRIMACEN), Mont-Saint Aignan Cedex, France
[2] University of Rouen/UMR6634/CNRS, Institut des Matériaux/
Faculté des Sciences et Techniques, St Etienne du Rouvray Cedex, France
[3] University of Porto/Departamento de Biologia/ Faculdade de Ciências/
Rua do Campo Alegre, Porto, Portugal

In this chapter, we describe a number of sample preparation and immunostaining procedures devoted to i) localization of cell wall heteropolysaccharides and proteoglycans within the cell wall domains and endomembrane system (Golgi stacks and associated vesicles) in various plant cell types and to ii) localization/mapping of Golgi-enzymes (glycosyltransferases) responsible for complex polysaccharide assembly within Golgi cisternae. We present procedures and examples of epitope localization using epifluorescence/confocal microscopy as well as immunogold electron microscopy.

2. Material

2.1 Plant materials

2.1.1 Arabidopsis and flax cultures

Arabidopsis (*Arabidopsis thaliana*) and flax (*Linum usitatissimum*) seedlings were grown as described by Durand et al. (2009). Briefly, for root tip or border-like cell studies, seeds were surface sterilized with 35% (v/v) ethanol and with 35% (v/v) bleach. After several washes in sterile distilled water, the seeds were sown on agar-solidified nutrient medium (Baskin et al., 1992). Growth conditions were identical to those described by Vicré et al. (2005). Seeds were grown in vertically orientated square Petri dishes in 16-h-day/8-h-night cycles at 24°C for 7 to 15 days. For Arabidopsis pollen and pistil studies, seeds were spread on the surface of sterile soil and cultured in a growth chamber as described by Dardelle et al. (2010) and Coimbra et al., (2007). Pollen was collected and grown *in vitro* in a liquid medium in the dark at 22°C (Boavida & McCormick, 2007). The flowers were collected at different stages of pistil development, according to the stages of early flower development set by Smyth *et al.* (1990), starting at stage 8, up to stage 13, when buds open and anthesis occurs.

2.1.2 Tobacco suspension-cultured cells

Wild type or transformed suspension-cultured cells of *Nicotiana tabacum* cv. Bright Yellow 2 (BY-2) expressing Arabidopsis protein fused to the Green Fluorescent Protein (GFP) (Chevalier et al., 2010) were cultured in the Murashige and Skoog (MS) liquid medium. Cells were subcultured every week by transferring 10 ml of suspension cells in 150 ml of fresh medium.

2.2 Monoclonal antibodies

A set of monoclonal antibodies (MAb) directed against the carbohydrate moiety of cell wall polysaccharides (Table 1) such as xyloglucan (XyG) and pectin domains including homogalacturonan (HG), xylogalacturonan, and side chains of rhamnogalacturonan-I (RG-I) were used. In addition, the cell surface glycoproteins such as arabinogalactan proteins (AGP) were localized with four MAbs (JIM8, JIM13, LM2 and MAC207).

2.3 Microscopes

For fluorescence studies, samples were examined using a confocal laser-scanning microscope (Leica TCS SP2 AOBS). Tetramethylrhodamine-5-(and 6)-isothiocyanate (TRITC) and Fluorescein isothiocyanate (FITC) fluorescences were imaged using an excitation wavelength of 543 nm or 488 nm, with the emission wavelength at 565–585 nm or 500–600

MAb	Epitope recognized	Polymer	References
CCRC-M1	α-Fuc-(1,2)-β-Gal	FucogalactoXyG	Puhlmann et al., (1994)
CCRC-M86	unknown	XyG	Pattathil et al., (2010)
JIM5	MeGalA-(1,4)-[(GalA)]$_4$-(1,4)-MeGalA	Weakly methylesterified HG	Clausen et al., (2003)
JIM7	GalA-(1,4)-[(MeGalA)]$_4$-(1,4)-GalA	methylesterified HG	Clausen et al., (2003)
JIM8	unknown	AGP	Pennell et al., (1991)
LM2	β-linked GlcA	AGP	Yates et al., (1996)
LM5	[(1-4)-β-D-Gal]$_{>3}$	Galactan of RG-I	Jones et al., (1997)
LM6	[(1-5)-α-L-Ara]$_{5/6}$	Arabinan of RG-I	Willats et al., (1998)
LM8	unknown	Xylogalacturonan	Willats et al., (2004)
LM13	linear (1-5)-α-L-arabinan	RG-I	Moller et al., (2008)
LM15	XXXG	XyG	Marcus et al., (2008)
MAC207	β-GlcA-(1,3)-α-GalA-(1,2)-Rha	AGP	Yates et al., (1996)

Table 1. List of monoclonal antibodies directed against cell wall carbohydrate epitopes used in this report. Fuc, fucose; Gal, galactose; MeGalA, 6-O-methyl-galacturonic acid; GalA, galacturonic acid; GlcA, glucuronic acid; Ara, arabinose; Rha, rhamnose. AGP, arabinogalactan protein; HG, homogalacturonan; MAb, monoclonal antibody; RG-I, rhamnogalacturonnan-I; XyG, xyloglucan. For more information see www.ccrc.uga.edu/~mao/wallMAb/Antibodies/antib and www.plantprobes.net

nm, respectively. Arabidopsis pollen tubes and pistils were observed under fluorescence illumination on a Leica DLMB microscope or a Zeiss Axio Imager Z1 inverted epifluorescence microscope equipped with FITC (absorption, 485–520 nm; emission, 520–560 nm wavelength) or filter UV filter (365/445 nm) for callose and β-glucan detection with aniline blue and calcofluor white, respectively. Images were acquired with a Leica DFC300FX camera or an Axiocam MR in automatic exposure mode, and processed with Axiovision 4.4 software.

For immunogold observations, grids were observed with a transmission electron microscope (TEM) apparatus (Tecnai 12 Bio-Twin; Philips) at 80 kV and images were acquired with an Erlangshen ES500W camera.

For all samples, control experiments were performed by omission of the primary antibody.

3. Spatial distribution of non cellulosic polysaccharides and proteoglycans in the cell wall and endomembrane compartments by fluorescence microscopy

These immunostaining methods were applied to various cell types including flax root cap and associated border-like cells, arabidopsis pollen tube or female sporophyte and tobacco suspension-cultured cells.

3.1 Cell surface immunostaining

Cell surface imaging was carried out in order to have an overview of plant cell wall composition. While the plant surface is the first structure involved during various biotic or abiotic stresses, the plant surface tissues are also implicated in plant growth and development. Therefore information on the composition of cell wall is important and can be obtained by immunocytochemistry and Fourier transform infrared microspectroscopy (Durand et al., 2009). We describe below several protocols dedicated to polysaccharides immuno-detection and cytochemical localization at the cell surface of flax root border-like cells (BLC), Arabidopsis pollen tube and pistil.

3.1.1 Root border-like cells

Root tips of most plant species produce a large number of cells programmed to separate from the root cap and to be released into the external environment named the border cells (Hawes et al., 2003). As described in Brassicaceae family plants (Driouich et al., 2007), flax doesn't produce isolated border cells but it does produce and release cells that remain attached to each other, forming a block of several cell layers called border-like cells (Vicré et al., 2005). We investigate the role of cell wall pectins in cell attachment and the organization of border-like cells in flax.

BLC from 5 days-old flax seedlings were put down gently on the clean surface of 10 wells glass slides (teflon printed diagnostic slide, Fischer Scientific). BLC were then fixed for 1 hour in 4% (w/v) paraformaldehyde (PAF) and 1% (v/v) glutaraldehyde in 50 mM piperazine-N,N'-bis(2-ethanesulfonic) acid buffer (PIPES), pH 7, and 1 mM $CaCl_2$ (adapted from Willats et al., 2001). BLC were washed in 50 mM PIPES, 1 mM $CaCl_2$, pH 7, and incubated for 30 min in a blocking solution of 3% (w/v) low-fat dried milk in phosphate-buffered saline (PBS), pH 7.2. After being carefully rinsed in PBS containing 0.05% (v/v) Tween 20 (PBST), BLC were incubated overnight at 4°C with the primary antibody (LM8 dilution 1:5 in 0.1% [v/v] PBST; Willats et al., 2004). After five washes with 0.05% PBST, roots were incubated with the secondary antibody, goat IgG anti-rat FITC conjugated (dilution 1:50 in 0.1% PBST) for 1 h at 37°C in the dark. BLC were rinsed in 0.05% PBST, mounted in anti-fade solution (Citifluor AF2; Agar Scientific). Between each immunostaining step, incubation medium was cautiously removed with filter paper to prevent the loss of BLC.

Labelling with the MAb LM8 specific for xylogalacturonan occurred in the outer surface of the border-like cells (Fig. 1). These results are similar to those already obtained for Arabidopsis (Vicré et al., 2005; Durand et al., 2009) that have revealed the importance of pectins in the organization of BLC.

3.1.2 Arabidopsis pollen tubes

Pollen tubes are fast growing tip-polarized cells carrying the two sperm cells. After landing on compatible stigmatic cells, pollen germinates and the tube grows deep inside the pistil, penetrating different tissues following guidance cues from both the sporophytic tissues and from the female gametophyte to arrive precisely at the micropylar end of the ovule and deliver sperm. Pollen grains can germinate *in vitro* and produce pollen tubes in liquid and on solid media (Fig. 2A). Arabidopsis pollen tubes are small (5 μm diameter) but can reach over 1mm long.

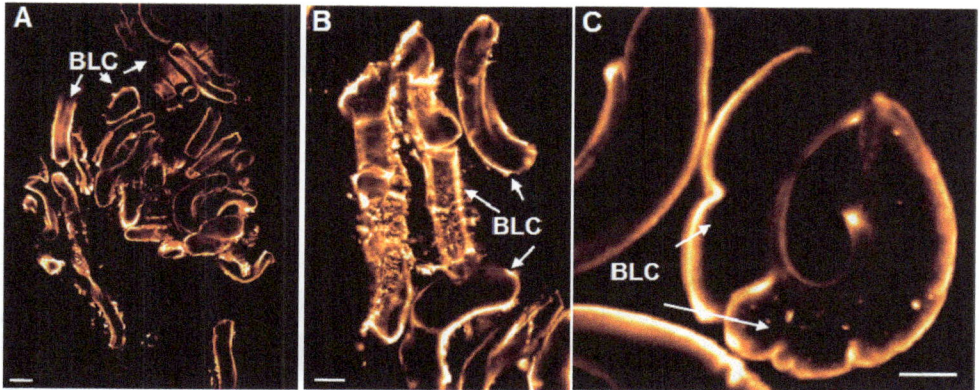

Fig. 1. Immunofluorescence labelling of flax root border-like cells (BLC) with the MAb LM8 specific for xylogalacturonan. Bars = 20 μm.

Pollen tubes grown in liquid germination medium were mixed (v/v) with a fixation medium containing 100 mM PIPES buffer, pH 6.9, 4 mM $MgSO_4$ $7H_2O$, 4 mM ethylene glycol tetraacetic acid (EGTA), 10% (w/v) Suc, and 5% (w/v) PAF and incubated for 90 min at room temperature. Pollen tubes were rinsed three times by centrifugation with 50 mM PIPES buffer, pH 6.9, 2 mM $MgSO_4$ $7H_2O$, and 2 mM EGTA and three times with phosphate-buffered saline (100 mM potassium phosphate, 138 mM NaCl, and 2.7 mM KCl, pH 7.4) supplemented or not with 3% fat-free milk. Primary antibodies were diluted at 1:5 or 1:10 as described previously (Mollet et al., 2000) with PBS (with or without 3% milk). Pollen tubes were rinsed with the buffer and incubated overnight at 4°C in the dark with the secondary antibody combined with FITC (Sigma) diluted at 1:50 with the appropriate buffer for 3 h at 30°C. For JIM, LM, and MAC antibody detection, goat anti-rat IgG (whole molecule)-FITC was used; for CCRC antibody detection, sheep anti-mouse IgG (whole molecule)-FITC was used.

Arabidopsis pollen tubes display specific cell wall organization with a strong detection of epitopes associated with arabinogalactan proteins in the tip region (Fig. 2B-C) as observed in other species such as lily (Mollet et al., 2002) or tobacco (Li et al., 1992). Moreover, pectic domains such as the the the arabinan side chains of RG-I (Fig. 2D-E) and methylesterified HGs (Fig. 2F) are also present in the cell wall of the pollen tubes. The cell surface immunolabeling gives important information on the accessibility of the epitopes. The enzyme or chemical treatments of the sample can sometimes be informative on the presence or not of a set of epitopes. As observed in Arabidopsis pollen tube, lily pollen tube was strongly labelled at the tip with the MAb directed against AGPs. Removal of the pectin with 0.1% pectinase extended the labelling for AGP to the entire pollen tube cell wall (Jauh & Lord, 1996).

In Arabidopsis, the female sporophyte is composed of a stigma with papillae, the receptive surface of the pollen grains, a short style and the ovary containing the ovules (Fig. 3B). Within the style and ovary, the pollen tubes grow in a specialized tissue, the transmitting tract guided toward the ovules.

Pollen tubes can also be observed *in vivo* (i.e. within the female tissues) by staining callose, a specific polysaccharide found in the intine wall of the pollen grain and pollen tube cell wall and within the pollen tube as callose plugs deposited periodically by the vegetative cell to maintain the expanding cell in the tip region (Fig. 3D-E). This method is widely used to assess the pollen tube growth between wild type pollen and mutant. Pollinated pistils were fixed with a mixture of acetic acid: ethanol (1:3 vol/vol) at least 12h. Rinsed 5 times in water, the pistils were softened by incubation with 8M NaOH for 12h at room temperature. The pistil were rinsed 5 times with water, placed on a glass slide and stained with 0.1% buffered decolorized aniline blue, pH 11 according to Johnson-Brousseau & McCormick (2004) for 4h and observed.

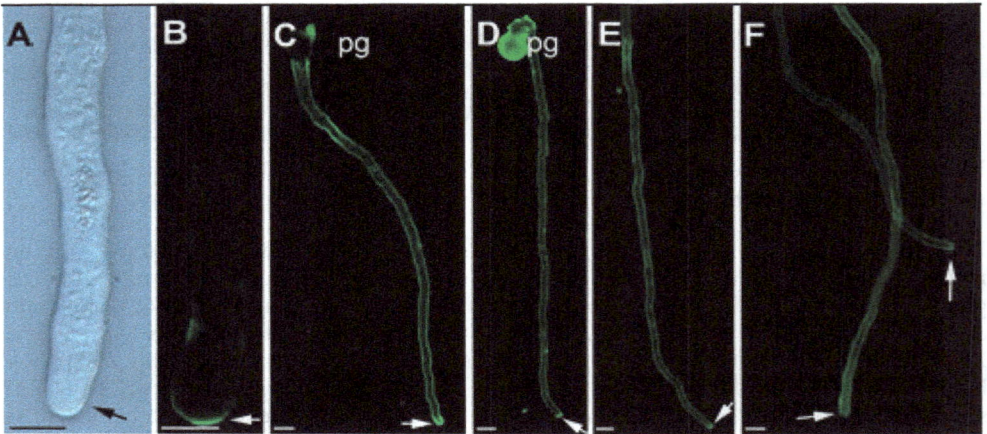

Fig. 2. Cell surface immunolocalization of cell wall epitopes in *Arabidopsis thaliana* pollen tube. A. Transmitted light of 6h-old *in vitro* grown pollen tubes, B-C. Localization of AGP epitopes at the tip of the pollen tubes using MAC207 and LM2 MAbs, respectively. D-E. Localization of arabinan epitopes found in the pectic polysaccharide RG-I in the entire pollen tube cell wall using LM6 and LM13 MAbs, respectively. F. Localization of highly methylesterified HG epitopes using JIM7 MAb. Arrow, pollen tube tip; pg, pollen grain. Bars = 5 µm.

After pollination, pollen grains germinate on the stigma and produce pollen tubes that grow within the cell wall of the papillae and in the transmitting tract of the style and ovary. Pollen grains are clearly detected on the stigma surface and the pollen tubes in the transmitting tract of the style and the ovary (Fig. 3D). Callose plugs are also visible. With higher magnification, the pollen tubes were observed in the ovary, emerging from the transmitting tract, growing on the funiculus toward the micropyle of the ovules (Fig. 3E).

3.2 Immunostaining of cell wall components on sections

The localization of cell wall polymers on sections is very instructive in particular to identify specific labelling of a given cell or tissue. When possible, the use of embedding medium is often avoided in order to increase the sugar epitope accessibility (Willats et al., 2001; Harholt et al., 2006). However, given the small sizes of Arabidopsis root tip, pollen tube or pistil, it is impossible to prepare hand sections. We currently use a procedure including fixation, dehydration and embedding to perform the sectioning step. We use methylmetacrylate

(Baskin, 1992) or London Resin White (LR White) resins as embedding media which do not disturb the detection of polysaccharides.

Pistils and anthers were fixed either with 4% PAF and 1% glutaraldehyde in 50 mM PIPES pH 7 for 2h or with 2% PAF and 2.5% glutaraldehyde in phosphate buffer (0.025 M, pH 7, with one micro drop of Tween 80), placed under vacuum for 1 h and then at 4 °C overnight. After dehydration in a graded ethanol series, the material was embedded in methylmetacrylate (Fig. 3A-C) or LR White (Fig. 3F-I). Sections (2 to 5 μm) were obtained with the Leica Reichart Supernova or EM UC6 microtome and placed on coated glass slides. Cytochemical staining with 0.05% toluidine blue showed that the cell wall of the transmitting tract cells is enriched in acidic polymers such as HG (Fig. 3B). The immunolocalization was performed as described in Coimbra et al., (2007) and Lehner al., (2010). Slides can be further counterstained with calcofluor white (fluorescent brightener; Sigma) for β-glucans (e.g., cellulose) detection as shown in Fig. 3F-I.

The arabinan epitopes associated with RG-I are strongly labelled in the style but very weakly in the cell wall of the papillae in the stigma (Fig. 3A). Weakly methylesterified HGs are also present in the cell wall of the ovary cells including the transmitting tract tissue (Fig. 3C). Glycosylated proteins, such as AGP, prevail in many stigma exudates, style transmitting tissues and pollen itself, and are believed to provide recognition signals and directional guidance for the pollen tube (Cheung et al., 1995; Wu et al., 2001). AGP represent a large group of highly glycosylated cell surface proteins, many of which contain glycosylphosphatidylinositol (GPI) anchors. They are suggested to act in signalling as soluble signals or as modulators and co-receptors of apoplastic morphogens. These immunolocalization studies, have shown a specific labelling of the central transmitting tissue of the style and also of the tracheary elements present all over the ovary, with MAbs JIM8 and JIM13 (Fig. 3F).

At the beginning of ovule development, when sections of Arabidopsis ovules were treated with MAbs JIM8 or JIM13, the specific labelling of the entire megaspore surface was striking. This labelling was specific to the first cell with a haploid constitution, marking the beginning of the gametophytic generation (Coimbra et al., 2007). As the development continues, the labelling with the same two MAbs was observed in the embryo sac wall, and was particularly intense in the synergid cells and in their filiform apparatus (Fig. 3H). At this more advanced stage of development, the labelling obtained with JIM8 and JIM13 extended to the integument micropylar cells and to the micropylar nucellus (Fig. 3H). The prevalence of AGPs in these reproductive tissues along with their possible role in signalling makes them excellent candidate molecules involved in guided pollen tube growth. This type of labelling was also shown for Nicotiana tabacum (Qin & Zhao, 2006) for Amaranthus hypochondriacus (Coimbra & Salema, 1997) and for Actinidia deliciosa (Coimbra & Duarte, 2003). The labelling obtained with the MAbs MAC207 and LM2, was found to be extensive and scattered throughout most cell types as shown in the style (Fig. 3G) and in the ovule, although excluded from the embryo sac (Fig. 3I).

The results obtained with this immunolabeling work were important not only from a developmental point of view, but also because they lay down the basis for the characterization of the expression of individual AGP genes in each of the developmental stages considered. As shown with cell surface labelling, removal of pectic polymers by enzymes (pectate lyase) or chemical treatments of the sections can effectively unmask other

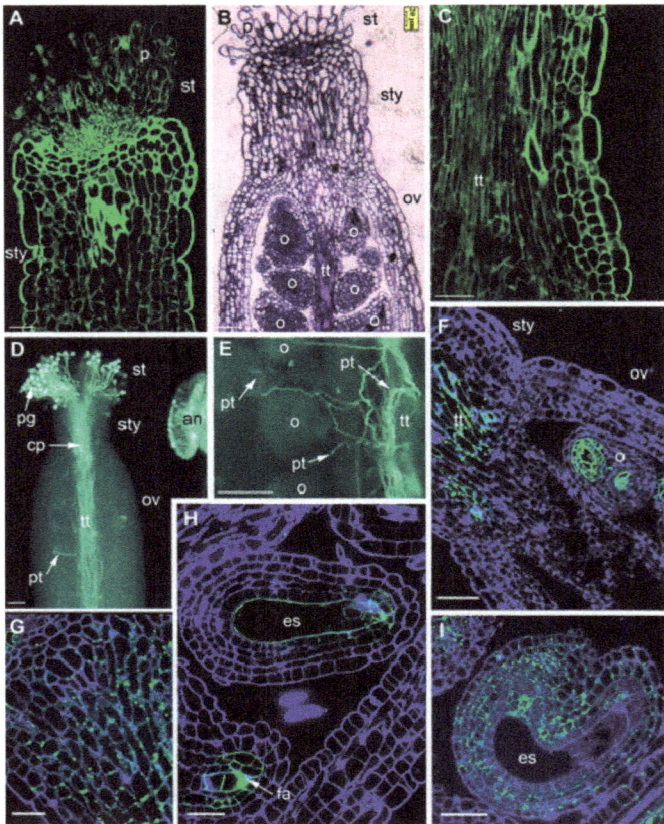

Fig. 3. Cytochemical staining and immunolocalization of cell wall epitopes in *Arabidopsis thaliana* pistil. A. Immunolocalization using the LM6 MAb specific for arabinan epitopes associated with RG-I in the cell wall of the stigma, style and ovary. B. Cytochemical staining of sectioned pistil with toluidine blue showing the acidic cell wall polysaccharides including the HG. C. Immunolocalization using the JIM5 MAb of epitopes associated with weakly methylesterified HG in the epidermal layer of the ovary and in the transmitting tract tissue. D-E. Cytochemical staining of callose with decolorized aniline blue on whole mount softened pollinated pistil showing the strong fluorescence of the pollen grains on the stigma surface, the pollen tube cell wall and the callose plugs within the female transmitting tissue in the style and ovary. E. A close up of D shows the pollen tubes emerging from the transmitting tract directed toward the ovules for the release of the sperm cells. F. The labelling with JIM8 is present in the centrally located transmitting tissue of the style, in the ovules embryo sac wall and extends into the integument micropylar cells. G. LM2 labelling all the style cells. H. The specific labelling by JIM8 of the embryo sac wall, the filiform apparatus and micropylar integuments are evident. I. This well developed ovule is showing all cell walls labelled by LM2, except the embryo sac cells and wall. an, anther; cp, callose plug; es, embryo sac; fa, filiform apparatus; o, ovule; ov, ovary; p, papillae; pg pollen grain; pt, pollen tube; st, stigma; Sty, style; tt, transmitting tract. Bars = 20 µm.

cell wall epitopes such those found on XyG as shown on tobacco stem sections by Marcus et al., (2008) and recently reviewed by Lee et al., (2011).

3.3 Immunostaining of cell wall polymers within the endomembrane compartment using protoplast

In higher plants, the endomembrane system and more precisely the Golgi apparatus plays a major role in the synthesis of the cell wall (Staehelin et al., 1990; Driouich et al., 1993). Involvement of plant Golgi stacks in the cell wall biosynthesis was revealed by cytochemical staining (Harris & Northcote, 1971) and biochemical experiments after fractionation (Ray, 1980; Gibeaut & Carpita, 1994). Hemicellulosic and pectic polymers are synthesized in the endomembrane system before their final assembly and deposition into the cell wall. Tobacco suspension-cultured cells (*Nicotiana tabacum*, Bright yellow-2 cells) are widely used in the study of endomembrane system (Pagny et al., 2003; Follet-Gueye et al., 2003; Toyooka et al., 2009; Langhans et al., 2011). Immunolabeling of N-glycan epitopes associated with Golgi membranes was easily achieved after a slight cell wall permeabilization (Fitchette et al., 1999). Here, we show that the production of protoplasts is necessary for performing immunodetection of cell wall polymer precursors in Golgi stacks.

3.3.1 Cell permeabilization and protoplast preparation

For tobacco cells immunolabeling, in 1.5 ml ependorf vials, four-day-old BY-2 cells (300 µl) were fixed with 800µl of 4% PAF in PBS (PBS: 8 g L^{-1} NaCl, 0.29 g L^{-1} KCl, 1.45 g L^{-1} Na_2HPO_4, 0.24 g L^{-1} KH_2PO_4) for 1 h at room temperature. After six washes for 5 min with PBS buffer, fixed tobacco cells were incubated 30 min in 1 ml enzymatic mixture of 1% cellulase (Sigma) and 1% pectinase (Sigma) diluted in PBS buffer. Cells were then washed four times in PBS. For the protoplast preparation, four-day-old BY-2 cells (400 µl) were put in 1.5 ml ependorf vials, and plasmolyzed in 800 µl plasmolysis medium (20 mM MES buffer pH 5.6; 0.6 M mannitol; 0.5 mM $CaCl_2$; 0.25 mM $MgCl_2$) for 15–20 min. Then the cell wall was digested overnight at 25°C by incubation in 1 ml enzymatic mixture of 1% cellulase (Sigma) and 1% pectinase (Sigma) diluted in plasmolysis medium containing 0.1% bovine serum albumin (BSA). After three washes with plasmolysis medium, protoplasts were fixed 1 h with 4% PAF in PBS at room temperature. Then protoplasts were washed four times with PBS. Then, 25 µl of cells or protoplasts were squashed onto 1:10 poly-L-Lysine (sigma) coated 10 wells slides (Fischer Scientific) and air dried.

3.3.2 Immunostaining

Protoplasts and tobacco cells were sequentially incubated 10 min with 0.5% Triton X100 in PBS, with 1% BSA and 3% half-milk in PBS, and washed three times with 3% BSA in PBS. Samples were incubated overnight at 4°C, with the primary antibody CCRC-M86 diluted at 1/10 in PBS containing 1% BSA, 3% normal goat serum (NGS). After six washes in PBS containing 1% BSA, samples were incubated 1 h at 25°C with a secondary antibody (goat IgG anti-rat TRITC conjugate from Sigma) diluted at 1/50 in PBS containing 1% BSA, 3% NGS. Finally cells or protoplasts were washed three times in PBS containing 1% BSA.

When the cell wall is only partially digested (Fig. 4A), the density of XyG epitope remaining in the cell wall limits the MAbs penetration inside the cell. So even under confocal microscopy observations, labelling of Golgi units is hardly detectable. In contrast, the complete removal of the cell wall in protoplasts allows the access of the MAb CCRC-M86 to the Golgi-associated epitopes (Fig. 4B). Moreover, it is well recognized that cell wall polymer synthesis is stimulated in protoplasts thus contributing to the increase of labelling of the endomembranes. With this procedure, we have clearly shown that sugar motifs associated with XyG (Fig. 4) or pectins (data not shown) are present in the Golgi units. These are numerous and dispersed throughout the cytoplasm (see numerous fluorescent spots on Fig. 4B).

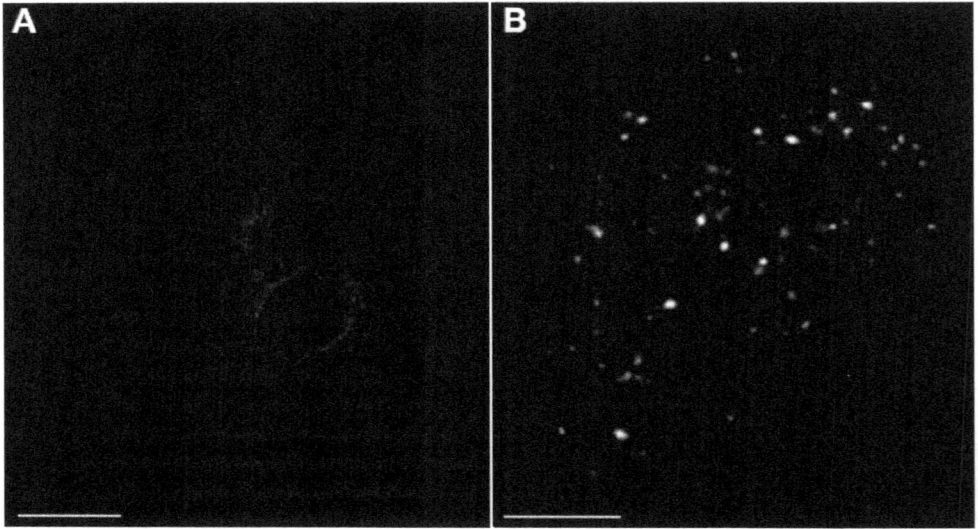

Fig. 4. Immunocytochemical localization of XyG epitopes in partially-digested tobacco BY-2 suspension-cultured cells and protoplasts with the MAb CCRC-M86. A- Confocal microscope images showing fluorescent staining of the cell surface and B- Golgi bodies. CCRC-M86 epitopes localize to both the cell wall and Golgi stacks of tobacco cell or protoplasts, respectively. Bars = 8 μm.

4. Spatial organization of non-cellulosic cell wall polysaccharides at the ultrastructural level

Due to the unability to subfractionate the plant Golgi stacks into *cis*, medial and *trans* cisternae, our knowledge, concerning the assembly of complex polysaccharides within the Golgi subcompartments, remains scarce. Progress towards the understanding of compartmentalization of matrix cell wall polysaccharide biosynthesis has come from immuno-electron microscopical analyses (Staehelin et al., 1990, Zhang & Staehelin, 1992, Driouich et al., 1993). These immunolabeling studies have been performed on samples prepared by HPF, a cryofixation technique that provides excellent preservation of Golgi stacks thereby allowing different cisternal subtypes to be easily distinguished. For ultrastructural and immunocytochemical investigations, we used the HPF coupled to freeze substitution (FS) technologies to prepare Arabidopsis root tips and pollen tubes, or tobacco suspension-cultured cells. We present and discuss here protocols allowing epitopes detection associated with polysaccharides and membrane proteins (e.g., glycosyltransferases responsible for polysaccharide biosynthesis).

4.1 TEM sample preparation

4.1.1 High-Pressure Freezing/Freeze Substitution (HPF/FS)

4.1.1.1 Pre-cooling treatment

Prior to freezing, 3 day-old tobacco (BY-2) suspension-cultured cells were concentrated on 50 μm nylon membrane and incubated successively in MS + 5% glycerol, then in MS+10% glycerol for 1h each, at +4°C and under light stirring. Cells were then transferred to MS+20% glycerol and immediately frozen (see § 4.1.1.2). Root tips of seven day-old arabidopsis seedlings were first immersed in 2-(N-morpholino) ethanesulfonic acid (MES) buffer (20mM, $CaCl_2.2H_2O$ 2mM, KCl 2mM, Sucrose 0.2M; pH 5.5) containing a small drop of red ruthenium (to allow visualisation of the roots) for 5 min. Roots were then incubated with 0.2M sucrose in MES buffer for 5 min before freezing. Six hour-old pollen tubes were kept in the germination medium (GM) before freezing.

4.1.1.2 High pressure freezing

High-pressure freezing was performed with the freezer HPF–EM PACT I Leica-microsystems (Studer et al., 2001). Tobacco cells were concentrated on 50μm nylon membrane and transferred into the cavity of a copper ring used for cryofracture which is 100μm in depth and 1.2 mm in diameter. Excess medium was lightly absorbed with filter paper. Using a horizontal loading station, the specimen carriers were tightened securely into the pod of specimen holder. After fixation on the loading device, specimens were frozen according to a maximum cooling rate of 13000°C/s, incoming pressure of 7.5 bars and working pressure of 4.8 bars. Rings containing frozen samples were stored in liquid nitrogen until the freeze substitution procedure was initiated.

Root tips of *Arabidopsis thaliana* (1 to 2 mm) were excised with a fine razor blade in freezing media (0.2M sucrose in MES buffer). Three or four roots were placed in a gold platelet (200 mm in depth and 1.2 mm in diameter) carrier pre-filled with the freezing medium and coated with soybean lecithin (100 mg mL^{-1} in chloroform). Specimens were immediately frozen in the EM PACT freezer under similar conditions of pressure and cooling rate as for the tobacco cells (see above). Time between excision of the root tips and freezing was <1 min.

Like root tips, centrifuged 6-h-old pollen tubes were transferred into the cavity of gold cupules coated with soybean lecithin (100 mg mL^{-1} in chloroform). Excess medium was removed using a filter paper and were frozen as describe above.

4.1.1.3 Freeze substitution and resin embedding

After HPF, pollen tubes were transferred to a freeze substitution automate (EM-AFS; Leica) precooled to -140°C. Freeze substitution conditions followed a modified procedure from D. Studer (personal communication). Substitution media were composed of 2% osmium in anhydrous acetone. Samples were substituted at -90°C for 72 h. The temperature was gradually raised (2°C h^{-1}) to -60°C and stabilized during 12 h, then gradually raised (2°C $^{-1}$)

to -30°C (12 h) and gradually raised again (2°C h⁻¹) to 0°C for 2 h. Samples were washed at room temperature with fresh anhydrous acetone. Infiltration was done at +4°C in acetone-Spurr resin (2:1, 1:1, 1:2, 8 h each step) and with pure resin for at least 2 days. Polymerization was performed at 60°C for 16 h. Samples destined for protein immunocytochemistry (Arabidopsis root tips and BY-2 cells) were substituted in anhydrous acetone + 0.5% uranyl acetate (UA) using similar program as that described above except that the final substitution step was performed at -15°C. Samples were rinsed twice with anhydrous ethanol. Infiltration was processed at -15°C in a solution containing Ethanol : LR White (ratio 2:1; 1:1; 1:2, 8h each step) and with pure resin for 2x24 h. Polymerisation was performed into the AFS apparatus at -15°C under ultra violet light for 48h. Using an ultracut EM-UC6 (Leica), thin sections (90 nm) were mounted on formvar-coated nickel grids.

4.2 Immunogold staining

4.2.1 Single immunogold labelling

4.2.1.1 Cell wall immunogold labeling on epoxy sections

Grids, with pollen tube sections, were rehydrated in a Tris-buffered saline buffer (TBS) + 0.2% BSA and blocked in a TBS/BSA 0.2%/milk 3% solution for 30 min. After three brief rinses in TBS/BSA 0.2% solution, grids were incubated 3 h at 25°C in primary antibodies: non diluted for LM15 and CCRC-M1 or diluted (LM6, 1:2; anti-callose, 1:100) in TBS/BSA 0.2% buffer. Then, grids were washed (six times for 5 min each) in TBS/BSA 0.2% and incubated for 1 h at 25°C in a 1:20 secondary antibody (goat anti-rat for LM6 and LM15, goat anti-mouse for CCRC-M1 and anti-callose) conjugated to 10-nm gold particles (British Biocell International). Finally, grids were washed (six times for 5 min each) in a TBS + BSA 0.2% buffer, 1 min in TBS, 10 min in TBS + glutaraldehyde 2%, 5 min in TBS, and then two times for 5 min each in double deionized water. The sections were stained with 0.5% (w/v) UA in methanol for 10 min in the dark, rapidly rinsed 10 times with water and stained with lead citrate for 10 min, and briefly rinsed 10 times with water. In this experiment, epoxy resin was used as embedding medium. It is generally well admitted that Epoxy resin, which allow good cellular structure preservation, is not recommended for immunocytochemical studies. In the case of XyG and callose epitopes studied in Arabidopsis pollen grain and pollen tube, the detection was done with success (Fig. 5). Subcellular observation of Arabidopsis pollen grains showed that the intine cell wall contains the hemicellulose epitopes associated with fucogalactoXyG (Fig. 5A) and XyG (Fig. 5B). On the exine wall, pollen coat material is also visible. In the pollen tube, a double cell wall can be detected back from the tip region with the inner cell wall containing callose (Fig. 5C).

4.2.1.2 Immunogold labeling of polysaccharides and synthesizing-enzymes in Golgi stacks on acrylic sections

Ultrathin sections of tobacco cells or Arabidopsis root tips were blocked in TBS (Tris-buffered saline: Tris–HCl 20 mM pH 7.2, 200 mM NaCl) supplemented with 0.2% of BSA and milk 3% for 30 min. Sections were then incubated with monoclonal antibodies CCRC-M86 at a 1/5 dilution in TBS + BSA and normal goat serum (NGS; British Biocell International) for 3 h at room temperature. After washing in TBS + BSA 0.2% buffer, grids were incubated for 1 h at room temperature in the rat secondary antibody conjugated to 10-nm gold particles (Tebu-British Biocell International, http://www.tebu-bio.com). After

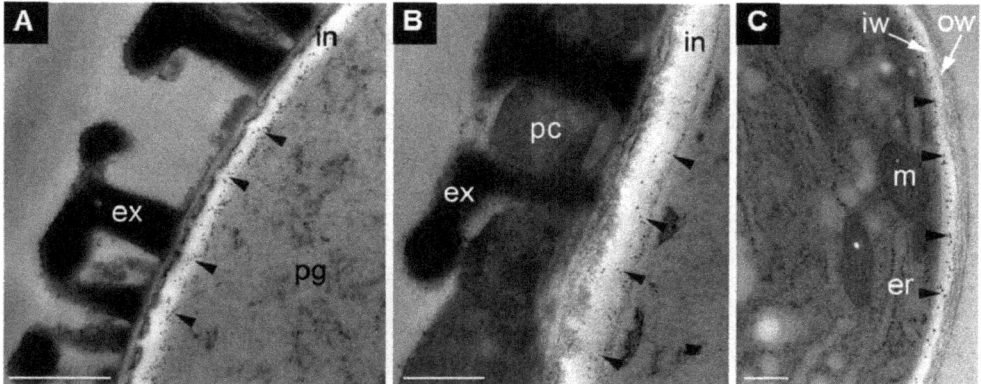

Fig. 5. A. Localization of fucosylated XyG in the intine wall of the pollen grain using the
CCRC-M1 MAb. B. Localization of the motif XXXG found in XyG in the intine wall of the
pollen grain using the LM15 MAb. C. Localization of callose in the inner cell wall of the
pollen tube using the anti-callose antibody. Arrowhead, gold particles; er, endoplasmic
reticulum; ex, exine; in, intine; iw, inner cell wall; m, mitochondria; ow, outer cell wall; pc,
pollen coat; pg, pollen grain. Bars = 0.2 μm.

washing and fixation in TBS + 2% glutaraldehyde, sections were first stained with 4%
vapour osmium for 3 h on a hotplate warmed to 40°C placed under a fume hood. For this
step, the grids were put closed to 4% osmium droplets on a parafilm paper. This system was
put in a small glass Petri dish who was kept hermetically closed until the total osmium
evaporation. Then sections were classically stained 10 min with 0.5% UA diluted in
methanol and 10 min in lead citrate.

Ultrathin sections of frozen cells expressing AtXT1–Green Fluorescent GFP (GFP) or
AtMURUS 3 (MUR3)–GFP fusion proteins (Chevalier et al., 2010) were blocked in TBS (Tris–
HCl 20 mM pH 7.2, 200 mM NaCl) supplemented with 0.2% of BSA and Tween 20 0.05% for
15 min. Sections were then incubated with the polyclonal anti-GFP antibody (BD living
colors A.v peptide antibody – Clontech) at a 1/5 dilution in TBS + BSA and NGS for 2 h at
room temperature. After washing in TBS + BSA 0.2% buffer, grids were incubated for 1 h at
room temperature in the rabbit secondary antibody conjugated to 10 nm gold particles
(Tebu-British Biocell International). After washing and fixation in TBS + 2% GA, sections
were stained as described above with osmium vapours, uranyl acetate and lead citrate.

The combined use of HPF-FS and cryo-embedding in acrylic resin is particularly adapted to
immunodetection of either membrane proteins or cell wall polysaccharides (Fig. 6 & Fig. 7).
The cryo-immobilisation by HPF ensures the sample vitrification without modification of
molecular structures. While chemical fixatives (such as aldehydes or osmium tetroxyde) are
relatively inert with sugar molecules, proteins are often covalently cross linked with
fixatives, which can disturb the epitope structure. During the cryo-dehydration step, we
have lightly fixed the cells by adding 0.5% uranyl acetate to avoid the modification of the
protein epitope. To enhance the electron micrograph contrast, the contact with osmium
vapour prior to classical uranyl acetate and lead citrate staining is efficient (Fig. 6, Fig. 7,
Fig.8 & 9) to enhance Golgi stack morphology. The non-fucosylated sugar motifs associated

with XyG, already detected in Golgi units in tobacco protoplasts (Fig. 4B), are mainly localized to the secretory vesicles of *trans* Golgi network (TGN) (Fig. 6A). With the same preparation procedure, it is also possible to immunolocalize XyG-synthesizing enzymes fused to GFP (e.g., AtXT1-GFP), within Golgi cisternae in transformed BY-2 cells (Fig. 6B). AtXT1-GFP was shown to localize mainly to the cis Golgi compartment (Chevalier et al., 2010).

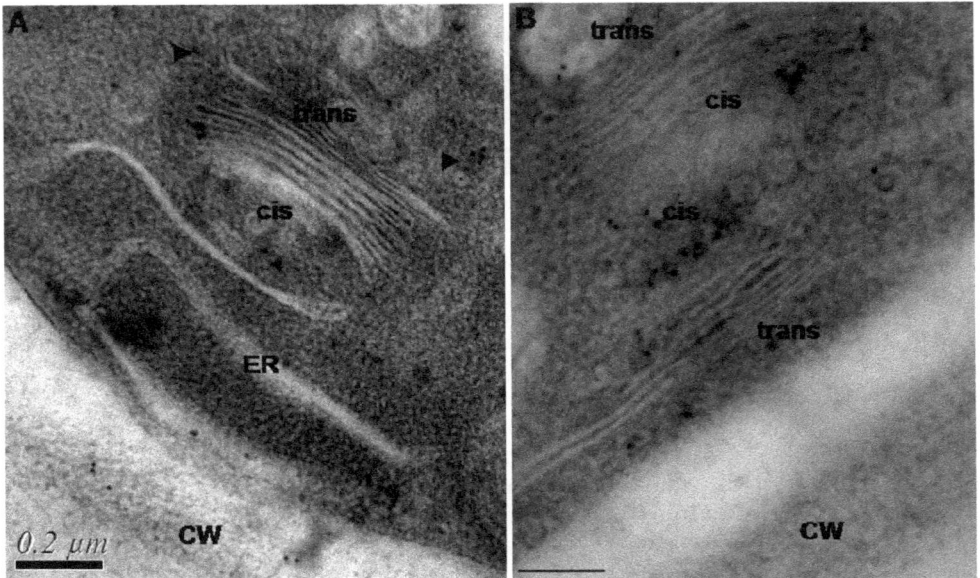

Fig. 6. Electron micrographs of HPF/FS and LR White embedded tobacco cells illustrating immunogold labelling of Golgi stacks and the cell wall. A- Non fucosylated motifs of XyG were immunodetected with the MAb CCRC-M86. B- AtXT1–GFP fusion protein was labelled with anti-GFP Ab, in Golgi stacks of transformed tobacco BY-2 suspension-cultured cells. The black head arrows indicate the secretory vesicles labelled. CW, cell wall; ER, endoplasmic reticulum; bars = 200 nm.

4.2.2 Double immunogold labeling

Information on the distribution of Golgi localized enzymes are currently obtained by using GFP technologies and confocal laser microscopy (Brandizzi et al., 2004). Immuno-electron microscopy is the only approach to study the distribution of GFP-labelling within a given organelle. Therefore, a central issue for immunogold at TEM level is to preserve epitopes of interest while minimizing the loss of ultrastructural details. Here, we study the localization of QUASIMODO2 protein (QUA2) in *qua2* mutant transgenic plants expressing the QUA2-GFP fusion protein (Mouille et al., 2007). QUA2 is a putative methyltransferase involved in the methylesterifcation of HG. It is well admitted that HG are synthesized in their methylesterified form and then possibly de-esterified by pectin methylesterases in the cell wall (Micheli, 2001).

With the sample substituted in 0.5% AU medium and embedded in LR White resin, we have been able to perform a double immunolabeling to localize Golgi enzymes fused to GFP and cell-wall polysaccharides. Ultrathin sections of Arabidopsis root tips of *qua2*-QUA2-GFP (Mouille et al., 2007) were blocked as describe above and incubated 2h at 25°C with the polyclonal anti-GFP antibody (BD living colors A.v peptide antibody – Clontech) at a 1/5 dilution in TBS + BSA and normal goat serum. After washing in TBS + BSA 0.2% buffer, sections were again saturated with 3% milk, 0.2% BSA in TBS buffer for 30 min. Incubation with the MAb JIM7 diluted to 1/5 in TBS + BSA and normal goat serum was done for 2h at 25°C. After six rinsing for 5 min with TBS buffer containing 0.2% BSA, grids were incubated for 1 h at room temperature in the rabbit secondary antibody conjugated to 10 nm gold particles diluted to 1/20 in TBS + BSA 0.2% + NGS 3%. Sections were then washed three times in TBS buffer + BSA 0.2%, and incubated 1 h at 25°C in the goat anti-rat secondary antibodies conjugated to 20-nm gold particles (Tebu-British Biocell International, http://www.tebu-bio.com). Sections were rinsed six times for 5 min with TBS buffer containing BSA 0.2%, and were followed by one washing in TBS buffer. After fixation in TBS + 2% GA, sections were stained as described above with osmium vapours, uranyl acetate and lead citrate.

The fusion protein is detected in the *trans* Golgi cisternae (Fig. 7C) while the methylesterified HG are immunolocalized not only in the *trans* cisternae but also in the TGN as previously described (Zhang & Steahelin, 1992; Toyooka et al., 2009). The lack of strong fixative in the substitution medium like osmium tetroxide or glutaraldehyde, allows the combined detection of pectic epitopes and its Golgi-associated biosynthetic enzyme. The cell wall labelling of the methylesterified HG epitopes (Fig. 7A & 7B) shows that the sample preparation did not alter the polymer localization. Given that the protein epitope is not impaired by strong fixatives, the double labelling of the product (HG) and the Golgi-localized methyltransferase is now possible. Our results suggest that HG methylation is likely to start in trans Golgi cisternae.

4.3 Quantitative mapping of polysaccharide-synthesizing enzymes within Golgi stacks

Statistical analysis of immunogold labelling was used to map XyG-glycosyltransferases within Golgi cisternae (Chevalier et al., 2010). As no antibodies directed against these enzymes are currently available, we have chosen to investigate their localization indirectly by using anti-GFP antibodies in transgenic tobacco cells expressing the glycosyltransferases fused to GFP. Immunogold labelling was quantified over each Golgi cisternae (fifty Golgi stacks with distinguishable cisternal subtypes) and subjected to chi-squared (χ^2) test to check for labelling specificity in transformed BY-2 cells versus wild type. The relative labelling index (RLI) was also determined to evaluate specific gold particles distribution within different subtypes of Golgi cisternae (Mayhew et al., 2002; Mayhew, 2011).

4.3.1 Labelling specificity of signal in transformed BY-2 cells

To evaluate the distribution of gold particles in Golgi stacks of transformed cells as compared with wild type cells, we used a chi-squared test (χ^2) as described by Mayhew et al., (2002). We considered the null hypothesis as 'no difference of gold distribution between wild type and transformed cells'. We analyzed gold labelling in different compartments of Golgi stacks: *cis*, medial, *trans* and the TGN (Fig. 8).

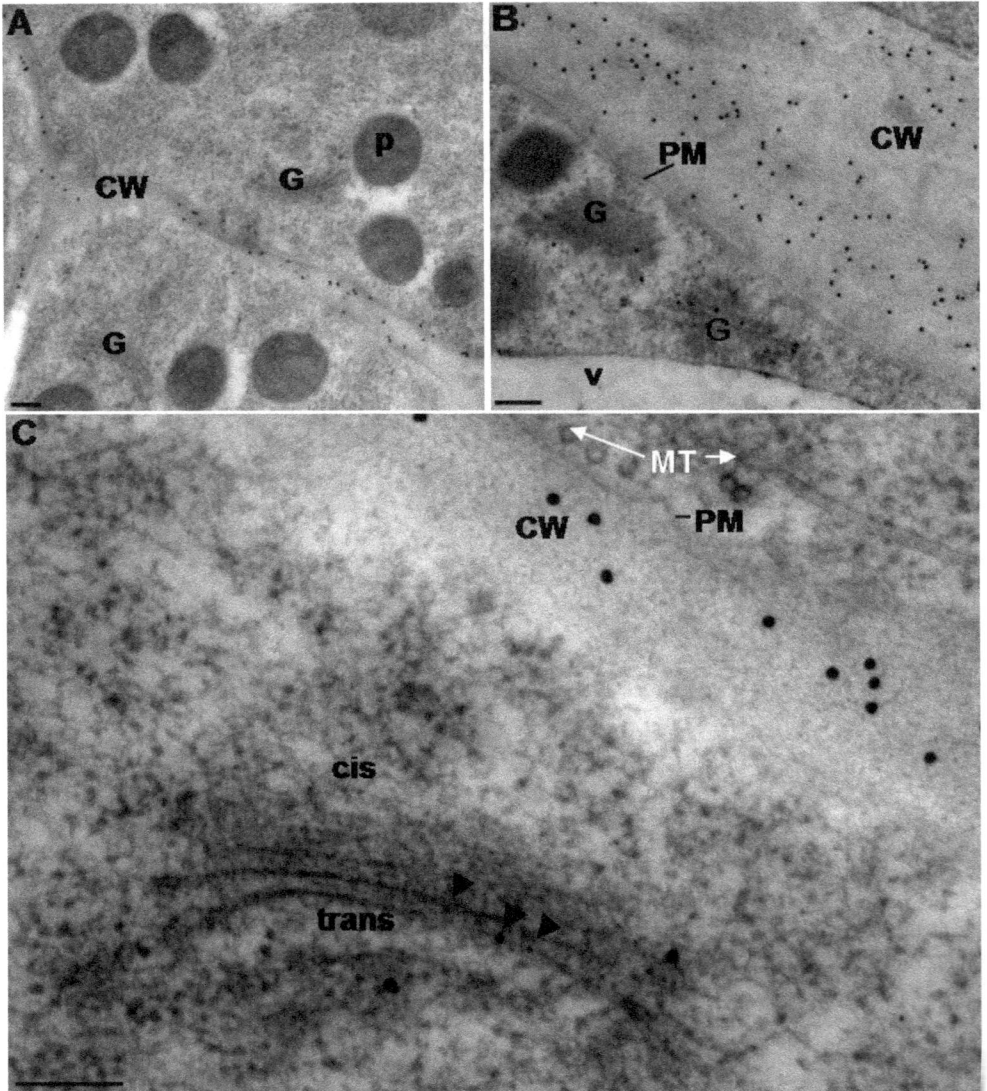

Fig. 7. Electron micrographs of HPF/FS and LR White-embedded ultrathin sections of
Arabidopsis root tips of *qua2*-QUA2-GFP. Double localization of AtQUA2-GFP with anti-
GFP antibody (10 nm gold particles) and of highly methylesterified HG with MAb JIM7 (20
nm gold particles) in Arabidopsis root meristematic cells. A-B: Note abundant labelling of
HG epitopes in the cell wall and in the last Golgi compartments. C, the black arrow heads
indicate the 10 nm gold particles corresponding to GFP labelling in the *trans* Golgi cisternae
CW, cell wall; G, Golgi ; MT, cortical microtubules ; P, plastid; PM, plasma membrane ; V,
vacuole. Bars = 100 nm.

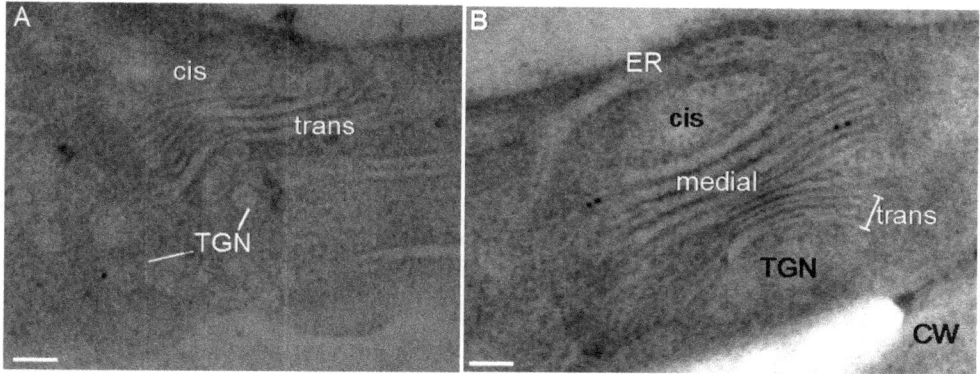

Fig. 8. Electron micrographs of GFP immunodetection on A- 3 day-old wild type BY-2 cells
or B- transformed cells expressing the fusion protein AtMUR3-GFP. 50 similar images are
used to determine the number of gold particles observed in each Golgi compartment: *cis*,
medial, *trans* cisternae and the *trans* Golgi network (TGN). CW, cell wall; ER, Endoplasmic
reticulum; At MUR3 is a XyG-galactosyltransferase, bars = 100 nm.

To compare the distribution of gold particles in these compartments between the two
tobacco lines, we counted the gold particles observed over 50 Golgi units both in non-
transformed cells (Wild type cells, termed WT cells) and transformed tobacco cells
(expressing Golgi-enzyme fused to GFP, termed GFP-cells), immunolabelled with anti-GFP
antibodies (Fig. 8). We prepared a table of gold particles counted for each tobacco lines and
their repartition in different Golgi compartments (Table 2).

For each Golgi compartment, we calculated the expected gold particle number (Eij)
according to the null hypothesis as 'no difference of gold distribution between wild type
and transformed cells and that the gold particles are randomly distributed across the Golgi
units'. The expected gold count (Eij) at two line cells for each selected Golgi compartment
are calculated by using the table 2 values1, from the product of column (total gold particles
observed in all Golgi stacks for wild type or transformed tobacco lines) and row totals (total
gold particles observed in all BY-2 lines for the selected compartment) divided by the big
total (total gold particles observed in four Golgi compartments, in wild type and
transformed BY-2 cells).

	cis (j_1)	**medial** (j_2)	*trans* (j_3)	**TGN** (j_4)	**Row Total** (j_5)
WT cells (i_1)	O_{11}	O_{12}	O_{13}	O_{14}	O_{15}
GFP cells (i_2)	O_{21}	O_{22}	O_{23}	O_{24}	O_{25}
Column Total (i_3)	O_{31}	O_{32}	O_{33}	O_{34}	**O_{35}**

Table 2. Gold particles counted (Oij) over 50 Golgi stacks from Wild type tobacco cells (WT
cells) or transformed BY-2 cells by Golgi localized enzymes fused to GFP (GFP cells). Oij
designates the gold particle number indicated in the row (i) and the column (j). O_{15} and O_{25}
designate the total gold particle number in WT cells and GFP cells, respectively. O_{31} to O_{34}
correspond to the total gold count in the four Golgi compartments of the two tobacco lines.
O_{35} is the big total (four Golgi compartments, in WT and GFP cells).

Example for the *cis* compartment of WT cells, expected gold number (E_{11}) is calculated as described below[1]:

$$E_{11}=[(O_{31} \times O_{15})/O_{35}] \tag{1}$$

To compare the distribution of gold particles between the two tobacco lines in these compartments, we created a contingency table (two columns; four rows) as described in Mayhew et al. (2002) and also in table 3. In this contingency table we report the observed (O_{ij}) and expected (E_{ij}) numbers of gold particles. For each tobacco line, total column should be equal ($O_{15}=E_{15}$; $O_{25}=E_{25}$). If expected number is superior to 5, the χ^2 test can then be applied. The corresponding partial χ^2 value is calculated as the square of the difference between observed and expected gold particle numbers divided by the expected gold number.

Example for the *cis* compartment of WT cells, χ^2 value (χ_{11}^2) is calculated as described below[2]:

$$\chi_{11}^2=[(O_{11}-E_{11})^2/ E_{11}] \tag{2}$$

The total χ^2 value ($\chi_{15}^2 + \chi_{25}^2$) is compared with χ^2 value in contingency table analysis with a degree of freedom equal to 3 in our experiment and corresponding to[3]:

$$\text{Degree of freedom}=[(\text{row number} -1)\times(\text{column number} -1)] \tag{3}$$

When the total χ^2 value ($\chi_{15}^2 + \chi_{25}^2$) is found to be superior to that of the contingency table, $\chi_{table}^2 = 12.38$ (with a degree of freedom of 3 and an uncertainty degree $\alpha = 0.005$) indicating that labelling is specific in Golgi stacks of transformed BY-2 cells.

	WT cells			GFP cells			Total observed gold
	(Oij)	(Eij)	χ_{ij}^2	(Oij)	(Eij)	χ_{ij}^2	
cis	O_{11}	E_{11}	χ_{11}^2	O_{21}	E_{21}	χ_{21}^2	O_{31}
medial	O_{12}	E_{12}	χ_{12}^2	O_{22}	E_{22}	χ_{22}^2	O_{32}
trans	O_{13}	E_{13}	χ_{13}^2	O_{23}	E_{23}	χ_{23}^2	O_{33}
TGN	O_{14}	E_{14}	χ_{14}^2	O_{24}	E_{24}	χ_{24}^2	O_{34}
Column Total	O_{15}	E_{15}	χ_{15}^2	O_{25}	E_{25}	χ_{25}^2	O_{35}

Table 3. Model of contingency table use for quantitative analysis of anti-GFP labelling over Golgi stacks in tobacco BY-2 cells expressing Golgi enzyme–GFP. This table contains two columns (Oij and Eij) and four rows (related to the four Golgi compartments studied).

4.3.2 Determination of fusion protein distribution in Golgi stacks

Quantification of the labelling was done according to Mayhew et al. (2002), on a total of 50 Golgi stacks with clearly distinguishable subtypes of cisternae (Fig. 9A) from transformed BY-2 cells expressing enzyme-GFP fused construct (noted GFP cells). First, gold particles were counted in each Golgi compartment (*cis*, medial, *trans* and the TGN) in 50 Golgi stacks from WT or GFP cells as described in § 4.3.1. The *'observed gold particles'* (n_0) in a given Golgi compartment in GFP cells is determined by subtracting the number of gold particles found in this compartment (Oij, GFP cells, 50 Golgi stacks) from the one counted in WT cells (O⁻j,

WT)[4]. Example for the *cis* compartment of WT cells, n_0 value is determined as described
below[4]:

$$n_{011} = O_{21} - O_{11} \tag{4}$$

Second, each electron micrograph is calibrated and a lattice of test points was superimposed
on each Golgi stack (so as to be random in position) by the plungins function of Image J
computer software (Fig. 9B); hit points were counted (*'Observed points'*, P) in each
compartment of the Golgi stack.

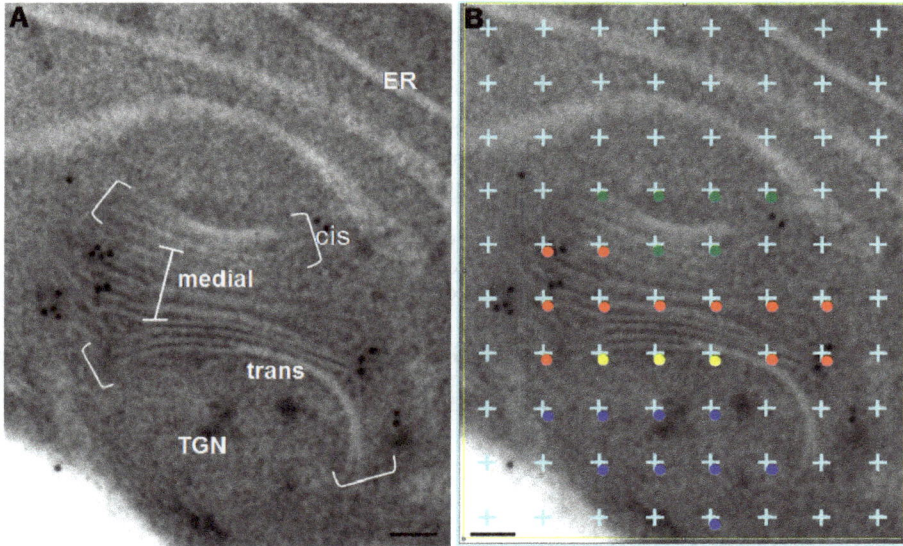

Fig. 9. Electron micrograph of a Golgi unit from At-MUR3-GFP transformed BY-2 cells.
(A) *cis*, medial, *trans* and *trans* Golgi network (TGN) are clearly distinguishable. (B) Lattice
(1015,43 area per point) was superimposed on the same electron micrograph calibrated
(1nm/1.01 pixels) by using Image J computer software. In this example, 6 *'Observed points*, P'
are present in the *cis* (green points), 11 in the medial (red points), 3 in the *trans* (yellow
points) and 9 in the TGN (blue points) compartments. Scale bars = 100 nm.

"*Observed gold particles*, n_0" and "*observed points*, P" are put in the relative labelling index
(RLI) table (Table 4). Then, the *'expected golds particles'* (n_e) (i.e. random distribution) is
calculated for each Golgi compartment[5] from the product of the observed points in the Golgi
compartments and column total of observed gold particles (total gold particles observed in
four Golgi compartments, $n_{0\ 51}$), divided by the column total of observed points (total points
observed in four Golgi compartments, $P_{\ 52}$). Example for the *cis* compartment of WT cells, n_e
value is determined as described[5]:

$$n_{e\ 13} = (P_{12} \times n_{0\ 51}) / P_{\ 52}. \tag{5}$$

Observed (n_0) and expected (n_e) gold particles distribution allowed calculation of the
relative labelling index[6] (RLI):

$$RLI = (n_0/n_e) \tag{6}$$

For a random distribution, the relative labelling index (RLI) is inferior or equal to 1, and for a compartment preferentially labelled RLI is superior to 1.

	Observed gold particles (n_0)	Observed points (P)	Expected gold particles (n_e)	RLI (n_0/n_e)
cis	$n_{0\,11}$	P_{12}	$n_{e\,13}$	$n_{0\,11}/\,n_{e\,13}$
medial	$n_{0\,21}$	P_{22}	$n_{e\,23}$	$n_{0\,21}/\,n_{e\,23}$
trans	$n_{0\,31}$	P_{32}	$n_{e\,33}$	$n_{0\,31}/\,n_{e\,33}$
TGN	$n_{0\,41}$	P_{42}	$n_{e\,43}$	$n_{0\,41}/\,n_{e\,43}$
Column total	$n_{0\,51}$	$P_{\,52}$	$n_{e\,53}$	

Table 4. Model of relative labelling index (RLI) table comprising 3 columns (n_0, P and n_e) and four rows (related to four Golgi compartments studied). Total observed gold particles and total of expected gold particles must be equal ($n_{0\,51} = n_{e\,53}$).

In a recent published study, we have applied this quantitative analysis to determine the Golgi mapping of three XyG synthesis enzymes in transformed tobacco cell lines (Chevalier et al., 2010). We show that the labelling of AtXT1–GFP (Fig. 6B) was mostly associated with *cis* and medial cisternae, whereas the labelling of AtMUR3–GFP (Fig. 8B & Fig.9) was mainly detected in medial and *trans* cisternae of Golgi stacks. In addition to HPF-FS protocol described here for BY-2 cell preparation, Tokuyasu cryo-sectioning approach is also amenable to this plant cell model (Chevalier et al., 2011). A new approach named the "rehydration method" was proposed by Van Donselaar et al. (2007). It combines HPF/FS procedure followed by a rehydration step, infiltration in sucrose and cryo-sectioning, and is particularly efficient to investigate the localization of epitopes from polysaccharides that are often lost during Tokuyasu process and was recently used to immunolocalize epitopes associated with XyG (Viotti et al., 2010). It could be of interest to adapt the rehydration method to plant suspension cells like tobacco BY-2 cells for quantitative study of cell wall polymers such as hemicelluloses or pectins.

5. Conclusion

In conclusion, many immuno-microscopical methods are available to investigate plant cell walls. A number of antibodies against cell wall carbohydrate epitopes are also available and can be used to examine the distribution of polysaccharides. It is however important to perform experiments with varying parameters to determine the best conditions for a given sample. Optical imaging instrumentation, based on confocal laser scanning microscopy (CLSM), offers the ability to determine the spatial organisation of polymers at the cell surface, across the cell wall compartment as well as within the endomembrane system. The CLSM minimizes interference of innate plant autofluorescence from chlorophylls or vacuole content and provides a quite precise picture of polysaccharides distribution. Moreover, cryo-sample preparation for immunogold EM studies has significantly improved the immunodetection at the subcellular level of the cell wall components and their synthesizing enzymes. Finally, the preservation quality of both the cell ultrastructure and epitopes allows the quantification of the labelling within specific membrane-bound compartments (e.g; Golgi stacks). A careful combination of microscopical and quantification methodologies is very promising for future studies devoted to imaging of cell wall glycomolecules and synthesizing enzymes.

6. Acknowledgments

Work in authors laboratories is/was supported by the Centre National de la Recherche Scientifique (CNRS), l'Université de Rouen, le Conseil Régional de Haute Normandie, the Plant Research Network VATA of Haute Normandie, the French ministry of Research and the FCT (Fundação para a Ciência e Tecnologia, Portugal) within the project PTDC/AGR-GPL/67971/2006.

7. References

Baskin T.I., Busby C.H., Fowke L.C., Saturant M. & Gubler F. (1992) Improvements in immunostaining samples embedded in methacrylate: localization of microtubules and other antigens throughout developing organs in plants of diverse taxa. *Planta*, Vol. 187, pp. 405-413

Boavida, L.C. & McCormick, S. (2007) Temperature as a determinant factor for increased and reproducible in vitro pollen germination in Arabidopsis thaliana. *The Plant Journal*, Vol.52, pp. 570–582

Brandizzi F., Irons S.L., Johansen J., Kotzer A. & Neumann U. (2004) GFP is the way to glow: Bioimaging of the plantendomembrane system. *Journal of Microscopy*, Vol. 214, pp.: 138-158.

Cheung, A.Y., Wang, H. & Wu, H.M. (1995) A floral transmitting tissue-specific glycoprotein attracts pollen tubes and stimulates their growth. *Cell*, Vol.82, pp. 383-393

Chevalier, L., Bernard, S., Ramdani, Y., Lamour, R., Bardor, M., Lerouge, P., Follet-Gueye, M.L. & Driouich, A. (2010) Subcompartment localization of the side chain xyloglucan-synthesizing enzymes within Golgi stacks of tobacco suspension-cultured cells. *The Plant Journal*, Vol.64, pp. 977-989

Chevalier L., Bernard S., Vicré-Gibouin M., Driouich A. & Follet-Gueye M.L. (2011) Cryopreparation of tobacco Bright Yellow-2 (BY-2) suspensioncultured cells and application to immunogold localization of Green Fluorescent Protein-tagged-membrane-proteins. *Journal of Biological Research*, Vol.16, pp. 202-216

Clausen, M.H., Willats, W.G.T. & Knox, J.P. (2003) Synthetic methyl hexagalacturonate hapten inhibitors of anti-homogalacturonan monoclonal antibodies LM7, JIM5 and JIM7. *Carbohydrate Research*, Vol.338, pp. 1797-1800

Coimbra, S. & Salema, R. (1997). Immunolocalization of arabinogalactan proteins in *Amaranthus hypochondriacus* L. ovules. *Protoplasma*, Vol.199, pp. 75-82.

Coimbra, S. & Duarte, C. (2003). Arabinogalactan proteins may facilitate the movement of pollen tubes from the stigma to the ovules in *Actinidia deliciosa* and *Amaranthus hypochondriacus*. *Euphytica*, Vol.133, pp. 171-178

Coimbra, S., Almeida, J., Junqueira, V., Costa, M.L. & Pereira, L.G. (2007) Arabinogalactan proteins as molecular markers in Arabidopsis thaliana sexual reproduction. *Journal of Experimental Botany*, Vol.58, pp. 4027-4035

Dardelle, F., Lehner, A., Ramdani, Y., Bardor, M., Lerouge, P., Driouich, A. & Mollet JC (2010) Biochemical and immunocytological characterizations of Arabidopsis thaliana pollen tube cell wall. *Plant Physiology*, Vol.153, pp. 1563-1576

Driouich A., Faye L. & Staehelin L.A. (1993) The plant Golgi apparatus: a factory for complex polysaccharides and glycoproteins. *Trends in Biochemistry. Science.*, Vol. 18, pp. 210-214

Driouich A., Durand C. & Vicré-Gibouin M. (2007) Formation and separation of root border cells. *Trends in Plant Science*, Vol. 12, pp.14–19

Driouich A. & Baskin T. (2008). Intercourse between cell wall and cytoplasm exemplified by arabinogalactan-proteins and cortical microtubules in the root of *Arabidopsis thaliana*. *Am. J. Botany.* 95: 1491-1497.

Durand C., Vicré-Gibouin M., Follet-Gueye M.L., Duponchel L., Moreau M., Lerouge P. & Driouich A. (2009) The organization pattern of root border-like cells of Arabidopsis is dependent on cell wall homogalacturonan. *Plant Physiology*, Vol.3, pp 1411-1421

Ellis, M., Egelund, J., Schultz, C.J. & Bacic, A. (2010) Arabinogalactan-proteins: key regulators at the cell surface? *Plant Physiology*, Vol.153, pp. 403-419

Fitchette A.C., Cabanes-Macheteau M., Marvin L., Martin B., Satiat-Jeunemaitre B., Gomord V., Crooks K., Lerouge P., Faye L. & Hawes C. (1999) Biosynthesis and immunolocalization of Lewis a-containing N-glycans in the plant cell. *Plant Physiology*, Vol. 121, pp. 333-44

Follet-Gueye, M.L., Pagny, S., Faye, L., Gomord, V. & Driouich, A. (2003) An improved chemical fixation method suitable for immunogold localization of green fluorescent protein in the Golgi apparatus of tobacco bright yellow (BY-2) cells. *Journal of Histochemistry and Cytochemistry*, Vol. 51, pp.931–940

Gibeaut D.M. & Carpita N.C. (1994) Biosynthesis of plant cell wall polysaccharides. *FASEB J.*, Vol. 8, pp. 904-915

Gilbert, H.J. (2010) The biochemistry and structural biology of plant cell wall deconstruction. *Plant Physiology*, Vol.153, pp. 444-455

Harholt J., Jensen J.K., Sørensen S.O., Orfila C., Pauly M. & Scheller H.V. (2006) ARABINAN DEFICIENT 1 is a putative arabinosyltransferase involved in biosynthesis of pectic arabinan in Arabidopsis. *Plant Physiology*, Vol. 140, pp. 49–58

Harris P.J. & Northcote D.H., (1971). Polysaccharide formation in plant Golgi bodies. *Biochemistry and Biophysical Acta.*, Vol. 237, pp. 56-64

Hawes M.C., Bengough G., Cassab G. & Ponce G. (2003) Root caps and rhizosphere. *Journal of Plant Growth Regulation*, Vol. 21, pp.352–367

Hématy, K., Cherk, C. & Somerville, S. (2009) Host-pathogen warfare at the plant cell wall. *Current Opinion in Plant Biology*, Vol.12, pp. 406-413

Jauh, G.Y. & Lord E.M. (1996) Localization of pectins and arabinogalactan-proteins in lily (*Lilium longiflorum* L.) pollen tube and style, and their possible roles in pollination *Planta*, Vol.199, pp. 251-261

Johnson-Brousseau, S.A. & McCormick, S. (2004) A compendium of methods useful for characterizing Arabidopsis pollen mutants and gametophytically expressed genes. *The Plant Journal*, Vol. 39, pp. 761–775

Jones, L., Seymour, G.B., & Knox, J.P. (1997) Localization of pectic galactan in tomato cell walls using a monoclonal antibody specific to (1->4)-beta-D-galactan. *Plant Physiology*, Vol.113, pp. 1405-1412.

Langhans M., Förster S., Helmchen G. & Robinson D.G. (2011) Differential effects of the brefeldin A analogue (6R)-hydroxy-BFA in tobacco and Arabidopsis. *Journal of Experimental Botany*,. Vol. 62, pp. 2949-2957

Lee, K.J.D., Marcus S.E. & Knox J.P. (2011) Cell wall biology: Perspectives from cell wall imaging. *Molecular Plant*, Vol.4, pp. 212-219

Lehner, A., Dardelle, F., Soret-Morvan, O., Lerouge, P., Driouich, A., & Mollet, J.C. (2010) Pectins in the cell wall of *Arabidopsis thaliana* pollen tube and pistil. *Plant Signaling and Behavior*, Vol.5, pp. 1282-1285

Li, Y.Q., Bruun, L., Pierson, E.S. & Cresti, M. (1992) Periodic deposition of arabinogalactan epitopes in the cell wall of pollen tubes of *Nicotiana tabacum* L. *Planta*, Vol.188, pp. 532-538

Liepman, A.H., Wightman, R., Geshi, N., Turner, S.R. & Scheller, H.V. (2010) Arabidopsis - a powerful model system for plant cell wall research. *The Plant Journal*, Vol.61, pp. 1107-1121

Marcus, S.E., Verhertbruggen, Y., Herve, C., Ordaz-Ortiz, J.J., Farkas, V., Pedersen, H.L., Willats, W.G.T. & Knox, J.P. (2008) Pectic homogalacturonan masks abundant sets of xyloglucan epitopes in plant cell walls. *BioMedCentral Plant Biology*, Vol.8, pp. 60-71

Mayhew, T.M., Lucocq, J.M. & Griffiths, G. (2002) Relative labelling index: a novel stereological approach to test for non-random immunogold labelling of organelles and membranes on transmission electron microscopy thin sections. *Journal of Microscopy*, Vol. 205, pp.153–164

Mayhew T.M. (2011) Quantifying immunogold localization on electron microscopic thin sections: a compendium of new approaches for plant cell biologists. *Journal of Experimental Botany*, Vol. 62, pp. 4101-4113

Micheli F. (2001) Pectin methylesterases: cell wall enzymes with important roles in plant physiology. Trends in Plant Sciences, Vol. 6, pp.414-419

Moller, I., Marcus, S.E., Haeger, A., Verhertbruggen, Y., Verhoef, R., Schols, H., Mikklesen, J.D., Knox, J.P. & Willats, W. (2008) High-throughput screening of monoclonal antibodies against plant cell wall glycans by hierarchial clustering of their carbohydrate microarray binding profiles. *Glycoconjugate Journal* Vol. 25, pp. 37-48.

Mollet, J.C., Park, S.Y., Nothnagel, E.A. & Lord, E.M. (2000) A lily stylar pectin is necessary for pollen tube adhesion to an in vitro stylar matrix. *The Plant Cell*, Vol.12, 1737-1749

Mollet, J.C., Kim, S., Jauh, G.Y. & Lord, E.M. (2002) AGPs, pollen tube growth and the reversible effects of Yariv phenylglycoside. *Protoplasma*, Vol.219, pp. 89-98

Mollet, J.C., Faugeron, C. & Morvan, H. (2007) Cell adhesion, separation and guidance in compatible plant reproduction, In: *Plant Cell separation and adhesion*, Roberts, J. & Gonzalez-Carranza, Z. (Ed.), *Annual Plant Reviews*, Vol.25, pp. 69-90

Mouille, G., Ralet, M.C., Cavelier, C., Eland, C., Effroy, D., Hématy, K., McCartney, L., Truong, H.N., Gaudon, V., Thibault, J.F., Marchant, A. & Höfte, H. (2007) Homogalacturonan synthesis in Arabidopsis thaliana requires a Golgi-localized protein with a putative methyltransferase domain. *The Plant Journal*, Vol.50, pp. 605–614

Pagny S., Bouissonnie F., Sarkar M., Follet-Gueye M.L., Driouich A., Schachter H., Faye L. & Gomord V. (2003) Structural requirements for Arabidopsis beta1,2-xylosyltransferase activity and targeting to the Golgi. *The Plant Journal*, Vol. 33, pp. 189-203

Pattathil, S., Avci, U., Baldwin, D., Swennes, A.G., McGill, J.A., Popper, Z., Bootten, T., Albert, A., Davis, R.H., Chennareddy, C., Dong, R., O'Shea, B., Rossi, R., Leoff, C., Freshour, G., Narra, R., O'Neil, M., York, W.S. & Hahn, M.G. (2010) A comprehensive toolkit of Plant cell wall glycan-directed monoclonal antibodies. *Plant Physiology*, Vol.153, pp. 514-525.

Pennell, R.I., Janniche, L., Kjellbom, P., Scofield, G.N., Peart, J.M. & Roberts, K. (1991) Developmental regulation of a plasma membrane arabinogalactan protein epitope in oilseed rape flowers. *The Plant Cell*, Vol.3, pp. 1317-1326.

Puhlmann, J., Bucheli, E., Swain, M.J., Dunning, N., Albersheim, P., Darvill, A.G. & Hahn, M.G. (1994) Generation of monoclonal antibodies against plant cell wall polysaccharides. I. Characterization of a monoclonal antibody to a terminal alpha-(1,2)-linked fucosyl-containing epitope. *Plant Physiology*, Vol.104, pp. 699-710

Qin, Y. & Zhao, J. (2006). Localization of arabinogalactan proteins in egg cells, zygotes, and two-celled proembryos and effects of β-D-glucosyl Yariv reagent on egg cell fertilization and zygote division in *Nicotiana tabacum* L. *Journal of Experimental Botany*, Vol.57, pp. 2061-2074

Ray P.M. (1980) Cooperative action of beta-glucan synthetase and UDP-xylose xylosyl transferase of Golgi membranes in the synthesis of xyloglucan-like polysaccharide. *Biochimistry and Biophysical Acta*, Vol. 629, pp. 431-444

Sandhu, A.P.S., Randhawa, G.S. & Dhugga, K.S. (2009) Plant cell wall matrix polysaccharide biosynthesis. *Molecular Plant*, Vol. 2, pp. 840-850

Seifert, G.J. & Blaukopf, C. (2010) Irritable walls: the plant extracellular matrix and signaling. *Plant Physiology*, Vol.153, pp. 467-478

Smyth, D.R., Bowman, J.L. & Meyerowitz, E.M. (1990) *Early flower development in* Arabidopsis. The Plant Cell, *Vol.2, pp. 755-767*

Staehelin L.A., Giddings T.H. Jr, Kiss J.Z. & Sack F.D. (1990) Macromolecular differentiation of Golgi stacks in root tips of Arabidopsis and Nicotiana seedlings as visualized in high pressure frozen and freeze-substituted samples. *Protoplasma*, Vol. 157, pp. 75-91

Studer, D., Graber, W., Al-Amoudi, A. & Eggli, P. (2001) A new approach for cryofixation by high-pressure freezing. *Journal of Microscopy*, Vol. 203, pp.285-294

Toyooka K., Goto Y., Asatsuma S., Koizumi M., Mitsui T. & Matsuoka K. (2009) A mobile secretory vesicle cluster involved in mass transport from the Golgi to the plant cell exterior. *The Plant Cell*, Vol. 4, pp. 1212-1229

Van Donselaar E., Posthuma G., Zeuschner D., Humbel B.M. & Slot J.W. (2007). Immunogold labeling of cryosections from high-pressure frozen cells. *Traffic*, 8: 471-485

Vicré, M., Santaella, C., Blanchet, S., Gateau, A. & Driouich, A. (2005) Root border-like cells of Arabidopsis. Microscopical characterization and role in the interaction with rhizobacteria. *Plant Physiology*, Vol.138, pp. 998-1008

Viotti C., Bubeck J., Stierhof Y.D., Krebs M., Langhans M., van den Berg W., van Dongen W., Richter S., Geldner N., Takano J., Jürgens G., de Vries S.C., Robinson D.G. & Schumacher K. (2010) Endocytic and secretory traffic in Arabidopsis merge in the trans-Golgi network/early endosome, an independent and highly dynamic organelle. *The Plant Cell*, Vol. 22, pp. 1344-1357

Willats, W.G.T., Marcus, S.E. & Knox, J.P. (1998) Generation of a monoclonal antibody specific to (1->5)-alpha-L-arabinan. *Carbohydrate Research*, Vol.308, pp. 149-152.

Willats W.G., McCartney L. & Knox J.P. (2001) In-situ analysis of pectic polysaccharides in seed mucilage and at the root surface of Arabidopsis thaliana. *Planta*, Vol. 213, pp. 37-44

Willats W.G., McCartney L., Steele-King C.G., Marcus S.E., Mort A., Huisman M., van Alebeek G.J., Schols H.A., Voragen A.G., Le Goff A., Bonnin E., Thibault J.F. & Knox J.P. (2004) A xylogalacturonan epitope is specifically associated with plant cell detachment. *Planta*, Vol. 218, pp.673-81

Wu, H., de Graaf, B., Mariani, C. & Cheung, A. Y. (2001). Hydroxyproline rich glycoproteins in plant reproductive tissues: structure, functions and regulation. *Cell and Molecular Life Sciences*, Vol.58, pp. 1418-1429

Yates, E.A., Valdor, J.-F., Haslam, S.M., Morris, H.R., Dell, A., Mackie, W. & Knox, J.P. (1996) Characterization of carbohydrate structural features recognized by anti-arabinogalactan-protein monoclonal antibodies. *Glycobiology*, Vol.6, pp. 131-139.

Zhang G.F. & Staehelin L.A. (1992) Functional compartmentalization of the Golgi apparatus of plant cells: an immunochemical analysis of high pressure frozen/freeze substituted sycamore suspension-cultured cells. *Plant Physiology*, Vol. 99, pp. 1070-1083

14

Immunocytochemistry of Proteases in the Study of *Leishmania* Physiology and Host-Parasite Interaction

Raquel Elisa da Silva-López
Department of Natural Products, Farmaguinhos
Oswaldo Cruz Foundation, Rio de Janeiro,
Brazil

1. Introduction

Leishmaniasis is a chronic disease caused by parasites from *Leishmania* genus and still represents a severe public health problem in the world and the incidence is increasing (Desjeux, 2004). There is no effective vaccine for prevention of any form of leishmaniasis and programs of prevention and drug therapy are the main mechanisms for disease control. On the other hand, current chemotherapy is the only way to treat cases of leishmaniasis. Since the 1940s, the pentavalent antimony compounds (e.g., Glucantime, Pentostam, or branded pentavalent formulations) have been the mainstays of antileishmanial therapy (Aït-Oudhia et al., 2011). Although these drugs are usually effective, they produce serious side effects, present difficulties of administration and high cost, the parasite persists in the scars of clinically cured patients (Schubach et al., 1998), and drug resistance has been observed (Castillo et al., 2010). Second-line drugs are used in areas with high rates of unresponsiveness to antimonial treatment or when it was not possible to administrate it. However, these drugs are even more toxic than antimony compounds and expensive, and these include pentamidine, amphotericin B, anti-fungal, allopurinol, and more recently, miltefosine, paramomicine and sitamaquine. Furthermore, they have low therapeutic index when compared to antimonial compounds (Almeida and Santos, 2011). Instead of determining treatment based on rational therapeutic indications, treatment of choice is frequently dictated by economic considerations and in a large majority of countries, chemotherapeutic approaches for all forms of leishmaniasis rely on the use of pentavalent antimonial compounds (Aït-Oudhia et al., 2011). The mechanism of pentavalent antimony compounds action is the inhibition of glycolytic pathway and β-oxidation enzymes of the parasites (Baiocco et al., 2009), but being a heavy metal it is non-selective and it is believed to interfere with other metabolic pathways of parasites and hosts. Furthermore, these drugs can interact with the zinc finger domain of proteins, and many proteins have this motif in their tridimensional structures (Demicheli et al., 2008).

Attempts to develop vaccines against *Leishmania* and drugs to treat cases of leishmaniasis are a continuing effort in search for novel parasite antigens. Various candidate molecules have also been tested and some degrees of protection against different species of *Leishmania* infection were observed (Chawla and Madhubala, 2010). In order to develop a rational drug for leishmaniasis chemotherapy, the biochemistry of *Leishmania* parasites needs to be better understood for the identification of these strategic targets. Immunocytochemistry strategies have localized these targets in *Leishmania*, providing valuable information about the roles of these molecules in the parasite life cycle and in the pathogenesis of leishmaniasis. So, the purpose of this chapter is to focus on the employment of immunolabeling and electron microscopy in order to localize proteases which are critical for the survival of the *Leishmania* parasites. Furthermore, this chapter will cover aspects of leishmaniasis epidemiology, *Leishmania* morphology, potential drug targets in *Leishmania* and proteases as targets in *Leishmania*, highlighting the function and subcellular localization of cysteine proteases, gp 63 metalloprotease and serine proteases.

2. Leishmaniasis and *Leishmania*

Leishmaniasis is one of the most significant neglected diseases and occurs in the tropical and subtropical regions of America, Asia, Africa and Europe. This disease is considered to be endemic in 88 countries, 72 of which are developing countries (Kaye and Scott, 2011). About 350 million people are at risk of *Leishmania* infection and as many as 12 million people in the world are believed to be currently infected. Approximately 1–2 million estimated new cases every year with the annual mortality rate of about 60,000 (Okwor and Uzonna, 2009). Leishmaniasis is a disease associated with the poverty, environmental changes, such as deforestation, building of dams, urbanization, and the accompanying migration of non-immune people to endemic areas. However, due to underreporting - notification of leishmaniasis is compulsory in only 32 of the 88 affected countries - and misdiagnosis, actual case numbers are expected to be higher. Furthermore, most affected people are hidden because the social stigma associated with deformities and disfigurement scars and due to they live in remote areas. Leishmaniasis-related disabilities impose a great social burden, and reduce economic productivity (WHO, accessed in August 15th, 2011). Over the past 20 years, leishmaniases have increasingly been recognized as an opportunistic infection in HIV-infected patients, with *Leishmania*–HIV co-infection common in areas where both diseases are endemic. The highest prevalence of co-infection cases occurs mostly in Spain and southwestern Europe, among injectable drug users. The presence of both pathogens concomitantly in the same host cell (macrophage) influences the expression and multiplication of both pathogens. HIV-1 infection increases the risk of developing visceral leishmaniasis by 100 to 2,300 times in endemic areas, reduces the likelihood of a therapeutic response and greatly increases the probability of relapse. Moreover, *Leishmania* promotes an increment in viral load and a more rapid progression to AIDS, which reduces life expectancy of infected patients (Ezra et al., 2010).

Clinical manifestations of leishmaniasis range from self-healing cutaneous, mucocutaneous skin ulcers and a long-lasting diffuse cutaneous in cellular-mediated immune response deficient hosts to a lethal visceral form (i.e., visceral leishmaniasis or kala-azar) and post-kala-azar dermal leishmaniasis. The clinical spectrum of this disease is associated with the species of *Leishmania* involved (Desjeux, 2004). Today, about 30 species of protozoan of the *Leishmania* genus (Order Kinetoplastida and Family Tripanosomatidae) are known and approximately 20 are pathogenic for humans and are the causative agents of the "Old" and "New Worlds" leishmaniasis. All members of the genus *Leishmania* Ross, 1903 are parasites of mammals. The two subgenera, *Leishmania* and *Viannia*, are separated on the basis of their location in the vector's intestine and isoenzyme analysis was used to define species complexes within the subgenera. These species generally present different epidemiological and clinical characteristics related to different genetic and phenotypic profiles. All species of the subgenus *Viannia* were isolated in the 'New World', while those of the subgenus *Leishmania* were isolated from the "Old World", except for species of the *L. (L.) mexicana* complex, *L. (L.) hertigi*, *L. (L.) deanei* — which are found in the 'New World' only — and *L. (L.) infantum/chagasi* and *L. (L.) major*, which are found in both the "New" and "Old Worlds" (Bañuls et al., 2007).

These parasites can be transmitted by female sandflies via anthroponotic or zoonotic cycles, transplacental, blood transfusion and through contaminated needles by injecting drug users. Vector transmission is the commonest way of parasite dissemination (Molina et al., 2003). *Leishmania* parasites have a dimorphic life-cycle: The flagellated, motile forms of *Leishmania* spp. are called promastigotes (Figure 1 A). They are found into the very alkaline digestive tract of the sandfly and progress through various morphologically distinct stages of differentiation to ultimately become the non-dividing, infectious 'metacyclic' promastigotes that are transmitted during a sandfly bite. These 'metacyclic' promastigotes are phagocyted by professional phagocytes such as macrophages and, inside these cells parasites survive and multiply as amastigotes (Figure 1B), a smaller form of *Leishmania* with non-exteriorized flagellum and very metabolic active (Seifert, 2011).

All members of the genus *Leishmania* are obligated intracellular parasites of several mammalian cells and survive under very acid, oxidant and hostile conditions into parasitophorous vacuoles environment, and they have evolved several mechanisms to avoid their degradation (Mougneu et al., 2011). These mechanisms include specific organelles and molecules, such as proteases, that are secreted or are intracellular expressed (Silva-López et al., 2005; Yao, 2003). Special organelles found in trypanosomatid *Leishmania* include mitochondrion and kinetoplast (De Souza et al., 2009a), megasomes (De Souza et al. 2009b) and glycosomes (Michels et al., 2006). Some of these organelles and the evolutive forms of *Leishmania* are schematically represented in the figure 1.

3. Potential drug targets in *Leishmania*

One of the features in the process of drug development is target identification in a biological pathway. In theory, during this identification in a pathogen, it is important that the putative target should be either absent in the host or substantially different from the host homolog so that it can be exploited as a drug target. Trypanosomatids, phylogenetically, branch out quite early from the higher eukaryotes. In fact, their cell organization is significantly different from the mammalian cells and thus, it is possible to find targets that are unique to these pathogens. Secondly, the target selected should be absolutely necessary for the survival of the pathogen. It is also important to consider the stage of the life-cycle of the pathogen in which the target protein is expressed. So, the most important targets are enzymes, since they regulate a specific biochemical pathway and their active sites can bind specific inhibitors that can be designed or found in the nature. A good enzyme target means that its inhibition should lead to loss in cell viability. Furthermore, it is important that the target selected should be assayable (Shukla et al., 2010). Many enzymes have been investigated in their capacity to control or regulate essential *Leishmania* biochemical pathways or some mechanisms that guarantee the parasite survival and proliferation for infection maintenance, such as the enzymes that regulate or participate in sterol biosynthesis,

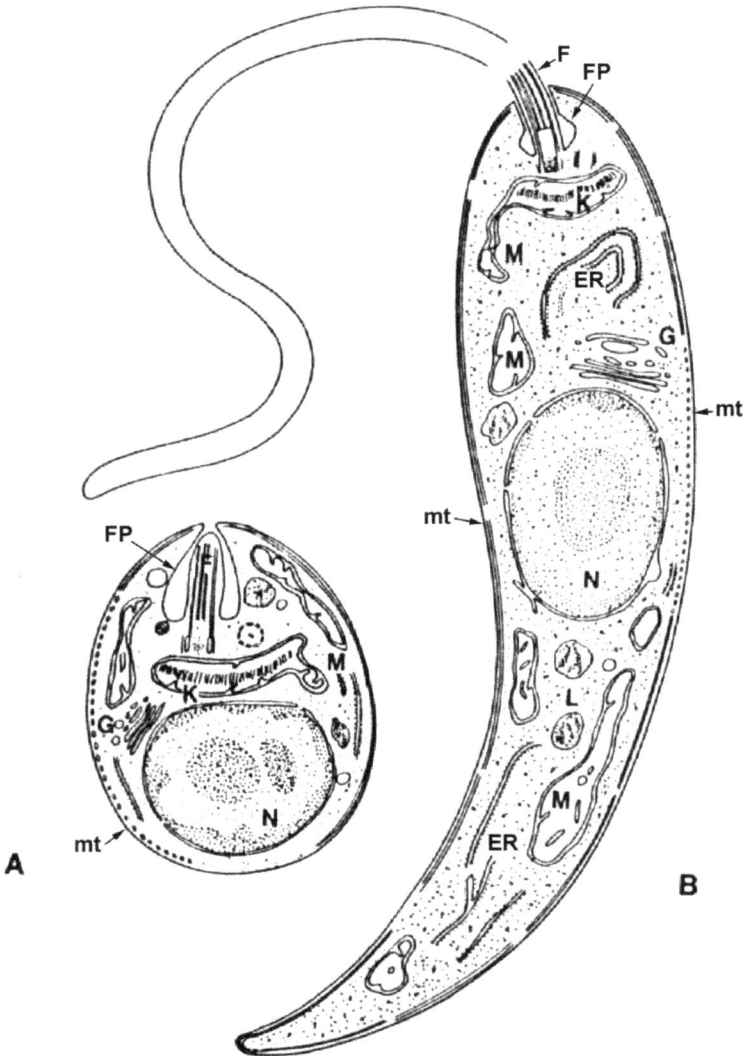

Fig. 1. Schematic representation of *Leishmania* sp forms and their organelles. A) Amastigotes. B) Promastigotes. F, flagellum; K, kinetoplast; N, nucleus; FP, flagellar pocket; ER, endoplasmic reticulum; M, mitochondrion; mt, microtubules; G, Golgi; L, lysosome (adapted by the author from Rey, 1991).

hypusine biosynthesis, glycolysis, purine salvage, glycosylphosphatidylinositol biosynthesis, folate biosynthesis and glyoxalase and trypanothione system or special enzymes such as protein kinases, topoisomerases and proteases (Chandra et al., 2010; Chawla and Madhubala, 2010).

Before discussing about the proteases, which is the purpose of this chapter, it is worth to mention about some enzymatic systems that are considered good targets in *Leishmania* (Table 1). Sterols are important components of the cell membrane and are essential to the cellular function. The sterol biosynthesis of trypanosomatids is very different from humans, because these parasites synthesize ergosterol and 24-methyl sterols, instead of cholesterol. The extensively studied squalene synthase and D24,25-sterol methyltransferase enzymes, only present in trypanosomatids, perform crucial roles in regulation of sterol metabolism and their specific inhibitors have showed anti-proliferative and growth inhibition effects of *Leishmania in vitro* (Goto et al., 2009).

Leishmania compartmentalize several important metabolic systems in special peroxisomes which are designated glycosomes. These organelles contain seven of the ten enzymes of glycolytic pathway, the pentose-phosphate pathway, β-oxidation of fatty acids, purine salvage, biosynthetic pathways for pyrimidines, ether-lipids and squalenes (Moyersoen et al., 2004). The glycolytic enzymes of *Leishmania* exhibited large phylogenetic distance with the mammalian hosts and, for this reasons, specific inhibitors have been designed for the most important regulator enzyme, the glyceraldehyde-3-phosphate dehydrogenase. These compounds inhibited growth of L. (L.) mexicana by blocking the energy production, since glycolysis is the most important source of energy for these parasites (Saunders et al., 2010). Furthermore, the biogenesis of these organelles occurs via peroxins self-interactions and the reduction of peroxin expression or their self-interaction inhibition induced the *Leishmania* death (Michels et al., 2006).

Protozoans of *Leishmania* genus lack the metabolic machinery to the synthesis of purine nucleotides and the parasites have to depend upon the purine salvage system to utilize purine from their hosts. Three phosphoribosyltransferases were identified in *Leishmania*, and the hypoxanthine-guanine phosphoribosyl transferase is the most important enzyme involved in purine salvage (Carter et al., 2008). Various inhibitors have been designed to target this enzyme due to its difference in substrate specificity with the host enzyme. Allopurinol is the most common inhibitor that is phosphorylated by the enzyme and incorporated into nucleic acid, leading to selective death of the parasite (Loiseau and Bories, 2006). Allopurinol has been shown to be effective against cutaneous and visceral leishmaniasis, but when used with other anti-leishmanial drugs was found to be even more effective (Castillo et al., 2011; Almeida and Santos, 2011). Besides, purines are transported through the parasite cell surface by nucleoside transporters and these transporters also uptake toxic nucleoside analogs which inhibits the parasite growth (Al-Salabi and Koning, 2005). So, these transporters represent an alternative strategy for interfering in *Leishmania* purine metabolism and develop novel drugs to leishmaniasis treatment. Some of these synthetic and natural products that inhibited specifically crucial steps in a metabolic pathway of *Leishmania* inducing the parasite death and reducing leishmaniatic lesion progression in susceptible animal models have been tested in controlled clinical trials. These compounds have showed different degrees of efficacy, therapeutic index and in general caused less adverse effects than that observed in patients treated with antimonials or with the second-line drugs that is currently being used for leishmaniasis treatment (Fernandes Rodrigues et al., 2008; Valdivieso et al., 2010; Almeida and Santos, 2011; Pereira et al., 2011).

Target enzymes	Methods of location	Subcellular location	Biological functions
Squalene synthase and *D24,25-sterol methyl-transferase.*	Subcellular fractionation	Membrane, glycosomes and mitochondrial/ microsomal vesicles.	Sterol biosynthesis, which are components of the cell membrane and *Leishmania* signaling (Goto et al., 2009).
Deoxyhypusine synthase and *deoxyhypusine hydroxylase*	not applied	not determined	Hypusine biosynthesis that are involved in *Leishmania* proliferation, differentiation, and biosynthesis of macromolecules (Chawla et al., 2010).
Glyceraldehyde-3-phosphate dehydrogenase	Subcellular fractionation	Glycosomes	Regulates the glycolysis that is the most important metabolic pathway in *Leishmania* ATP supply (Saunders et al., 2010).
Hypoxanthine-guanine phosphoribosyl transferase	Subcellular fractionation	Glycosomes	Purine salvage for nucleotides and nucleosides (Carter et al., 2008).
Glycosylphosphatidylinositol biosynthetic enzymes	Subcellular fractionation immunofluorescence	Tubular subdomain of the endoplasmic reticulum	Glycosylphosphatidyl-inositol acts as a membrane anchor for many cell-surface proteins of eukaryotes (Ilgoutz et al., 1999).
methylene-tetrahydrofolate dehydrogenase/cyclohydrolase and formyl-tetrahydrofolate ligase.	Subcellular fractionation western blotting immunofluorescence	Cytosol	Formyl-tetrahydrofolate biosynthesis. Folates are used in purine biosynthesis and mitochondrial initiator methionyl-tRNA^{Met}formylation (Vickers et al., 2009)
Glyoxalase system	immunofluorescence	kinetoplast	catalyzes the formation of d-lactate from methylglyoxal, a toxic product of glycolysis, using trypanothione as substrate (Chauhan and Madhubala, 2009)

Target enzymes	Methods of location	Subcellular location	Biological functions
Trypanothione system	Subcellular fractionation immunofluorescence	mitochondria and cytosol	H_2O_2-detoxification (Krauth-Siegel et al., 2007)
Protein kinases	immunofluorescence immunocytochemistry	cytosol	Enzymatic activity regulation by addition of phosphate (Baqui et al., 2000).
Topoisomerases	immunofluorescence	Kinetoplast mitochondria	Topoisomerases I catalyze the cleavage of one strand of DNA, whereas topoisomerases II catalyze the cleavage of a double-stranded DNA, requiring ATP as cofactor (Banerjee et al, 2011)
Proteases	Subcellular fractionation immunofluorescence immunocytochemistry	Membrane, flagellar pocket, megasomes and endocytic/ exocytic vesicles	Hydrolysis of peptide bonds in proteins and peptides. They are crucial in *Leishmania* life cycle, in host-parasite relationship, and in leishmaniasis pathogenesis (Silva-López et al., 2010a, b).

Table 1. Main protein targets in *Leishmania* sp.

4. Proteases as targets in *Leishmania*

Proteases, also known as peptidases, are hydrolytic enzymes that cleave peptide bounds in proteins and peptides, releasing peptides with variable sizes and free amino acids. Unlike most enzymes, proteases lack specificity toward a substrate, i.e., a specific protein. Instead, they are very specific for a peptide containing the scissile peptide bond and the amino acids involved in the neighborhood of the peptide bonds instead of the whole molecule (Garcia-Carreno and Del Toro, 1997). They are ubiquitously found in all living beings from unicellular to higher organisms and are the most studied class of enzyme (Shinoda and Miyoshi, 2011). They participate in various physiological processes through the modification of proteins, such as digestion of food proteins, tissue remodeling, neuropeptides, hormones, and proenzyme processing, cellular metabolism by proteasomes, blood clotting, complement activation cascade reactions, metabolism regulation and a vast number of other biological phenomena as well as they are encoded by 2% of all genes in an organism (López-Otín and Bond, 2008). These enzymes are grossly classified as: exopeptidases which act on the ends of protein substrates and are designated as amino-or carboxypeptidases, and endopeptidases acting on the interior of protein substrates. Endopeptidases classification rests on the type of residue at the active site: the hydroxyl group of serine proteases (EC 3.4.21) and the sulfidryl group of cysteine proteases (EC 3.4.22) are the nucleophile during catalysis, while activated water is the nucleophile for aspartic (EC 3.4.23), and metalloproteases (EC. 3.4.24) (Barret, 1994). Proteases are further classified according to their structure in families and Clans. A family is a set of homologous proteolytic enzymes that show significant similarity in amino acid sequence, and homologous families that have arisen from a single evolutionary origin of peptidases are grouped together in a Clan (Rawlings et al., 2010). The different types of proteases have particular characteristics that allow their proteolysis to function in a huge diversity of environments. They are very well adapted to surrounding conditions and can act in acid or basic pH, hypo and hyper osmolarity, higher or lower temperatures, and for these reasons they can be found in all cellular compartments and in all organs of the higher organisms.

Besides their physiological necessity, proteases are potentially hazardous to their proteinaceous content and their activity must be precisely controlled by the respective cell or organism. However if this activity is unregulated it can destroy cells, tissues and organs and can kill organisms. So, the control has to be very efficient and is normally achieved by regulated expression/secretion, zymogen production, enzyme activation, degradation of the mature enzymes and most important by protease inhibitors, that can react with the active site or other domain of the enzyme, impeding its capacity to bond and hydrolyze the substrate. In general, there are two types of inhibitors: (a) small non-proteinaceous compounds, secreted by microorganisms that irreversibly modify the amino acid residue of the enzyme active site, and (b) the huge number of natural inhibitors, which are pseudo substrates isolated from various cells, tissues and organisms often that accumulate in high quantities in plant seeds, and various body fluids. Inhibitors of different types occur commonly among living organisms and viruses, which stresses their ultimate role in physiological processes (Otlewski et al., 1999; Krowarsch et al., 2003; Silva-López, 2009). These inhibitors are valuable tools for investigation of the biochemical properties and the biological functions of proteases, besides they are employed in the treatment of many diseases and are under investigation as chemotheraphic in the treatment of leishmaniasis (Silva-López et al., 2007; Valdivieso et al., 2010; Olivier and Hassani, 2010; Pereira et al., 2011).

Many studies have focused their attention on the crucial roles of proteases in the *Leishmania* life cycle, in the host-parasite relationship, and in the pathogenesis of leishmaniasis. These enzymes are important virulence factors and they have been implicated in a wide variety of adaptation mechanisms for in-host parasite survival, which include modulation of the host immune system, invasion and destruction of host connective tissues, enabling parasites to migrate to specific sites for growth and development and/or acquire essential nutrients that guarantee survival and proliferation for infection maintenance (Mottran et al., 2004; McKerrow et al., 2006; Matos Guedes et al., 2010; Gómez and Olivier, 2010; Silva-López, 2010 b; Yao, 2010; Swenerton et al, 2010; Swenerton et al, 2011) and their importance has been confirmed through findings that specific protease inhibitors kills *Leishmania* parasites (Silva-López et al., 2007).

It is important to point out that the subcellular location of *Leishmania* proteases provides knowledge about the function of these enzymes in the parasite physiology, and consequently if they are potential targets to develop new chemotherapy for leishmaniasis treatment. There are some strategies to determine the proteases localization and immunocytochemistry techniques are the most capable for specifically detecting and localizing macromolecules within thin sections of any cell type, preserving the structure of most antigenic sites and retaining the antigenicity (Herman, 1988). Using these strategies, many proteases were successfully localized in both forms of Leishmania and their importance in the parasite physiology and infection was elucidated. Cysteine proteases, gp63 metalloprotease and, more recently serine proteases are the most investigated proteolytic enzymes of the genus Leishmania and some considerations about their functions and subcellular localization using immunocytochemistry studies will be focused in this chapter.

4.1 Cysteine proteases

North and Coombs in 1981 reported for the first time the proteolytic activity in both promastigote and amastigote forms of Leishmania. They demonstrated that the highest activity was found in amastigotes and belonged to the cysteine protease class (North and Coombs, 1981), and since then these enzymes were extensively studied. So, at this time proteases were also considered virulence factors in Leishmania, because as in other parasites these enzymes have a recognized role in the mechanisms of invasion, survival and migration in host tissues (Kozar, 1961). The immunocytochemistry constitute an universal assay capable of detecting and localizing macromolecules in any cell type using specific unlabeled primary antibody directed at the antigen of interest and then indirectly localizing the primary antibody with a second label consisting of antibodies conjugated to an electron-dense material (Lunedo et al., 2011) and considerations about this valuable technique will be done at the end of this chapter. These cysteine proteases were localized for the first time in amastigotes of L. (L.) mexicana, a member of mexicana complex, by the post-embedding immunocytochemistry assays using as the primary antibody IgG fraction purified from a rabbit antiserum against L. (L.) mexicana amastigote cysteine protease and the anti-rabbit IgG immunospecificity complexed with gold colloidal (5-10 nm diameter) as the secondary antibody. The acid ester hydrolase arylsulfatase (EC 3.1.6.1) was also employed as a marker of lysosomes using 4-nitrocatechol sulphate as substrate and barium chloride as capture agent, forming electron dense barium deposits at reaction sites (the enzyme location), that is visualized by electron microscopy. These enzymes, cysteine protease and arylsulfatase, were found into larges organelles that contain putative lysosomal enzymes that was previously named "megasomes" (Pupkins et al., 1986). Megasomes are large lysosome-like structures, previously described in amastigote forms of Leishmania belonging to the mexicana complex, whose major constituents are the cysteine proteinases (Ueda-Nakamura et al., 2001). It is important to point out is that L. (L.) mexicana has the ability to cause both a cutaneous and a diffused cutaneous leishmaniasis in South and Central America and, is a member of the mexicana complex that is formed by L. (L.) amazonensis, L. (L.) pifanoi, L. (L.) garnhami, L. (L.) venezuelensis and L. (L.) forattinni. Besides, L. (L.) amazonensis is one of the most important etiologic agent that causes cutaneous and diffuse cutaneous leishmaniasis in Latin America, L. (L.) donovani causes cutaneous and mainly visceral diseases in the "Old World" and L. (L.) major is associated with cutaneous and mucocutaneous leishmaniasis in the "Old World" (Bañuls et al., 2007). Comparative studies with amastigotes of L. (L.) amazonensis, L. (L.)

donovani, revealed that *L. (L.) amazonensis* was similar to *L. (L.) mexicana* in possessing both high content of cysteine protease activity in amastigotes and a large numbers of megasomes, whereas the other two species lacked both of these features (González et al., 2008). The presence of numerous megasomes in the amastigote is a characteristic of the *Leishmania* subgenus, particularly of mexicana complex (Pupkis et al., 1986).

Further studies also employing light microscopy and post-embedding immunocytochemistry and the weak base 3-(2,4 dinitroanilino)-3'-amino-N-methyldipropylamine (DAMP) as a probe, localized acidic compartments of *L. (L.) amazonensis* amastigotes since it is known that DAMP concentrates in acidic compartments of cultured cells. This probe was mainly accumulated within megasomes and in dense inclusion vacuoles, proving that megasomes have a low pH maintained by an active process, besides suggesting that these organelles may be targets for amino acid derivatives (Antoine, 1988). It is important to consider some immunocytochemical aspects to localize the DAMP probe in *Leishmania*. In this assay the anti-DNP (2,4-Dinitrophenol) immune serum was prepared in rabbit by injecting human IgG-DNP mixed with a complete Freund´s adjuvant. After fixation and reaction with an antibody against-DNP it can be visualized by anti-rabbit IgG conjugated to gold (7 nm) particles as secondary antibody (Antoine, 1988).

The immunocytochemistry is a powerful technique for investigating the location of specific proteins in a cell. The biosynthesis, enzymatic processing, and immunocytochemical localization of a major cysteine protease of *L. (L.) pifanoi*, were investigated employing *L. (L.) pifanoi* axenic cultured amastigotes and *L. (L.) amazonensis* lesion-derived and, both polyclonal antisera and monoclonal antibodies specifically recognized either the mature cysteine protease or the carboxyl-terminal extension domain (Duboise et al., 1994) using post-embedding protocols. *L. (L.) pifanoi* is another member of the mexicana complex that causes cutaneous leishmaniasis in humans (Colemares et al., 2002). It is important to emphasize that all proteases are synthetized as a high molecular weight zymogen without proteolytic activity and the formation of mature protease involves the processing of the zymogen by cleavage of specific peptides bounds and removal of C or N-terminal domains (Neurath, 1984). Electron microscopic immunolocalization of both catalytic and C-terminal domains showed intense labeling of megasomes, indicating that this cleavage occurs in this organelle. Furthermore, specific cysteine proteinase inhibitors blocked the processing of cysteine protease *in vivo* and also inhibited parasite cell division. Moreover, a low level of the mature protease was also associated with the flagellar pocket and plasma membrane. Consistent with this observation, a low level secretion of this cysteine protease into the culture medium was detected (Duboise et al., 1994). It is known that the flagellar pocket is a secluded extracellular compartment in the anterior portion of trypanosomatids, formed by an invagination of the plasma membrane at the base of the flagellum and is the only part of the cell surface that supports exocytosis and endocytic traffic in *Leishmania* and other trypanosomatids because of its lack of attached microtubules (De Souza, 2006; Bonhivers et al., 2008). The flagellar pocket membrane is an obligatory intermediary station for membrane-bound molecules trafficking between intracellular membranes and the cell surface and vice-versa (Overath et al., 1997).

The processing and trafficking of cysteine proteases were also studied in *L. mexicana* (L) using axenic amastigotes and anti-cysteine protease B antiserum and the study showed that the proteolytic processing of zymogen and maturation of the enzyme is redundant and required other types of cysteine proteases and is not easy to paralyze it with inhibitors. However, cysteine proteases are addressed to megasomes via the flagellar pocket and therefore differs from trafficking in mammalian cells (Brooks et al., 2000), making this pathway an important target to block the *Leishmania* growth.

Post-embedding immunocytochemistry assays, using antibody against *Leishmania* cysteine protease, was also employed to prove the importance of these proteases in the pathogenesis of leishmaniasis. This *Leishmania* protease was detected in the lesion sections from infected mice by *L. (L.) mexicana* amastigotes, possibly due to amastigotes lysis and releasing of megasomes content into the parasitophorous vacuole of infected macrophages. These proteases were also found extracellularly in the host mice tissue presumably as a result of macrophage rupture and appear to persist within the lesion, where they may damage host cells and the extracellular matrix proteins (Ilg et al., 1994). Additionally, it was demonstrated that metacyclic promastigotes of *L. (L.) mexicana*, the infective form of parasite, exhibited higher proteolytic activity than multiplicative promastigotes and amastigotes, expressing quantitatively more and with a distinct pattern composed of multiple proteases (Bates et al., 1994). Suggesting that the expression of proteases varies according to evolutive forms of parasite and are important both to survival within the host, and infection of the mammalian cells.

The information about the cellular location and distribution of cysteine proteases in *Leishmania* parasites draws attention to the importance of these enzymes in the parasite physiology. Further studies have clarified the biochemical properties and functions of these enzymes, as well as their gene expression (Ramos et al., 2004; Hide and Bañuls, 2008). Cysteine proteases are the major proteolytic activity in *Leishmania* and they are required for survival and growth of protozoan into fagolisosomes and leishmaniasis pathogenesis (Marín-Villa et al., 2008), because of this, they are considered the most important virulence factors of *Leishmania* since they influence the interaction between the parasite and mammalian host (Mottram, et al., 2004). Genome analysis has revealed the great diversity of cysteine proteases of *Leishmania* (Hide and Bañuls, 2008) and they are distributed in eight families within clan CA. Family C1 contains cysteine protease A and cysteine protease B, which are both cathepsin L-like, and cysteine protease C, which is cathepsin B-like. Cysteine protease B is unusual as it has a 100-amino acid C-terminal extension in comparison with most cysteine proteases of the group, and exists as multiple isoenzymes, which are encoded by a tandem array of similar cysteine protease B genes located in a single locus (the arrays comprise eight genes in *L. major*) (Saffari and Mohabatkar, 2009).

Although the exact roles of cysteine proteases in *Leishmania* pathogenesis are unclear, it has been demonstrated that *Leishmania* cannot grow within macrophages in the presence of specific protease inhibitors (Duboise et al., 1994). Besides, it was demonstrated that *L. (L.) chagasi* cathepsin L-like cysteine protease (Ldccys2) are specifically expressed in amastigote and is necessary for macrophage infection and for survival of the parasite within macrophage cells (Mundondi et al., 2005). The role of the same enzyme, Ldccys2, was investigated in *L. (L.) pifanoi* and *L. (L.) amazonensis* employing post-embedding immunocytochemistry using antibodies against recombinant C-terminal extension of Ldccys2 and anti-Ldccys2 catalytic domain both of *L. (L.) pifanoi* and 10nm-gold labeled goat anti-rabbit as a secondary antibody (Marín-Villa et al, 2008). The polyclonal antibody specific to Ldccys2 C-terminal extension recognized cysteine proteases from both parasites and also detected a predominant location of this peptide in the lysosome and flagellar pocket of cultured axenic amastigotes of both parasite species. However, its location was shifted towards the surface of the parasites during macrophage infection. This same antibody significantly reduced macrophage infection in both *L. (L.) pifanoi* and *L. (L.) amazonensis*, confirming that Ldccys2 C-terminal domain is essential for macrophage infection. Importantly, the entrance into macrophages is mediated by the endocytosis of opsonized parasites, which are recognized by membrane receptors present on the macrophage surface (Rittig and Bogdan, 2000). Besides, confirming previous reports that C-terminal extensions of proteases are highly immunogenic in *T. cruzi*, antibodies against this peptide in sera of leishmaniasis patients was detected. This study suggests an essential role for *Leishmania* cysteine proteinases C-terminal extensions at early stages of infection (Marín-Villa et al, 2008). Other studies demonstrated that when *Leishmania* parasites are exposed to various stress conditions, such as heat shock and oxidant agents, they release cysteine protease C, a cathepsin B-like enzyme, which is involved in the cell death cascade of the parasite (El-Fadili et al., 2010).

Beside the roles described above, *Leishmania* cysteine proteases, specifically of type B, can modulate the immune response of mammalian hosts to favor parasite survival and proliferation. They are themselves immunogenic, since *L. (L.) mexicana* cysteine proteases are T cell immunogen, resulting in the development of potentially protective Th1 cell lines. This finding suggests that the cysteine proteases could also be a vaccine candidate and that homologous enzymes in other parasites species may also be so (Khoshgoo et al., 2008; Saffari and Mohabatkar, 2009; Fedelli et al., 2010, Doroud et al, 2011).

Many evidences indicate that *Leishmania* cysteine proteases could be targets to develop rational drugs to treat leishmaniasis, so specific inhibitors were produced by combinatorial synthetic chemistry optimization using models of both *L. (L.) major* cathepsin B and L, through a structure based drug design screen (Scheidt et al, 1998). These compounds were tested against *Leishmania* cysteine proteases and both amastigotes and promastigotes (Selzer et al., 1999; Schurigt et al., 2010). The electron microscopy and post-embedding immuno-cytochemical localization strategies were also used to study the effect of hydrazine derivatives in parasite and to confirm target protease localization at the site of inhibitor-induced abnormalities (Selzer et al., 1999). After 24 h of treatment, ultrastructural alterations included autophagic vacuoles, undigested cell debris, and multivesicular bodies into dilated megasomes and flagellar pocket were observed. These abnormalities resemble alterations seen in lysosomal storage diseases caused by lysosomal hydrolases deficiency. Using a polyclonal antiserum raised against the native *L. (L.) major* cysteine protease B and a secondary goat antibody to rabbit IgG conjugated with 10-nm gold particles it was possible to observe, only in treated promastigotes, heavily labeled in dilated megasomes and in flagellar pocket, confirming the specific effect of inhibitors in the site of cysteine proteases location (Selzer et al., 1999). Other cysteine protease inhibitors from natural resources, such as plant cystatins, or chemically synthetized, such as pseudopeptide substrate analogues, derivatives of aziridine, triazoles, α-ketoheterocycles and NO-donors, were assayed against *Leishmania*. These compounds provided different degrees of inhibition in promastigote growth and viability, amastigote survival and reduction in the macrophages infection rate (Duboise et al., 1994; Pral et al., 1996; Alves et al., 2001; Tornøe et al., 2004; Ascenzi et al.,

2004; Ordóñez-Gutiérrez et al., 2009, Schurigt et al., 2010; Steert et al., 2010). Although cysteine proteases inhibitors look promising, the activity of the three cysteine proteases families would need to be blocked to completely prevent parasite invasion or replication in the host cells and lesion development, and non-selective inhibitors can also affect the host cysteine proteases. An alternative to this problem would be to develop inhibitors which prevent the cysteine proteases precursor processing resulting in a retrograde accumulation of unprocessed proteases and proteins in organelles of endocytic/exocytic pathway which lead to the parasite´s death.

4.2 The major surface protein or gp 63 metalloprotease

After the discovery of cysteine proteases in both amastigote and promastigote forms of *Leishmania*, a protease of 63 kDa was purified and characterized as the major surface protein of promastigotes also called glycoprotein 63 (gp63) (Etges et al., 1986). This protease was identified in different species of *Leishmania*, including *L. (L.) major*, *L. (L.) donovani*, *L. (L.) infantum*, *L. (L.) tropica*, *L. (L.) mexicana*, *L. (L.) amazonensis*, *L. (V.) braziliensis*, and *L. (L.) enriettii*, and was proved to be structurally and functionally conserved in Old and New World *Leishmania* species (González et al., 2008). It is important to note that *L. (L.) major*, *L. (L.) donovani*, *L. (L.) infantum* and *L. (L.) tropica* cause cutaneous and mucocutaneous leishmaniasis in the "Old World" while *L. (L.) enriettii* is non-pathogenic for humans (Bouvier et al., 1987). In amastigote forms of *L. (L.) mexicana* the surface glycoprotein gp63 was localized by post-embedding immunocytochemistry strategies using a monoclonal antibody against promastigote gp63 of *L. (L.) mexicana* and goat anti-mouse IgG conjugated with 5-nm gold particles. This protease was located in amastigote surface, however the label was more intense within the flagellar pocket of the parasite, which is also involved with endocytosis and secretion of molecules (Overath et al., 1997), and is primarily associated with dense material in the lumen of this pocket (Medina-Acosta et al., 1989). The isolation and analysis of surface proteins from lesion amastigotes indicated that gp63 is also the most abundant protein on the amastigote surface (Medina-Acosta et al., 1989).

Gp63 (EC 3.4.24.36) is a zinc metalloprotease that accounts for about 1% of the total protein in promastigotes of *Leishmania* and is also termed as major surface protease, surface acid protease, promastigote surface protease, and leishmanolysin. This metalloproteases belongs to the M8 family of endopeptidases, sharing several characteristics with mammalian matrix metalloproteases (Yao et al., 2003). This enzyme hydrolyzes only proteins, but not peptides, at various pH values (acid, neutral or basic) depending on the protein substrate (Tzinia and Soteriadou, 1991). These observations suggest that gp63 can catalysis in different environment conditions, such as in the sandfly midgut and macrophage parasitophorous vacuoles, because this enzyme is present in both promastigotes and amastigotes *Leishmania* surfaces. Since then, the functions of this strategic protease were extensively investigated. Gp63 plays several important roles in the pathogenesis of leishmaniasis, including (i) evasion of complement-mediated lysis, (ii) facilitation of macrophage infection by promastigotes, (iii) interaction with the extracellular matrix, (iv) inhibition of natural killer cellular functions, (v) resistance to antimicrobial peptide killing, (vi) degradation of macrophage and fibroblast cytosolic proteins, and (vii) promotion of survival of intracellular amastigotes (Yao, 2010). The gp63 overexpression caused increased host infection and intracellular parasite survival, on the other hand, gp63-deficient parasites infecting macrophages mice, showed a diminished infection and survival (Yao, 2003). Thus, gp63 contributes to parasite virulence by exerting a novel type of control over complement fixation. Organisms expressing gp63 can exploit the opsonic properties of complement while avoiding its lytic effects (Brittingham et al., 1995).

The major surface protease is also important in the *Leishmania*-sandfly interaction. The development and forward migration of *Leishmania* parasites in the sandfly gut was accompanied by morphological transformation to highly mobile, non-dividing 'metacyclic' forms. Metacyclogenesis is associated with developmentally regulated changes in expression of gp63 and its expression in L. *(L.) major* promastigotes surface was studied by post-embedding immunocytochemical analysis using the GP63-specific monoclonal antibodies which demonstrates a clear expression from 2 to 7 days post-blood feeding (Davies et al., 1990). This protease proved to be essential in the development and survival of parasite into sandfly gut, because it degrades hemoglobin and other proteins in the blood meals, thereby providing nutrients needed for the growth of promastigotes and protects promastigotes from degradation by the midgut digestive enzymes (Hajmová et al., 2004). Besides all the functions of Gp63 in *Leishmania*-hosts interactions, the native protein, recombinant or specific peptides from this cleavage was able to elicit a protective immunity to many species of *Leishmania* infection in a variety of animal models (Handman, et al., 1990; Abdelhak et al., 1995; Bhowmick et al., 2008; Mazumder et al., 2011). The immunogenicity and antigenicity of gp63 is very well known. Additionally, it was immunolocalized in the lumen of flagellar pocket indicating that this protease is secreted by *Leishmania* parasites and, and it explains why can be found antibodies against gp63 in patient sera with leishmaniasis (Sayal et al., 1994). For these reasons, gp63 is one of the major candidate molecules for vaccine development against leishmaniasis (Chawla and Madhubala, 2010). Unlike the *Leishmania* cysteine proteases inhibitors that are extensively investigated for drug development to leishmaniasis treatment, studies about gp63 inhibitors did not identify any compound that block the biological functions of this protease and, as the immunogenicity and antigenicity of gp63 has always been recognized many studies were conducted to develop vaccines against *Leishmania* using gp63 or its derivatives as immunogen.

4.3 Serine proteases

Although serine proteases are the most studied enzymes in all living organisms, the first studies about proteases of *Leishmania* identified important proteolytic activity belonged to cysteine protease class, as discussed before (North and Coombs, 1981). Almost two decades later the activity of a serine peptidase was purified and characterized from soluble extracts of L. *(L.) amazonensis* promastigotes (Andrade et al., 1998). Unlike other proteases described in *Leishmania*, it does not hydrolyze proteins or large peptides, but cleaves only small peptides substrates, at the carboxyl side of basic residues and aromatic residues preceding basic residues, which characterizes the enzyme as an oligopeptidase. This was the first study that reports the presence of serine peptidase activity in *Leishmania* and even more an oligopeptidase (Andrade et al., 1998). It is important to consider that *Trypanosoma* species do not express enzymes showing serine protease activities, but only serine oligopeptidases with specific functions in many steps of mammalian cell invasion (Silva-López et al., 2008; Alvarez et al., 2011).

Besides the oligopeptidase activity, *L. (L.) amazonensis* showed expressive activity of serine proteases (Silva-López et al., 2004; Silva- López and De Simone, 2004 a, b; Silva-López et al., 2005). This type of proteolytic enzymes were first obtained from a cell-free supernatant of axenic *L. (L.) amazonensis* promastigotes and was proven to be originated from the parasite despite having been purified from culture supernatant. The post-embedding immunocytochemistry strategy was critical to demonstrate the relationship between the extracellular serine protease and promastigotes, using a rabbit antiserum raised against a heat-inactivated 56-kDa serine protease obtained from culture supernatant and purified using aprotinin-agarose affinity chromatography, and anti-rabbit antibody labeled with 10-nm gold particles (Silva-López et al., 2004). In this study amastigotes from lesions of infected mice were also used in order to investigate the subcellular location of this serine protease and infer possible functions for this enzyme. It was possible to observe that the antibody reacted poorly with the parasite surface and moderately with internal structures in most samples (about 95%) of both forms of the parasite (Figure 2).

In promastigotes, gold particle labeling showed the serine protease to be predominantly located in the flagellar pocket and in vesicular structures which are morphologically similar to the compartments found in mammalian endocytic/exocytic pathways (Figure 2 A and B). It is worth noting, as mentioned previously, that the flagellar pocket is a secluded extracellular compartment in the anterior portion of *Leishmania* formed by an invagination of the plasma membrane at the base of the flagellum and is the only part of the cell surface that supports exocytosis and endocytic traffic of molecules (De Souza, 2006; Bonhivers et al., 2008). This pocket is an obligatory intermediary station for membrane-bound molecules trafficking between intracellular membranes and the cell surface and vice-versa (Overath et al., 1997). Both membrane-bound and secreted proteins appear on the cell surface, underscoring the role of this membrane in delivery of proteins to the cell surface and exterior (Bonhivers et al., 2008). In amastigotes, the enzyme was detected not only in subcellular structures similar to those of promastigotes, such as the flagellar pocket and cytoplasmic vesicles (Figure 2), but also in electron-dense structures corresponding to megasomes (Figure 2 C and D). As commented before, megasomes are large lysosome-like structures and are the main sites of proteolytic activity in *Leishmania* belonging to the mexicana complex, whose major constituents are the cysteine proteases, which results in differentiation process participation and in parasite intracellular survival (Ueda-Nakamura et al. 2002).

The processing and trafficking of cysteine proteases, the best studied lysosomal *Leishmania* proteases, has been reported in *L. pifanoi* and *L. (L.) mexicana* and is targeted to megasomes via the flagellar pocket and has been previously discussed (Duboise et al., 1994; Brooks et al. 2000). These results demonstrated that *L. (L.) amazonensis* secretes a 56-kDa serine protease into the culture supernatant through the flagellar pocket with the participation of different components that resemble mammalian endocytic/exocytic organelles. Furthermore, the fact that this enzyme is located in megasomes, where cysteine proteases are also found, indicate that the serine protease can contribute, in association with the cysteine proteases, to maintain the parasite life cycle and leishmaniasis pathogenesis and so, also represents a novel target in *Leishmania* parasite. This secreted serine protease was further purified and their biochemical characteristics and kinetics parameters were investigated. This enzyme is a dimeric protein of about 115 kDa, with subunits of 56kDa, very well adapted to the environment conditions and certainly contributes to survival and growth of the parasite

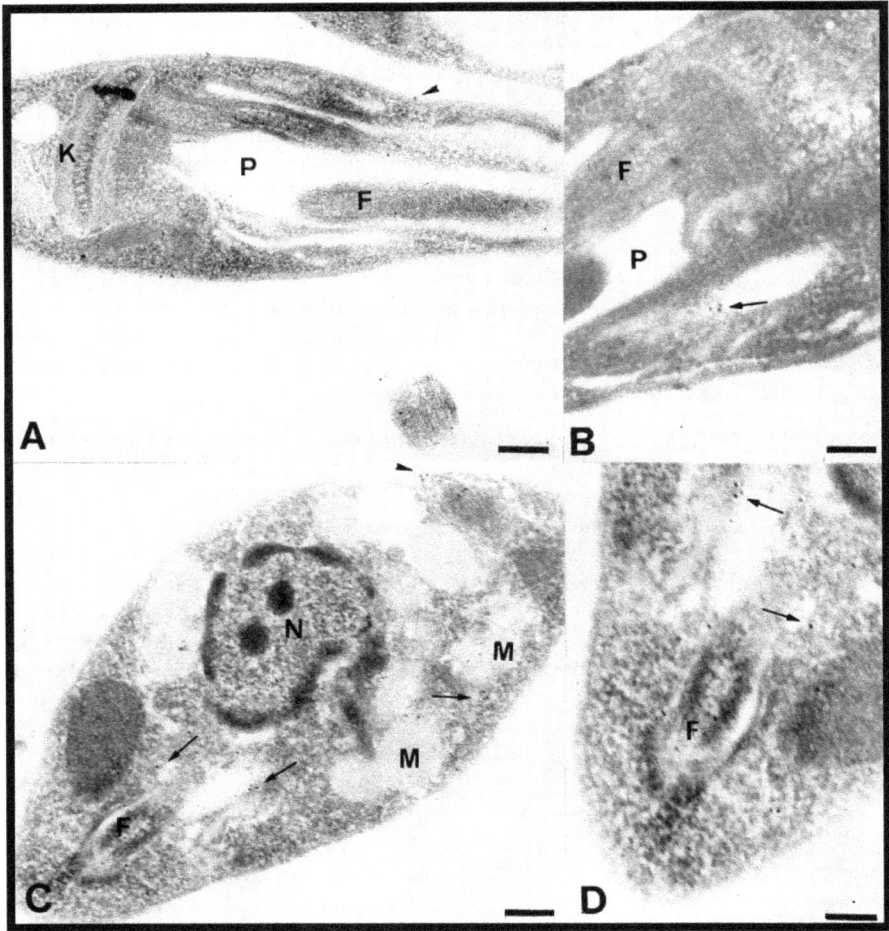

Fig. 2. Subcellular location of serine protease in *L. (L.) amazonensis* promastigotes. (A) Promastigote forms showing immunolabeling at flagellar pocket. In both forms, the cell surface was poorly labeled (Arrowhead). (C) Amastigotes displayed moderate labeling in flagellar pocket, cytoplasmic vesicles (Arrows) and in megasomes. Scale bars; A and C: 2.0 μm. High magnification images of the anterior region of (B) promastigote and (D) and amastigote showing immunolabeling at cytoplasmic vesicles (Arrows) that subtending the flagellar pocket in both forms of *L. (L.) amazonensis*. Scale bars; B and D: 3.2 μm. Flagellar pocket (P), flagellum (F), kinetoplast (K), megasome (M) and nucleus (N) (Silva-López et al., 2004).

inside their hosts, since it is found in promastigotes and amastigotes forms of *Leishmania* (Silva-López et al., 2005).

Two other serine proteases were also purified from water and detergent soluble intracellular extracts of *L. (L.) amazonensis* promastigotes and exhibited different properties and must perform distinct functions in the *Leishmania* metabolism and physiology (Silva- López and De Simone, 2004 a and b). The subcellular location of the 68-kDa serine protease, an enzyme purified from the water soluble intracellular extracts of promastigote parasites, was performed by electron microscopy post-embedding immunocytochemistry, using the antiserum raised in rabbit by injecting the 68-kDa serine protease purified as previously described (Silva-López and Giovanni De Simone 2004 a). As observed in the figure 3, the antibody reacted against the parasite cytoplasmic membrane and internal structures in the analyzed cells. Cytoplasmic gold particles are seen bound to the external surface and to the flagellar pocket membrane (Figure 3 A), and were localized predominantly in cytoplasmic vesicular structures morphologically similar to that of the endocytic/ exocytic pathways and tubulovesicular structures close to the flagellar pocket region (Figure 3 B and C) (Morgado-Díaz, et al., 2005).

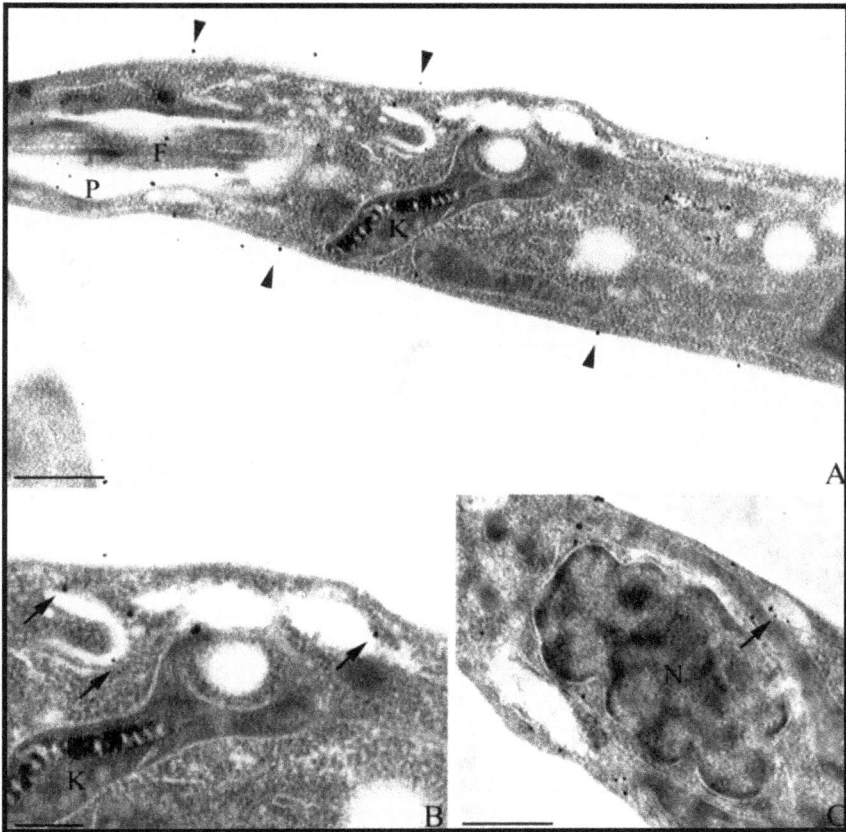

Fig. 3. Subcellular localization of 68 kDa serine protease in *L. (L.) amazonensis* promastigotes. A: gold particles are seen bound to the external surface (arrowheads) and to the flagellar pocket membrane, B: high magnification showing labeling in cytoplasmic vesicles and tubulovesicular structures close to the flagellar pocket (arrows). C: high magnification showing labeling (arrow) in cytoplasmic vesicles. P: flagellar pocket; F: flagellum; K: Kinetoplast; N: nucleus. Bar = 0.25 µm (Morgado-Díaz, et al., 2005).

It is important to compare the localization of both 68 kDa intracellular and 56 kDa secreted serine proteases: While the 68 kDa enzyme is mainly located in membranes of intracellular compartments and plasma membrane, the 56 kDa protease, previously described, reacted poorly with the parasite surface and moderately with internal structures. However, it was predominantly located in the flagellar pocket, megasomes and structures that are morphologically similar to the compartments that are found in mammalian endocytic/exocytic pathways (Silva-López et al. 2004), which justified the released into the extracellular environment.

The activity of serine proteases were also isolated from aqueous, detergent and extracellular extracts of *L. (V.) braziliensis* promastigotes employing aprotinin-agarose affinity chromatography (Matos Guedes at al., 2007). *L. (V.) braziliensis* is the major species of *Leishmania* associated with cutaneous and mucosal forms of leishmaniasis in Brazil and Latin America (Bañuls et al., 2007). These proteases display some biochemical similarities with *L. (L.) amazonensis* serine proteases, demonstrating a conservation of this class of proteolytic activity in the *Leishmania* genus and suggesting similar subcellular location and functions. This was the first study to report the purification of a serine protease from *L. (V.) braziliensis* (Matos Guedes at al., 2007).

The first evidences of the possible functions of these serine proteases in *Leishmania* were obtained using specific serine protease inhibitors. These compounds induced parasite death, with regard to time and doses dependence and, significant morphological alterations. These structural changes were observed in the region of the flagellar pocket and included the appearance of vesicles, which were accompanied by bleb formations of the membrane that covers this pocket which was importantly altered (Silva-López et al., 2007). These effects in the flagellar pocket (a structure of intense exocytic/endocytic activities) indicated that these compounds are endocyted through this structure, and inhibited the serine proteases which are located in this pocket, as previously described by immunocytochemical studies (Silva-López et al., 2004). In the cytoplasm the presence of vesicles that resemble autophagic vacuoles was also noted. The autophagy is a catabolic process involving the degradation of a cell's own components through the lysosomal machinery and is required for normal turnover, starvation, stress responses differentiation, development and in a certain type of cell death (Kiel, 2010). Serine proteases inhibitors caused cell death of *L. (L.) amazonensis* promastigotes inducing the formation of autophagic vacuoles, since the features associated with *Leishmania* apoptosis were not observed (Paris et al., 2004). No modification was found in any of the other cellular structures of the parasites treated with these inhibitors. Furthermore, all parasites exhibited shape alterations (Silva-López et al., 2007). Although the described results indicate that these enzymes are essential for parasite survival, their functions in *Leishmania* physiology are unclear. If these enzymes participate in the exocytosis/endocytosis pathway through the processing of intracellular proteins or even in the maintenance of morphological organization of *Leishmania* remains to be elucidated. However, these findings suggest that *Leishmania* serine proteases appear to be promising targets for the development of specific inhibitors for leishmaniasis chemotherapy.

Serine protease activities were also described for *L. (L.) chagasi*, the causative agent of visceral leishmaniasis in Latin America (Bañuls et al., 2007). These enzymes were isolated from aqueous, detergent soluble and culture supernatant of *L. (L.) chagasi* promastigote extracts and respectively named as LCSII, LCSI and LCSIII. The characterization of these enzymes employed similar strategies used for *L. (L.) amazonensis* serine proteases (Silva-López et al., 2010a). The same rabbit antiserum against the 56-kDa extracellular protease and anti-rabbit antibody labeled with 10-nm gold particles were used to determine the subcellular localization of the serine proteases in *L. (L.) chagasi* promastigotes employing post-embedding immunocytochemistry strategies. The reactivity of this antiserum was first assayed with all three serine proteases by immunoblotting proteases and indicated that serine proteases of both parasites are related proteins. The antibody did not react with the parasite surface but strongly with internal structures in most samples. The gold particles labeling confirmed that serine proteases are located in the flagellar pocket region and intracellular vesicles (Figure 4), demonstrating that LCSIII which was obtained from culture supernatant, follow the same route of secretion to the extracellular environment utilizing the flagellar pocket. *L. (L.) chagasi* serine proteases were also located in contractile vacuoles and in vesicles located at the posterior region of the parasite body, next to the nucleus. The contractile vacuoles are intracellular vesicles immediately adjacent to the plasma membrane of the flagellar pocket and are involved in fluid secretion via this pocket (Linder and Staehelin, 1979). Other serine peptidases, such as the extracellular serine peptidases from *L. (L.) amazonensis* (Silva-López et al., 2004) (Figure 2) and *T. cruzi* (Silva-López et al., 2008) employed the same route of secretion, since they were also evidenced in the flagellar pocket and contractile vacuoles.

Notably, the endocytic pathway of the *Leishmania* promastigotes comprises a network of tubular endosomes, multivesicular bodies and un unusual multivesicular tubule (MVT)–lysosome, originally observed in *L. (L.) mexicana* (Alberio et al., 2004), and are the main sites of proteolytic activity in *Leishmania*, as well as being crucial for the differentiation process and parasite intracellular survival (Ueda-Nakamura et al., 2007).

Recent studies demonstrated that *L. (L.) donovani* express a very similar secreted serine protease like *L. (L.) amazonensis* which was also located in the flagellar pocket of promastigotes by post-embedding immunogold labeling using anti-115-kDa serine protease antibody and a gold 10 nm particles conjugated secondary antibody. Besides, this enzyme is particularly expressed in virulent strains and is also associated with the metacyclic stage of *L. (L.) donovani* promastigotes. It is postulated that 115 kDa serine protease could be a potential vaccine candidate since it plays important roles in the macrophage infection and is secreted to extracellular environments (Choudhury et al., 2010).

Besides the expression of secreted 115-kDa serine protease, two other serine proteases were identified and characterized in *L. (L.) donovani* promastigotes, using biochemical and molecular strategies: subtilisin (Swenerton et al., 2010) and oligopeptidase B (Swenerton et al., 2011) serine proteases. The functions of *Leishmania* subtilisin (Clan SB, family S8) was studied in parasites with gene knock-out for this enzyme, which resulted in reduced ability to undergo promastigote to amastigote differentiation *in vitro* and amastigotes revealed abnormal membrane structures, retained flagella, and increased binucleation. These "knock-out" parasites displayed reduced virulence in both hamster and murine infection models. Furthermore, proteomic analysis indicated that *Leishmania* subtilisin is the maturase for tryparedoxin peroxidases that detoxifies reactive oxygen intermediates and maintain redox homeostasis that is essential for *Leishmania* virulence (Swenerton et al., 2010). Using similar proteomic strategies was demonstrated that *L. (L.) donovani* oligopeptidase B (Clan SC, family S9A) regulate the function of enolase, since parasites "knock-out" of this peptidase

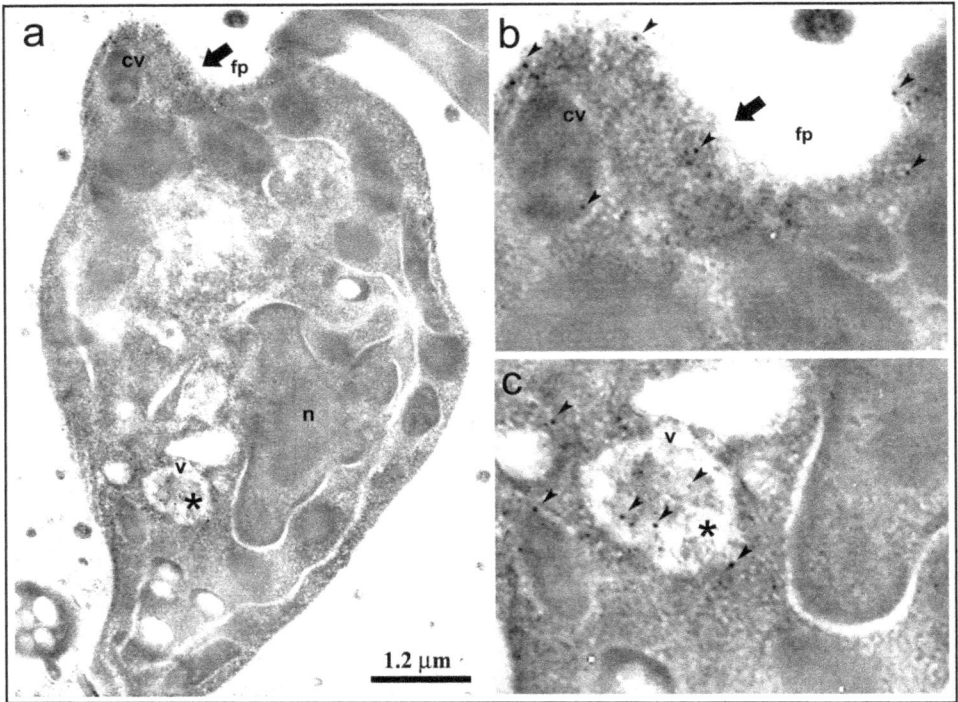

Fig. 4. Immunolocalization of serine proteases in promastigotes of *L. (L) chagasi*.
Representative ultrathin section of promastigotes labeled with polyclonal antiserum to 56-kDa *L. amazonensis* serine protease. (a) Gold particles are seen bound to the flagellar pocket membrane (arrow) and into intracellular vesicles (asterisk) of *L. chagasi* promastigote. (b-c) High magnification images showing labeling in the anterior region of promastigote, in flagellar pocket (arrow head), contractile vacuoles and into intracellular vesicles (arrow head and asterisk), respectively. Flagellar pocket (fp), contractile vacuole (cv), vesicle (v) and nucleus (n) (Silva-López et al., 2010a).

showed enolase abnormally increased but enzymatically inactive. Aside from its classic role in carbohydrate metabolism, enolase was found to localize in cytoplasmic membranes, where it binds host plasminogen and functions as a virulence factor for several pathogens. As expected, there was a striking alteration in macrophage responses to *Leishmania* when oligopeptidase B was deleted, so the enzyme interfered in parasite enolase activity and immune evasion. Besides, these "knock-out" parasites displayed decreased virulence in the murine footpad infection model (Swenerton et al., 2011).

It is also important to emphasize the roles of serine proteases in the host immune system modulation. Mice vaccination with soluble proteases isolated from *L. (L.) amazonensis* promastigote antigens directly activated IL-4, IL-10 and TGF-beta production by immune cells and primed mice to respond to parasite challenge with a strong Jones-Mote cutaneous hypersensitivity reaction, and increased susceptibility to infection. So, serine proteases are key components of *L. (L.) amazonensis* promastigote antigens responsible for disease-promoting immunity (Matos Guedes et al., 2010) and besides being important targets to drug development against *Leishmania*, they are also vaccine candidates for leishmaniasis.

In addition to the proteases already discussed, aspartic protease activity was identified and characterized in *L. (L.) mexicana* promastigotes (Valdivieso et al., 2007). This activity was target of anti proliferative effect on *Leishmania* sp. promastigotes and axenic amastigotes by HIV aspartyl-protease inhibitors, Ac-Leu-Val-Phenylalaninal, Saquinavir mesylate and Nelfinavir. The latter two compounds are currently used as part of antiretroviral therapy. This effect appears to be the result of cell division blockage. In addition, these drugs induced in culture a decrease in the percentage of co-infected HIV/*Leishmania* monocytes and amastigotes of *Leishmania* per macrophage. The finding of a dose-dependent inhibition of *Leishmania* aspartyl-protease activity by these drugs allows us to propose this activity as the drug parasite target. A direct action of these HIV aspartyl-protease inhibitors on *Leishmania* parasites would be correlated with the effect that highly active antiretroviral therapy has had in the decrease of HIV/*Leishmania* co-infection, opening an interesting perspective for new drugs research development based on this novel parasite protease family (Valdivieso et al., 2010).

It is very clear that the employment of specific protease inhibitors could block *Leishmania* proteolytic activity and interfere in the progression of leishmaniasis. A recent study demonstrated that host uncontrolled matrix metalloprotease activity in the cutaneous lesions caused by *L. (V.) braziliensis* may result in intense tissue degradation and, consequently, poor healing wounds, which were associated with unsatisfactory response to antimonials treatment (Maretti-Mira et al., 2011). Thus a pharmaceutical formulation containing protease inhibitors can inhibit both host and parasite proteases and helps heal leishmaniac lesions.

5. Technical considerations

All immunocytochemical experiments begin with tissue fixation, which serves the dual purpose of preserving the cellular structure and the *in vivo* distribution of antigens. However, antigens are chemically modified by fixation and further denatured by dehydration and embedment. Formaldehyde fixation preserves most antigenic sites but it is reversible and it does not maintain good ultrastructure. Osmium post fixation is essential to preserve membrane structure and ultrastructural detail; unfortunately, osmium often irreversibly destroys antigenic sites. In immunocytochemical protocols to localize serine proteases in *L. (L.) amazonensis* and *L. (L.) chagasi*, the parasites were fixed in 4% paraformaldehyde/1% glutaraldehyde in 0.1M sodium cacodylate buffer, pH 7.3 (Silva-López et al., 2004; Morgado-Díaz et al. 2005; Silva-López et al., 2010a). The samples were dehydrated in methanol and embedded at progressively lowered temperature in Lowicryl K4M resin. Lowicryl is a hydrophilic acrylic resin that tolerates partial dehydration and is processed and photopolymerized at subfreezing temperatures. The hydrophilic properties of Lowicryl result in excellent antigen retention and consequently in high label density, specificity, and low background (Herman, 1988). After embedment, thin sections were collected on 400 mesh uncoated nickel grids, incubated for 30 min at room temperature in phosphate buffered saline (PBS) containing 1.5% bovine serum albumin and 0.01% Tween 20, pH 8.0 (blocking buffer) in order to block unspecific bounds. The grids were incubated for 60 min in the presence of the primary antibodies: anti-56 kDa extracellular serine protease or anti-68 kDa intracellular serine protease diluted in blocking buffer. Both antibodies were produced using similar protocols. Serine proteases were purified using affinity chromatography on aprotinin-agarose columns (Silva-López and De Simone, 2004b; Silva-López et al., 2005). The antiserum was raised in rabbit by injecting the homogeneous heat inactivated purified serine proteases emulsified in complete (first booster) and incomplete (subsequent boosters) Freud's adjuvant. After the fourth injection the antibody reactivity was checked by immunoblotting. So, grids containing embedded parasites were finally incubated for 60 min with goat anti-rabbit antibody labeled with 10-nm gold particles. The grids were subsequently washed with PBS and distilled water, stained with uranyl acetate and lead citrate and observed in a Zeiss EM10C transmission electron microscope (Silva-López et al., 2004; Morgado-Díaz et al. 2005; Silva-López et al., 2010a). The quality of antibodies is essential to obtain reliable results in immunocytochemistry assays, since they must be specific and sensitive enough to bind the antigen of interest in the cellular structure without labeling other intracellular sites that do not contain the antigen. Furthermore, to localize antigens by electron microscopy, it is necessary to impart electron density to the bound antibodies. Colloidal gold probes have been extensively adopted for use in post-embedding and pre-embedding immunocytochemistry assays (Bendayan et al. 1987).

Immunocytochemical experiments may be accomplished with various procedures for pre- or post-embedding labeling. Each method offers distinct advantages and disadvantages regarding to specificity, density of antigen labeling, and structural preservation. Pre-embedding immunocytochemistry assays requires cryoprotected tissue and labeled with primary and indirect electron-dense labels which enter the tissue by diffusion. The labeled tissue is then embedded, sectioned, and examined. The primary advantage is the excellent antigen retention as the consequence of few pre-labeling and processing steps. The disadvantages result from the poor penetration of both primary antibodies and secondary labels into the tissue, limiting the label to a gradient in the superficial few micrometers, and require costly instrumentation. Post-embedding immunocytochemistry is the most employed technique in most electron microscopy subcellular location studies and was discussed above. In these procedures tissues are fixed, dehydrated, and embedded in plastic using protocols similar to those of conventional EM. Thin plastic sections are labeled by immersion in solutions of primary antibodies followed by electron-dense second label. The main advantages are that the skills and methods are similar to those employed in conventional EM, and no specialized equipment beyond that found in any EM laboratory is required (Herman, 1988).

6. Concluding remarks

The immunocytochemistry is a technique of choice that permits routine and reproducible localization of most moderately abundant antigens using specific antibodies against certain antigens. The subcellular location of enzymes suggests their function in the metabolism of specific organelle, cell or organism. The immunocytochemical localization of *Leishmania* proteases in megasomes, flagellar pocket, cytoplasmic membrane, contractile vacuoles, cytoplasmic vesicles, tubulovesicular structures and as secreted enzymes into the extracellular environment indicates the versatility of these proteases. They participate in exocytic/endocytic pathways, in processing of endogenous proteins or enzymes, in the digestion of exogenous proteins for parasite nutrition or signaling. The localization of *Leishmania* proteases in membrane or in the extracellular medium suggests that these enzymes could be important mediators with their hosts, and thus modulate a host immune response. Since *Leishmania* proteases perform crucial roles in parasite physiology and in the host-parasite interaction, they are absolutely necessary for the survival of the pathogen and the leishmaniasis progression and, in addition they are substantially different from the host homolog. So, they are considered important targets in *Leishmania*. Furthermore, specific protease inhibitors induced important alterations in parasite morphology, reduced the viability and growth of *Leishmania* and killed axenic promastigotes and amastigotes and intracellular amastigotes thus becoming a promising candidate for leishmaniasis treatment (Silva-López et al., 2007; Valdivieso et al., 2010; Olivier & Hassani, 2010; Pereira et al., 2011). In conclusion, immunocytochemical strategies contributed and continue to contribute for the specific identification of targets in *Leishmania* which is a rational approach for drug development in the leishmaniasis treatment.

7. References

Abdelhak S, Louzir H, Timm J, Blel L, Benlasfar Z, Lagranderie M, Gheorghiu M, Dellagi K, Gicquel B. (1995) Recombinant BCG expressing the leishmania surface antigen Gp63 induces protective immunity against *Leishmania major* infection in BALB/c mice. Microbiology. 141, 1585-1592

Aït-Oudhia K, Gazanion E, Vergnes B, Oury B, Sereno D. (2011) *Leishmania* antimony resistance: what we know what we can learn from the field. Parasitol Res. Jul 29 [e-pud].

Alberio SO, Dias SS, Faria FP, Mortara RA, Barbiéri CL, Freymüller Haapalainen E. (2004) Ultrastructural and cytochemical identification of megasome in *Leishmania (Leishmania) chagasi*. Parasitol Res. 92, 246-254.

Almeida OL, Santos JB. (2011) Advances in the treatment of cutaneous leishmaniasis in the new world in the last ten years: a systematic literature review. An Bras Dermatol. 86,497-506.

Al-Salabi MI, Koning HP. (2005) Purine nucleobase transport in amastigotes of *Leishmania mexicana*: involvement in allopurinol uptake. Antimicrob Agents Chemother. 49, 3682-3689.

Alvarez VE, Niemirowicz GT, Cazzulo JJ. (2011) The peptidases of *Trypanosoma cruzi*: Digestive enzymes, virulence factors, and mediators of autophagy and programmed cell death. Biochim Biophys Acta. May 19 [e-pud].

Alves LC, St Hilaire PM, Meldal M, Sanderson SJ, Mottram JC, Coombs GH, Juliano L, Juliano MA. (2001) Identification of peptides inhibitory to recombinant cysteine proteinase, CPB, of *Leishmania mexicana*. Mol Biochem Parasitol. 114, 81-88.

Andrade AS, Santoro MM, de Melo MN, Mares-Guia M. (1998) *Leishmania (Leishmania) amazonensis*: purification and enzymatic characterization of a soluble serine oligopeptidase from promastigotes. *Exp Parasitol.* 89, 153-160.

Antoine JC, Jouanne C, Ryter A, Benichou JC. (1988) *Leishmania amazonensis*: acidic organelles in amastigotes. Exp Parasitol. 67, 287-300.

Ascenzi P, Bocedi A, Gentile M, Visca P, Gradoni L. (2004) Inactivation of parasite cysteine proteinases by the NO-donor 4-(phenylsulfonyl)-3-(2-(dimethylamino)ethyl)thio)-furoxan oxalate. Biochim Biophys Acta. 1703, 69-77.

Baiocco P, Colotti G, Franceschini S, Ilari A. (2009) Molecular basis of antimony treatment in leishmaniasis. J Med Chem. 52, 2603-2612.

Banerjee B, Sen N, Majumder HK. (2011) Identification of a Functional Type IA Topoisomerase, LdTopIIIβ, from Kinetoplastid Parasite Leishmania donovani. Enzyme Res. 2011:230542.

Bañuls AL, Hide M, Prugnolle F. (2007) Leishmania and the Leishmaniases: A Parasite Genetic Update and Advances in Taxonomy, Epidemiology and Pathogenicity in Humans. Advances in Parasitol, 64, 1-113.

Baqui MM, Milder R, Mortara RA, Pudles J. (2000) In vivo and in vitro phosphorylation and subcellular localization of trypanosomatid cytoskeletal giant proteins. Cell Motil Cytoskeleton. 47, 25-37.

Barret, A. J., 1994. Classification of peptidases. Methods Enzymol. 244, 1-15.

Bates PA, Robertson CD, Coombs GH. (1994) Expression of cysteine proteinases by metacyclic promastigotes of Leishmania mexicana. J Eukaryot Microbiol.41, 199-203.

Bendayan M, Nanci A, Kan FWK 1987. Effect of tissue processing on colloidal gold cytochemistry. J Histochem Cytochem. 35, 483-496.

Bhowmick S, Ravindran R, Ali N. (2008) gp63 in stable cationic liposomes confers sustained vaccine immunity to susceptible BALB/c mice infected with Leishmania donovani. Infect Immun. 76, 1003-1015.

Bonhivers M, Nowacki S, Landrein N, Robinson DR. (2008) Biogenesis of the trypanosome endo-exocytotic organelle is cytoskeleton mediated. PLoS Biol. 6, e105.

Bouvier J, Etges R, Bordier C. (1987) Identification of the promastigote surface protease in seven species of Leishmania. Mol Biochem Parasitol. 24, 73-79.

Brittingham A, Morrison CJ, McMaster WR, McGwire BS, Chang KP, Mosser DM. (1995) Role of the Leishmania surface protease gp63 in complement fixation, cell adhesion, and resistance to complement-mediated lysis. J Immunol. 155, 3102-3111.

Brooks DR, Tetley L, Coombs GH, Mottram JC (2000) Processing and trafficking of cysteine proteases in Leishmania mexicana J Cell Sci. 113, 4035-4041.

Carter NS, Yates P, Arendt CS, Boitz JM, (2008) Ullman B.Purine and pyrimidine metabolism in Leishmania. Adv Exp Med Biol. 625, 41-54.

Castillo E, Dea-Ayuela MA, Bolás-Fernández F, Rangel M, González-Rosende ME (2010). The kinetoplastid chemotherapy revisited: current drugs, recent advances and future perspectives. Curr Med Chem. 17, 4027-4051.

Chandra S, Ruhela D, Deb A, Vishwakarma RA. (2010) Glycobiology of the Leishmania parasite and emerging targets for antileishmanial drug discovery.Expert Opin Ther Targets. 14,739-757.

Chauhan SC, Madhubala R. (2009) Glyoxalase I gene deletion mutants of Leishmania donovani exhibit reduced methylglyoxal detoxification. PLoS One. 4, e6805.

Chawla B, Jhingran A, Singh S, Tyagi N, Park MH, Srinivasan N, Roberts SC, Madhubala R. (2010) Identification and characterization of a novel deoxyhypusine synthase in Leishmania donovani. J Biol Chem. 285, 453-463.

Chawla B, Madhubala R. (2010) Drug targets in Leishmania. J Parasit Dis. 34, 1-13.

Choudhury R, Das P, Bhaumik SK, De T, Chakraborti T. (2010) In situ immunolocalization and stage-dependent expression of a secretory serine protease in Leishmania donovani and its role as a vaccine candidate. Clin Vaccine Immunol. 17, 660-667.

Colmenares M, Constant SL, Kima PE, McMahon-Pratt D. (2002) *Leishmania pifanoi* pathogenesis: Selective Lack of a Local Cutaneous Response in the Absence of Circulating Antibody. Infec Imm. 70, 6597-6605.

Davies CR, Cooper AM, Peacock C, Lane RP, Blackwell JM. (1990) Expression of LPG and GP63 by different developmental stages of *Leishmania major* in the sandfly *Phlebotomus papatasi*. Parasitology. 101, 337-343.

De Souza W, Attias M, Rodrigues JC. (2009a) Particularities of mitochondrial structure in parasitic protists (Apicomplexa and Kinetoplastida). Int J Biochem Cell Biol. 41, 2069-2080.

De Souza W, Sant'Anna C, Cunha-e-Silva NL. (2009b) Electron microscopy and cytochemistry analysis of the endocytic pathway of pathogenic protozoa. Prog Histochem Cytochem. 44, 67-124.

De Souza W. (2006) Secretory organelles of pathogenic protozoa. An Acad Bras Cienc. 78, 271-291.

Demicheli C, Frézard F, Mangrum JB, Farrell NP. (2008) Interaction of trivalent antimony with a CCHC zinc finger domain: potential relevance to the mechanism of action of antimonial drugs. Chem Commun (Camb). 39, 4828-4830.

Desjeux P. (2004) Leishmaniasis: current situation and new perspectives. Comp Immunol Microbiol Infec Dis. 27, 305-318.

Doroud D, Zahedifard F, Vatanara A, Najafabadi AR, Rafati S. (2011) Cysteine proteinase type I, encapsulated in solid lipid nanoparticles induces substantial protection against *Leishmania major* infection in C57BL/6 mice. Parasite Immunol. 33, 335-348.

Duboise SM, Vannier-Santos MA, Costa-Pinto D, Rivas L, Pan AA, Traub-Cseko Y, De Souza W, McMahon-Pratt D. (1994) The biosynthesis, processing, and immunolocalization of *Leishmania pifanoi* amastigote cysteine proteinases. Mol Biochem Parasitol. 68, 119-132.

El-Fadili AK, Zangger H, Desponds C, Gonzalez IJ, Zalila H, Schaff C, Ives A, Masina S, Mottram JC, Fasel N. (2010) Cathepsin B-like and cell death in the unicellular human pathogen *Leishmania*. Cell Death Dis. 2, 1:e71.

Etges R, Bouvier J, Bordier C. (1986) The major surface protein of *Leishmania* promastigotes is a protease. J Biol Chem. 261, 9098-9101.

Ezra N, Ochoa MT, Craft N. (2010) Human immunodeficiency virus and leishmaniasis. J Glob Infect Dis. 2, 248-257.

Fedeli CE, Ferreira JH, Mussalem JS, Longo-Maugéri IM, Gentil LG, dos Santos MR, Katz S, Barbiéri CL. (2010) Partial protective responses induced by a recombinant cysteine proteinase from *Leishmania (Leishmania) amazonensis* in a murine model of cutaneous leishmaniasis. Exp Parasitol. 124, 153-158

Fernandes Rodrigues JC, Concepcion JL, Rodrigues C, Caldera A, Urbina JA, de Souza W. (2008) In vitro activities of ER-119884 and E5700, two potent squalene synthase inhibitors, against *Leishmania amazonensis*: antiproliferative, biochemical, and ultrastructural effects. Antimicrob Agents Chemother. 52, 4098-4114.

Garcia-Carreno FL, Del Toro MAN. (1997) Classification of proteases without tears. Biochem. Education 25, 161-167.

Gómez MA, Olivier M. (2010) Proteases and phosphatases during *Leishmania*-macrophage interaction: paving the road for pathogenesis.Virulence. 1, 314-318.

González U, Pinart M, Reveiz L, Alvar J (2008) Interventions for Old World cutaneous leishmaniasis. The Cochrane Collaboration. Published by JohnWiley & Sons, Ltd.

Goto Y, Bhatia A, Raman VS, Vidal SE, Bertholet S, Coler RN, Howard RF, Reed SG. (2009) *Leishmania infantum* sterol 24-c-methyltransferase formulated with MPL-SE induces cross-protection against L. major infection. Vaccine. 27, 2884-2890.

Hajmová M, Chang, KP, Kolli, B, Volf P. (2004) Down-regulation of gp63 in *Leishmania amazonensis* reduces its early development in *Lutzomyia longipalpis* Microbes and Infection 6, 646–649

Handman E, Button LL, McMaster RW. (1990) *Leishmania major*: production of recombinant gp63, its antigenicity and immunogenicity in mice. Exp Parasitol. 70, 427-435.

Herman, E.M. (1988). Immunocytochemical localization of macromolecules with the electron microscope. Ann Rev Plant Physiol Plant Mol. Biol. 39, 139-155.

Hide M, Bañuls AL. (2008) Polymorphisms of *cpb* multicopy genes in the *Leishmania (Leishmania) donovani* complex. Trans Royal Soc Trop Med Hyg. 102, 105-106.

Ilg T, Fuchs M, Gnau V, Wolfram M, Harbecke D, Overath P. (1994) Distribution of parasite cysteine proteinases in lesions of mice infected with *Leishmania mexicana* amastigotes. Mol Biochem Parasitol. 67, 193-203.

Ilgoutz SC, Mullin KA, Southwell BR, McConville MJ. (1999) Glycosylphosphatidylinositol biosynthetic enzymes are localized to a stable tubular subcompartment of the endoplasmic reticulum in *Leishmania mexicana*. EMBO J. 18, 3643-3654.

Kaye P, Scott P. (2011) Leishmaniasis: complexity at the host–pathogen interface. Nat Rev. 9, 604-615.

Khoshgoo N, Zahedifard F, Azizi H, Taslimi Y, Alonso MJ, Rafati S. (2008) Cysteine proteinase type III is protective against *Leishmania infantum* infection in BALB/c mice and highly antigenic in visceral leishmaniasis individuals. Vaccine. 26, 5822-5829.

Kiel JA. (2010) Autophagy in unicellular eukaryotes. Philos Trans R Soc Lond B Biol Sci. 365, 819-830.

Kozar, Z. (1961) Mechanisms of invasion and migration of parasites in host's organism. World Wide Abstr Gen Med. 7,541-559.

Krauth-Siegel LR, Comini MA, Schlecker T. (2007) The trypanothione system. Subcell Biochem. 44, 31-51

Krowarsch D, Cierpicki T, Jelen F, Otlewski J (2003) Canonical protein inhibitors of serine proteases. Cell Mol Life Sci. 60, 2427–2444.

Linder, J.C., Staehelin, L.A. (1979) A novel model for fluid secretion by the trypanosomatid contractile vacuole apparatus. Cell Biol. 83, 371-382.

Loiseau PM, Bories C. (2006) Mechanisms of drug action and drug resistance in *Leishmania* as basis for therapeutic target identification and design of antileishmanial modulators. Curr Top Med Chem. 6, 539-550.

López-Otín C, Bond JS. (2008) Proteases: Multifunctional enzymes in life and disease. J. Biol Chem. 283, 30433-30437.

Lunedo SN, Thomaz-Soccol V, de Castro EA, Telles JE. (2011) Immunocytochemical and immunohistochemical methods as auxiliary techniques for histopathological diagnosis of cutaneous leishmaniasis. Acta Histochem. Jul 9 [e-pud].

Maretti-Mira AC, de Oliveira-Neto MP, Da-Cruz AM, de Oliveira MP, Craft N, Pirmez C. (2011) Therapeutic failure in American cutaneous leishmaniasis is associated with gelatinase activity and cytokine expression. Clin Exp Immunol. 163, 207-214.

Marín-Villa M, Vargas-Inchaustegui DA, Chaves SP, Tempone AJ, Dutra JM, Soares MJ, Ueda-Nakamura T, Mendonça SC, Rossi-Bergmann B, Soong L, Traub-Csekö YM. (2008) The C-terminal extension of Leishmania pifanoi amastigote-specific cysteine proteinase Lpcys2: a putative function in macrophage infection. Mol Biochem Parasitol. 162, 52-59.

Matos Guedes HL, Pinheiro RO, Chaves SP, De-Simone SG, Rossi-Bergmann B. (2010) Serine proteases of Leishmania amazonensis as immunomodulatory and disease-aggravating components of the crude LaAg vaccine. Vaccine. 28, 5491-546.

Matos Guedes HL, Rezende JM, Fonseca MA, Salles CM, Rossi-Bergmann B, De-Simone SG. (2007) Identification of serine proteases from Leishmania braziliensis. Z Naturforsch C. 62, 373-381.

Mazumder S, Maji M, Das A, Ali N. (2011) Potency, efficacy and durability of DNA/DNA, DNA/protein and protein/protein based vaccination using gp63 against Leishmania donovani in BALB/c mice. PLoS One. 6, e14644

McKerrow JH, Caffrey C, Kelly B, Loke P, Sajid M. (2006) Proteases in Parasitic diseases. Annu Rev Pathol Mech Dis. 1, 497–536.

Medina-Acosta E, Karess RE, Schwartz H, Russell DG. (1989) The promastigote surface protease (gp63) of Leishmania is expressed but differentially processed and localized in the amastigote stage. Mol Biochem Parasitol. 37, 263-73.

Michels PA, Bringaud F, Herman M, Hannaert V. (2006) Metabolic functions of glycosomes in trypanosomatids. Biochim Biophys Acta. 1763, 1463-1477.

Molina R, Gradoni L, Alvar J (2003) HIV and the transmission of Leishmania. Ann Trop Med Parasitol. 97, 29-45.

Mottram JC, Coombs GH, Alexander J. (2004) Cysteine peptidases as virulence factors of Leishmania. Curr Opin Microbiol. 7, 375-81.

Mougneau E, Bihl F, Glaichenhaus N. (2011) Cell biology and immunology of Leishmania. Immunol Rev. 240, 286-296.

Moyersoen J, Choe J, Fan E, Hol WG, Michels PA. (2004) Biogenesis of peroxisomes and glycosomes: trypanosomatid glycosome assembly is a promising new drug target. FEMS Microbiol Rev. 28, 603-643.

Neurath H. (1984) Evolution of proteolytic enzymes. Science. 224, 350-357.

North MJ, Coombs GH. (1981) Proteinases of Leishmania mexicana amastigotes and promastigotes: analysis by gel electrophoresis. Mol Biochem Parasitol. 3, 293-300.

Okwor I, Uzonna, JE. (2009) Immunotherapy as a strategy for treatment of leishmaniasis: a review of the literature. Immunotherapy. 1, 765-76.

Olivier M, Hassani K. (2010) Protease inhibitors as prophylaxis against leishmaniasis: new hope from the major surface protease gp63.Future Med Chem. 2,539-542.

Ordóñez-Gutiérrez L, Martínez M, Rubio-Somoza I, Díaz I, Mendez S, Alunda JM. (2009) Leishmania infantum: antiproliferative effect of recombinant plant cystatins on promastigotes and intracellular amastigotes estimated by direct counting and real-time PCR. Exp Parasitol. 123, 341-346.

Otlewski J, Krowarsch D, Apostoluk W (1999) Protein inhibitors of serine proteinases. Acta Biochim Pol. 46, 531–565.

Overath P, Stierhof Y-D, Wiese M (1997) Endocytosis and secretion in trypanosomatid parasites — tumultuou traffic in a pocket. Trends Cell Biol. 7, 27–33.

Paris C, Loiseau PM, Bories C, Bréard J (2004) Miltefosine Induces Apoptosis-Like Death in *Leishmania donovani* promastigotes. Antimicrob Agents Chemother. 48, 852-859.

Pereira IO, Assis DM, Juliano MA, Cunha RL, Barbieri CL, do Sacramento LV, Marques MJ, dos Santos MH. (2011) Natural products from *Garcinia brasiliensis* as *Leishmania* protease inhibitors. J Med Food. 14, 557-562.

Pral EM, Alfieri SC. (1996) Uptake of Z-Tyr[125I]-AlaCHN2, an irreversible cysteine proteinase inhibitor, by lesion amastigotes, axenic amastigotes and promastigotes of *Leishmania mexicana*. Braz J Med Biol Res. 29, 987-994.

Pupkis MF, Tetley L, Coombs GH. (1986) *Leishmania mexicana*: amastigote hydrolases in unusual lysosomes. Exp Parasitol. 62, 29-39.

Ramos CS, Franco FA, Smith DF, Uliana SR. Characterisation of a new *Leishmania* META gene and genomic analysis of the META cluster. (2004) FEMS Microbiol Lett. 238, 213-219.

Rawlings, N.D., Barrett, A.J., Bateman, A., 2010. MEROPS: the peptidase database. Nucleic Acids Res. 38, 227-33.

Rey, L. Parasitologia. 2 ed. Rio de Janeiro: Guanabara Koogan, 2a edição, 1991.

Rittig MG, Bogdan C. (2000) *Leishmania*–host-cell interaction: complexities and alternative views. Parasitol Today 16, 292- 297.

Saffari B, Mohabatkar H. (2009) Computational Analysis of csteine proteases (Clan CA, Family Cl) of *Leishmania major* to Find Potential Epitopic Regions. Genom Proteom Bioinformatic. 7, 87-95.

Sanyal T, Ghosh DK, Sarkar D. (1994) Immunoblotting identifies an antigen recognized by anti gp63 in the immune complexes of Indian kala-azar patient sera. Mol Cell Biochem. 130, 11-17.

Saunders EC, DE Souza DP, Naderer T, Sernee MF, Ralton JE, Doyle MA, Macrae JI, Chambers JL, Heng J, Nahid A, Likic VA, McConville MJ. (2010) Central carbon metabolism of *Leishmania* parasites. Parasitology. 137, 1303-1313.

Scheidt KA, Roush WR, McKerrow JH, Selzer PM, Hansell E, Rosenthal PJ. (1998) Structure-based design, synthesis and evaluation of conformationally constrained cysteine protease inhibitors. Bioorg Med Chem. 6, 2477-2494.

Schubach A, Marzochi MC, Cuzzi-Maya T, Oliveira AV, Araujo ML, Pacheco R, Momen H, Coutinho SG, Marzochi KB. (1998) Cutaneous scars in American tegumentary leishmaniasis patients: a site of *Leishmania (Viannia) braziliensis* persistence and viability eleven years after antimonial therapy and clinical cure. Am. J. Trop. Med. Hyg. 58, 824-827.

Schurigt U, Schad C, Glowa C, Baum U, Thomale K, Schnitzer JK, Schultheis M, Schaschke N, Schirmeister T, Moll H. (2010) Aziridine-2,3-dicarboxylate-based cysteine cathepsin inhibitors induce cell death in *Leishmania major* associated with accumulation of debris in autophagy-related lysosome-like vacuoles. Antimicrob Agents Chemother. 54, 5028-5041.

Seifert K. (2011) Structures, targets and recent approaches in anti-leishmanial drug discovery and development. Open Med Chem J. 5, 31-39.

Selzer PM, Pingel S, Hsieh I, Ugele B, Chan VJ, Engel JC, Bogyo M, Russell DG, Sakanari JA, McKerrow JH. (1999) Cysteine protease inhibitors as chemotherapy: lessons from a parasite target. Proc Natl Acad Sci U S A. 96, 11015-11022.

Shinoda S, Miyoshi S. (2011) Proteases produced by vibrios. Biocontrol Sci. 16, 1-11.

Shukla AK, Singh BK, Patra S, Dubey VK. (2010) Rational approaches for drug designing against leishmaniasis. Appl Biochem Biotechnol. 160, 2208-2218.

Silva-López RE, Morgado-Díaz JA, Alves CR, Côrte-Real S, De Simone SG. (2004) Subcellular localization of an extracellular serine protease in *Leishmania (Leishmania) amazonensis*. Parasitol Res. 93, 328-31.

Silva-López RE, De Simone SG. (2004a) *Leishmania (Leishmania) amazonensis*: purification and characterization of a promastigote serine protease. Exp Parasitol. 107, 173-182.

Silva-López RE, De Simone SG. (2004b) A serine protease from a detergent-soluble extract of *Leishmania (Leishmania) amazonensis*. Z Naturforsch C. 59, 590-598.

Silva-López RE, Coelho MG, De Simone SG. (2005) Characterization of an extracellular serine protease of *Leishmania (Leishmania) amazonensis*. Parasitology. 131, 85-96.

Silva-López RE, Morgado-Díaz JA, Chávez MA, De Simone SG. (2007) Effects of serine protease inhibitors on viability and morphology of *Leishmania (Leishmania) amazonensis* promastigotes. Parasitol Res. 101, 1627-35.

Silva-López RE, Morgado-Díaz JA, dos Santos PT, Giovanni-De-Simone S. (2008) Purification and subcellular localization of a secreted 75 kDa *Trypanosoma cruzi* serine oligopeptidase. Acta Trop. 107, 159-167.

Silva-López RE. (2009) Proteases Inhibitors Originated from Plants: Useful Approach for Development of New Drug, Revista Fitos 4, 108-119.

Silva-López RE, Santos TR, Morgado-Díaz JA, Tanaka MN, De Simone SG. (2010a) Serine protease activities in *Leishmania (Leishmania) chagasi* promastigotes. Parasitol Res. 107, 1151-1162.

Silva-López RE. (2010b) *Leishmania* proteases: new targets for rational drug development. Quim Nova. 33, 1541-1548.

Steert K, Berg M, Mottram JC, Westrop GD, Coombs GH, Cos P, Maes L, Joossens J, Van der Veken P, Haemers A, Augustyns K. (2010) α-ketoheterocycles as inhibitors of *Leishmania mexicana* cysteine protease CPB. Chem Med Chem. 5, 1734-1748.

Swenerton RK, Knudsen GM, Sajid M, Kelly BL, McKerrow JH. (2010) *Leishmania* subtilisin is a maturase for the trypanothione reductase system and contributes to disease pathology. J Biol Chem. 285, 31120-31129.

Swenerton RK, Zhang S, Sajid M, Medzihradszky KF, Craik CS, Kelly BL, McKerrow JH. (2011) The oligopeptidase B of *Leishmania* regulates parasite enolase and immune evasion. J Biol Chem. 286, 429-440.

Tornøe CW, Sanderson SJ, Mottram JC, Coombs GH, Meldal M. (2004) Combinatorial library of peptidotriazoles: identification of [1,2,3]-triazole inhibitors against a recombinant *Leishmania mexicana* cysteine protease. J Comb Chem. 6, 312-324.

Tzinia AK, Soteriadou KP. (1991) Substrate-dependent pH optima of gp63 purified from seven strains of *Leishmania*. Mol Biochem Parasitol. 47, 83-89.

Ueda-Nakamura T, Attias M, de Souza W. (2001) Megasome biogenesis in *Leishmania amazonensis*: a morphometric and cytochemical study. Parasitol Res. 87, 89-97.

Ueda-Nakamura, T., Attias, M., de Souza, W. (2007) Comparative analysis of megasomes in members of the *Leishmania mexicana* complex. Res Microbiol. 158, 456-462.

Valdivieso E, Dagger F, Rascón A. (2007) *Leishmania* mexicana: identification and characterization of an aspartyl proteinase activity. Exp Parasitol. 116, 77-82.

Valdivieso E, Rangel A, Moreno J, Saugar JM, Cañavate C, Alvar J, Dagger F. (2010) Effects of HIV aspartyl-proteinase inhibitors on *Leishmania* sp. Exp Parasitol. 126, 557-563.

Vickers TJ, Murta SM, Mandell MA, Beverley SM. (2009) The enzymes of the 10-formyl-tetrahydrofolate synthetic pathway are found exclusively in the cytosol of the trypanosomatid parasite *Leishmania major*. Mol Biochem Parasitol. 166, 142-152.

World Healthy Organization, http://www.who.int/leishmaniasis/en/index.html, accessed in August 15th, 2011.

Yao C, Donelson JE, Wilson ME. (2003) The major surface protease (MSP or GP63) of *Leishmania* sp. Biosynthesis, regulation of expression, and function. Mol Biochem Parasitol. 132, 1-16.

Yao C. (2010) Major surface protease of trypanosomatids: one size fits all? Infect Immun. 78, 22-31.

Permissions

The contributors of this book come from diverse backgrounds, making this book a truly international effort. This book will bring forth new frontiers with its revolutionizing research information and detailed analysis of the nascent developments around the world.

We would like to thank Hesam Dehghani, for lending his expertise to make the book truly unique. He has played a crucial role in the development of this book. Without his invaluable contribution this book wouldn't have been possible. He has made vital efforts to compile up to date information on the varied aspects of this subject to make this book a valuable addition to the collection of many professionals and students.

This book was conceptualized with the vision of imparting up-to-date information and advanced data in this field. To ensure the same, a matchless editorial board was set up. Every individual on the board went through rigorous rounds of assessment to prove their worth. After which they invested a large part of their time researching and compiling the most relevant data for our readers. Conferences and sessions were held from time to time between the editorial board and the contributing authors to present the data in the most comprehensible form. The editorial team has worked tirelessly to provide valuable and valid information to help people across the globe.

Every chapter published in this book has been scrutinized by our experts. Their significance has been extensively debated. The topics covered herein carry significant findings which will fuel the growth of the discipline. They may even be implemented as practical applications or may be referred to as a beginning point for another development. Chapters in this book were first published by InTech; hereby published with permission under the Creative Commons Attribution License or equivalent.

The editorial board has been involved in producing this book since its inception. They have spent rigorous hours researching and exploring the diverse topics which have resulted in the successful publishing of this book. They have passed on their knowledge of decades through this book. To expedite this challenging task, the publisher supported the team at every step. A small team of assistant editors was also appointed to further simplify the editing procedure and attain best results for the readers.

Our editorial team has been hand-picked from every corner of the world. Their multi-ethnicity adds dynamic inputs to the discussions which result in innovative outcomes. These outcomes are then further discussed with the researchers and contributors who give their valuable feedback and opinion regarding the same. The feedback is then collaborated with the researches and they are edited in a comprehensive manner to aid the understanding of the subject.

Apart from the editorial board, the designing team has also invested a significant amount of their time in understanding the subject and creating the most relevant covers. They scrutinized every image to scout for the most suitable representation of the subject and create an appropriate cover for the book.

The publishing team has been involved in this book since its early stages. They were actively engaged in every process, be it collecting the data, connecting with the contributors or procuring relevant information. The team has been an ardent support to the editorial, designing and production team. Their endless efforts to recruit the best for this project, has resulted in the accomplishment of this book. They are a veteran in the field of academics and their pool of knowledge is as vast as their experience in printing. Their expertise and guidance has proved useful at every step. Their uncompromising quality standards have made this book an exceptional effort. Their encouragement from time to time has been an inspiration for everyone.

The publisher and the editorial board hope that this book will prove to be a valuable piece of knowledge for researchers, students, practitioners and scholars across the globe.

List of Contributors

Elke Bocksteins
Department of Biomedical Sciences, University of Antwerp, Antwerp, Belgium Department of Pharmacology, University of Iowa Carver, College of Medicine, Iowa City, Iowa, USA

Andrew J. Shepherd and Durga P. Mohapatra
Department of Pharmacology, University of Iowa Carver, College of Medicine, Iowa City, Iowa, USA

Dirk J. Snyders
Department of Biomedical Sciences, University of Antwerp, Antwerp, Belgium

Araceli Diez-Fraile, Nico Van Hecke and Katharina D'Herde
Department of Basic Medical Sciences, Ghent University, Ghent, Belgium

Christopher J. Guérin
Department for Molecular Biomedical Research, VIB, Ghent, Belgium Department of Biomedical Molecular Biology, Ghent University, Ghent, Belgium

Ana L. De Paul, Jorge H. Mukdsi, Juan P. Petiti, Silvina Gutiérrez, Amado A. Quintar, Cristina A. Maldonado and Alicia I. Torres
Centro de Microscopía Electrónica, Facultad de Ciencias Médicas, Universidad Nacional de Córdoba, Argentina

Shyam Gajavelli, Amade Bregy, Markus Spurlock, Daniel Diaz, Stephen Burks, Christine Bomberger, Carlos J. Bidot, Shoji Yokobori, Julio Diaz, Jose Sanchez-Chavez and Ross Bullock
Lois Pope LIFE Center, University of Miami, Miller School of Medicine, Miami, FL, USA

Arzu Karabay, Şirin Korulu and Ayşegül Yıldız Ünal
Istanbul Technical University, Turkey

Alfonsa García-Ayala
Department of Cell Biology and Histology, Faculty of Biology, University of Murcia, Murcia, Spain

Elena Chaves-Pozo
Centro Oceanográfico de Murcia, Instituto Español de Oceanografía (IEO), Carretera de la Azohía s/n. Puerto de Mazarrón, Murcia, Spain

Chiharu Sogawa, Norio Sogawa and Shigeo Kitayama
Okayama University, Japan

Joan E. Rodríguez-Gil
Dept. Animal Medicine & Surgery, Autonomous University of Barcelona, Spain

Kittipong Maneechotesuwan and Adisak Wongkajornsilp
Faculty of Medicine Siriraj Hospital, Mahidol University, Thailand

Eugenia Mato
Networking Research Center on Bioengineering, Biomaterials and Nanomedicine (CIBER-BBN), EDUAB-HSP Hospital Santa Creu i Sant Pau, Barcelona, Spain

Maria Lucas and Anna Novials
Diabetes and Obesity Laboratory, CIBER de Diabetes y Enfermedades Metabólicas Asociadas (CIBERDEM), Institut d'Investigacions Biomèdiques August Pi i Sunyer (IDIBAPS) - Hospital Clínic, Universitat de Barcelona, Spain

Silvia Barceló
Proteomics Unit, IIS Aragón Instituto Aragonés de Ciencias de la Salud (ICS), Unidad Mixta de Investigación, C/Domingo Miral s/n, Zaragoza, Spain

Alejandra Kun
Department of Proteins & Nucleic Acids, Instituto de Investigaciones Biológicas Clemente Estable (IIBCE), Montevideo, Uruguay

Gonzalo Rosso, Lucía Canclini, Mariana Bresque, Carlos Romeo, Karina Cal, Alicia Hanuz and José Roberto Sotelo
Department of Proteins & Nucleic Acids, Instituto de Investigaciones Biológicas Clemente Estable (IIBCE), Montevideo, Uruguay

Aldo Calliari
Biophysics Area, School of Veterinary, (UdelaR), Montevideo, Uruguay Department of Proteins & Nucleic Acids, Instituto de Investigaciones Biológicas Clemente Estable (IIBCE), Montevideo, Uruguay

José Roberto Sotelo-Silveira
Cell and Molecular Biology Department, School of Sciences, (UdelaR), Montevideo, Uruguay Department of Genetics, Instituto de Investigaciones Biológicas Clemente Estable (IIBCE), Montevideo, Uruguay

Hesam Dehghani
Embryonic and Stem Cell Biology and Biotechnology Research Group, Research Institute of Biotechnology, and Department of Basic Science, Faculty of Veterinary Medicine, Ferdowsi University of Mashhad, Mashhad, Iran

Marie-Laure Follet-Gueye
University of Rouen/Laboratoire « Glycobiologie et Matrice Extracellulaire Végétale »/ UPRES EA 4358, Institut Fédératif de Recherche Multidisciplinaire sur les Peptides 23/ Plate-forme de Recherche en Imagerie Cellulaire de Haute Normandie (PRIMACEN), Mont-Saint Aignan Cedex, France

Mollet Jean-Claude, Vicré-Gibouin Maïté, Bernard Sophie, Plancot Barbara, Dardelle Flavien, Ramdani Yasmina and Driouich Azeddine
University of Rouen/Laboratoire « Glycobiologie et Matrice Extracellulaire Végétale »/ UPRES EA 4358, Institut Fédératif de Recherche Multidisciplinaire sur les Peptides 23/ Plate-forme de Recherche en Imagerie, Cellulaire de Haute Normandie (PRIMACEN), Mont-Saint Aignan Cedex, France

Chevalier Laurence
University of Rouen/UMR6634/CNRS, Institut des Matériaux/Faculté des Sciences et Techniques, St Etienne du Rouvray Cedex, France

Coimbra Silvia
University of Porto/Departamento de Biologia/ Faculdade de Ciências/ Rua do Campo Alegre, Porto, Portugal

Raquel Elisa da Silva-López
Department of Natural Products, Farmaguinhos Oswaldo Cruz Foundation, Rio de Janeiro, Brazil

www.ingramcontent.com/pod-product-compliance
Lightning Source LLC
Chambersburg PA
CBHW070728190326
41458CB00004B/1084